AF690662

AI, Numerical Optimization, IoT and Blockchain for Healthcare 4.0

Other related titles:

You may also like

- PBHE063 | Imoize | Cybersecurity in Emerging Healthcare Systems – Technologies for Healthcare 4.0: From AI and IoT to blockchain | 2024
- PBHE058 | Ramamurthy | Technologies for Healthcare 4.0: From AI and IoT to blockchain | 2023
- PBHE061 | Imoize | Artificial Intelligence and Blockchain Technology in Modern Telehealth Systems | 2023
- PBPC053 | Arya | Nature-inspired Optimization Algorithms and Soft Computing: Methods, technology and applications for IoTs, smart cities, healthcare and industrial automation | 2023
- PBHE043 | Kumar | Evolving Predictive Analytics in Healthcare: New AI techniques for real-time interventions | 2022
- PBSE016 | Tyagi | Machine Learning, Blockchain Technologies and Big Data Analytics for IoTs: Methods, technologies and applications | 2022

We also publish a wide range of books on the following topics:
Computing and Networks
Control, Robotics and Sensors
Electrical Regulations
Electromagnetics and Radar
Energy Engineering
Healthcare Technologies
History and Management of Technology
IET Codes and Guidance
Materials, Circuits and Devices
Model Forms
Nanomaterials and Nanotechnologies
Optics, Photonics and Lasers
Production, Design and Manufacturing
Security
Telecommunications
Transportation

All books are available in print via https://shop.theiet.org or as eBooks via our Digital Library https://digital-library.theiet.org.

IET HEALTHCARE TECHNOLGIES SERIES 065

AI, Numerical Optimization, IoT and Blockchain for Healthcare 4.0

Edited by
Saurav Mallik, Ben Othman Soufiane, Junichi Iwata, Subramaniam Karuppiah Barathi Sangeetha, Venkatesan Muthukumaran and Priyanka Roy

The Institution of Engineering and Technology

About the IET

This book is published by the Institution of Engineering and Technology (The IET).

We inspire, inform and influence the global engineering community to engineer a better world. As a diverse home across engineering and technology, we share knowledge that helps make better sense of the world, to accelerate innovation and solve the global challenges that matter.

The IET is a not-for-profit organisation. The surplus we make from our books is used to support activities and products for the engineering community and promote the positive role of science, engineering and technology in the world. This includes education resources and outreach, scholarships and awards, events and courses, publications, professional development and mentoring, and advocacy to governments.

To discover more about the IET please visit https://www.theiet.org/.

About IET books

The IET publishes books across many engineering and technology disciplines. Our authors and editors offer fresh perspectives from universities and industry. Within our subject areas, we have several book series steered by editorial boards made up of leading subject experts.

We peer review each book at the proposal stage to ensure the quality and relevance of our publications.

Get involved

If you are interested in becoming an author, editor, series advisor, or peer reviewer please visit https://www.theiet.org/publishing/publishing-with-iet-books/ or contact author_support@theiet.org.

Discovering our electronic content

All of our books are available online via the IET's Digital Library. Our Digital Library is the home of technical documents, eBooks, conference publications, real-life case studies and journal articles. To find out more, please visit https://digital-library.theiet.org.

In collaboration with the United Nations and the International Publishers Association, the IET is a Signatory member of the SDG Publishers Compact. The Compact aims to accelerate progress to achieve the Sustainable Development Goals (SDGs) by 2030. Signatories aspire to develop sustainable practices and act as champions of the SDGs during the Decade of Action (2020–30), publishing books and journals that will help inform, develop, and inspire action in that direction.

In line with our sustainable goals, our UK printing partner has FSC accreditation, which is reducing our environmental impact to the planet. We use a print-on-demand model to further reduce our carbon footprint.

Published by The Institution of Engineering and Technology, London, United Kingdom

The Institution of Engineering and Technology (the "**Publisher**") is registered as a Charity in England & Wales (no. 211014) and Scotland (no. SC038698).

Copyright © The Institution of Engineering and Technology and its licensors 2026

First published 2025

All intellectual property rights (including copyright) in and to this publication are owned by the Publisher and/or its licensors. All such rights are hereby reserved by their owners and are protected under the Copyright, Designs and Patents Act 1988 ("**CDPA**"), the Berne Convention and the Universal Copyright Convention.

With the exception of:

(i) any use of the publication solely to the extent as permitted under:

  a. the CDPA (including fair dealing for the purposes of research, private study, criticism or review); or
  b. the terms of a licence granted by the Copyright Licensing Agency ("**CLA**") (only applicable where the publication is represented by the CLA); and/or

(ii) any use of those parts of the publication which are identified within this publication as being reproduced by the Publisher under a Creative Commons licence, Open Government Licence or other open source licence (if any) in accordance with the terms of such licence, no part of this publication, including any article, illustration, trade mark or other content whatsoever, may be used, reproduced, stored in a retrieval system, distributed or transmitted in any form or by any means (including electronically) without the prior permission in writing of the Publisher and/or its licensors (as applicable).

The commission of any unauthorised activity may give rise to civil or criminal liability.

Please visit https://digital-library.theiet.org/copyrights-and-permissions for information regarding seeking permission to reuse material from this and/or other publications published by the Publisher. Enquiries relating to the use, including any distribution, of this publication (or any part thereof) should be sent to the Publisher at the address below:

The Institution of Engineering and Technology
Futures Place,
Kings Way, Stevenage,
Herts, SG1 2UA,
United Kingdom

www.theiet.org

Whilst the Publisher and/or its licensors believe that the information and guidance given in this publication is correct, an individual must rely upon their own skill and judgement when performing any action or omitting to perform any action as a result of any statement, opinion or view expressed in the publication and neither the Publisher nor its licensors assume and hereby expressly disclaim any and all liability to anyone for any loss or damage caused by any action or omission of an action made in reliance on the publication and/or any error or omission in the publication, whether or not such an error or omission is the result of negligence or any other cause. Without limiting or otherwise affecting the generality of this statement and the disclaimer, whilst all URLs cited in the publication are correct at the time of press, the Publisher has no responsibility for the persistence or accuracy of URLs for external or third-party internet websites and does not guarantee that any content on such websites is, or will remain, accurate or appropriate.

Whilst every reasonable effort has been undertaken by the Publisher and its licensors to acknowledge copyright on material reproduced, if there has been an oversight, please contact the Publisher and we will endeavour to correct this upon a reprint.

Trade mark notice: Product or corporate names referred to within this publication may be trade marks or registered trade marks and are used only for identification and explanation without intent to infringe.

Where an author and/or contributor is identified in this publication by name, such author and/or contributor asserts their moral right under the CPDA to be identified as the author and/or contributor of this work.

British Library Cataloguing in Publication Data
A catalogue record for this product is available from the British Library

ISBN 978-1-83724-028-9 (hardback)
ISBN 978-1-83724-029-6 (PDF)

Typeset in India by MPS Limited

Cover image: Dan/Moment via Getty Images

Contents

Preface

The recent advancements in science and technology have revolutionized almost each aspect of our lives, while healthcare service is no other exception. With the advent of Healthcare 4.0, a paradigm shift has been occurred, shifting us toward the modern, smarter, more accurate, and patient-centric healthcare systems. The fusion of artificial intelligence (AI), numerical optimization, blockchain and the Internet of Things (IoT) is in the core of this transformation. This book basically explores these groundbreaking techniques and their potential applications to reshape the future of healthcare services.

AI's capability to analyze large volumes of datasets and make real-time record-breaking decisions is transforming disease diagnosis/detection, treatment/ therapeutic services, and patient management. Numerical optimization plays a critical role in fine-tuning of the healthcare systems, enabling improved decision-making, resource allocation, as well as personalized treatment plans. IoT utilizing the interconnected smart devices and sensors offers a real-time monitoring service of patients and medical equipment that can low down the response time in emergencies and enhance the overall nursing and healthcare management. Meanwhile, blockchain technology ensures secure, transparent, and decentralized data management that can resolve critical issues in privacy and data sharing in the healthcare.

This book brings together all of these technologies and offers better insights into how they converge to create an improved ecosystem that enhances the efficiency, accuracy, transparency, as well as the reliability of healthcare delivery. It is designed to serve mathematicians, biomedical researchers, academics, doctors, healthcare professionals, and technology enthusiasts who are really interested in the emerging trends and future scopes of the overall diagnosis and healthcare technologies.

The chapters here provide a detailed exploration of each technology, starting with the foundational concepts and gradually advancing to emerging complex applications in the real-world healthcare scenarios. Case studies, practical examples, and research findings are included to illustrate the transformative potential of these technologies. From AI-driven diagnostic techniques and optimized treatment paths to IoT-enabled smart nursing homes, hospitals, and blockchain-based patient data management systems, this book basically offers a comprehensive sitemap for the future of diagnosis and healthcare systems.

About the editors

Saurav Mallik is a research scientist in the Department of Pharmacology and Toxicology, R Ken Coit College of Pharmacy, The University of Arizona, USA. Previously, he worked as a postdoctoral fellow in the Harvard T H Chan School of Public Health, University of Texas Health Science Center at Houston, and University of Miami Miller School of Medicine, USA. He obtained his PhD degree from the Department of Computer Science and Engineering, Jadavpur University, Kolkata, India in 2017 while his PhD works carried out in Machine Intelligence Unit (MIU), Indian Statistical Institute (ISI), Kolkata, India as a junior research fellow in the DST (Department of Science and Technology, New Delhi, Government of India)-sponsored Swarnajayanti project. He also worked in the Department of Computer Science and Engineering, Jadavpur University, Kolkata, India as UGC (University Grant Commission, Government of India) research fellow. He is the recipient of research associate from the Council of Scientific and Industrial Research), MHRD, Government of India in 2017. He is also a recipient of "Emerging Researcher In Bioinformatics" Award from Bioclues & BIRD Award steering committee, India (2020). He received two times Travel Grant Award for *International Conference on Intelligent Biology and Medicine (ICIBM)*, June 2018 at Los Angeles, California and August 2021 at Philadelphia, PA, USA. Dr Mallik has coauthored more than 240 research papers in various peer-reviewed international journals, conferences, and book chapters. He also has more than 30 authored/edited book publication in Taylor & Francis, River publishers, IET, Elsevier, Springer, Bentham, etc. His papers are highly cited (google scholar citation >3200 and h-index=31). He attended many national and international conferences in USA and India. He is currently an active member of the Institute of Electrical and Electronics Engineers, American Association for Cancer Research, and Association for Computing Machinery, USA and life member of BIOCLUES, India. He is on the editorial board of journals including *Frontiers in Genetics*, *BMC Bioinformatics, Frontiers in Bioinformatics, Frontiers in Applied Mathematics and Statistics*, *Archives of Medical Sciences*, *Mathematics*, *Electronics, Bioengineered* (Taylor & Francis), *International Journal of Biomedical Imaging*, *Chemistry & Biodiversity* (Wiley), *International Journal of Molecular Sciences*, etc. He is also a member of international advisory committee of many reputed engineering colleges in India. His research areas include data mining, computational biology, bioinformatics, biostatistics, and machine learning.

Ben Othman Soufiane is an assistant professor of computer science at the Applied College, King Faisal University, Saudi Arabia, from 2025. He received his PhD degree in computer science from Manouba University, Tunisia in 2016 for his dissertation on "Secure data aggregation in wireless sensor networks." He also holds MS degrees from Monastir University in 2012. Dr Ben Othman has published more than 140 papers at reputed international journals, conferences, and book chapters. He is an editorial board member for various journals and conferences. He serves as an associate editor/academic editor for international journals including *IEEE Access*, *IEEE Sensors*, and *IEEE Internet of Things*. He is a technical program committee member for more than a dozen of international conferences. His research interests include the internet of medical things, wireless body sensor networks, wireless networks, artificial intelligence, machine learning and big data, software testing, and blockchain.

Junichi Iwata is a professor at the University of Michigan School of Dentistry, USA. Previously, he was a professor at the University of Texas Health Science Center, USA. He was previously a research associate at the Center for Craniofacial Molecular Biology, University of Southern California, USA. He has authored more than 110 peer-reviewed research articles. His research laboratory aims to comprehensively understand the cellular and molecular mechanisms in craniofacial congenital disabilities and diseases such as cleft lip and palate, tooth developmental defects, bone diseases, muscle disorders, and Sjögren's disease. He employs a comprehensive array of multidisciplinary approaches, including genetics, genomics, proteomics, bioinformatics, biochemistry, and molecular biology—to characterize the cell-signaling network and cellular metabolic processes related to membrane trafficking. His research areas include craniofacial development, genetics, noncoding RNAs, autophagy, and mouse models.

Subramaniam Karuppiah Barathi Sangeetha is an associate professor at the Manipal Institute of Technology Bengaluru, Manipal Academy of Higher Education, Manipal, India. She has 16 years of teaching experience and 10 years of research experience, and she has published various research papers in high-quality journals. She is a life member of the ISTE and IEI and has published more than 60 journal articles and conference papers, as well as 10 book chapters and 10 IPR patents in machine learning and deep learning. Her research areas include machine learning, data science, deep learning, and data visualization.

Venkatesan Muthukumaran is an assistant professor at the SRM Institute of Science and Technology, Kattankulathur, India. He is a fellow of the International Association for Cryptologic Research (IACR), India and a life member of the IEEE. He has published more than 100 journal articles and conference papers, 10 book chapters, and 10 IPR patents in algebraic with IoT applications. His research areas include machine learning, data science, blockchain, IoT, data mining, and algebraic cryptography.

Priyanka Roy is currently working as an assistant professor in the School of Advanced Sciences and Languages, VIT Bhopal University. She completed her masters and PhD from the Department of Mathematics, IIT Kharagpur. She was a recipient of DST INSPIRE SHE fellowship in 2009–14. She has published several articles in reputed international journals, conferences, and book series. She serves as reviewer in several international journals of operations research and computational mathematics. Her research interest mainly focuses on numerical optimization, generalized convexity, optimization under uncertainty, interval analysis, portfolio optimization, and game theory.

Introduction

Saurav Mallik[1,2], Priyanka Roy[3] and Ben Othman Soufiane[4]

In recent times, the rapid advancements in science and technology have revolutionized almost every sector of our lives including disease detection and healthcare services. With the advent of Healthcare 4.0, a paradigm shift has been happening, moving ourselves to faster, smarter, more technical, efficient, and patient-centric healthcare systems. The fusion of artificial intelligence (AI), numerical optimization, the Internet of Things (IoT), and blockchain are at the core of this transformation. This book explores various groundbreaking technologies, namely, AI, numerical optimization, IoT, and blockchain for Healthcare 4.0, and their potential to restructure the future of healthcare. This book brings together these technologies and also offers old and new insights into how they converge to evolve an ecosystem that enhances the efficiency, accuracy, fault tolerance, and reliability of healthcare delivery. It is basically designed to serve mathematicians, researchers, academics, hospital and healthcare professionals, and technology enthusiasts who are really interested in the emerging trends and future possibilities of the hospital and healthcare technology. The chapters provide a detailed exploration of each technology, starting with foundational concepts and gradually advancing to complex applications in real-world healthcare scenarios. Case studies, practical examples, and research findings are included to illustrate the transformative potential of these technologies.

Chapter 1 proposed a technique about medical image forgery detection. The transmission of breast cancer medical images over the internet is vulnerable to tampering, posing serious risks to diagnostic accuracy and patient safety. Subtle alterations in digital images may go unnoticed, particularly by less experienced medical personnel, leading to misdiagnoses and compromised patient outcomes. This research proposes a framework for detecting image forgery using stacked machine learning (ML) and deep learning models. Features are extracted via Weber local descriptors and TensorFlow Hub's CNN-based pre-trained models, with authenticity verification performed using extreme learning machines and optimized pipelines. The approach, tested on breast cancer images, integrates 17 supervised

[1]Department of Environmental Health, Harvard T. H. Chan School of Public Health, USA
[2]Department of Pharmacology and Toxicology, R Ken Coit College of Pharmacy, University of Arizona, USA
[3]School of Advanced Sciences and Languages, VIT Bhopal University, India
[4]Department of Computer Science, Higher Institute of Computer Sciences, Gouvernorat de Médenine, Tunisia

ML techniques and pre-trained models such as AlexNet, ResNet, ResNet-50, and VGG, optimized through Elephant Herding Optimization and Genetic Algorithm methods. The proposed method demonstrates high accuracy and outperforms existing studies in detecting forgery, ensuring reliable diagnostic outcomes. Chapter 2 focused on the overview of blockchain technology in healthcare. Blockchain technology is revolutionizing healthcare by addressing critical issues in patient-centered care, data security, and interoperability. This chapter explores blockchain's transformative potential through compelling use cases, emphasizing its role in creating tamper-proof health record ledgers that ensure data integrity and enhance patient privacy. Real-world examples illustrate how blockchain safeguards sensitive information, prevents unauthorized access, and facilitates secure data sharing among providers. Its decentralized architecture resolves interoperability challenges, while smart contracts enable seamless data exchange across systems, improving collaboration and patient outcomes. As healthcare digitizes, blockchain emerges as a powerful driver of change, reshaping the industry and redefining stakeholder relationships through innovative applications.

Chapter 3 proposes a novel approach to patient identity management in healthcare by integrating Python and Solidity within blockchain technology to enhance data security and integrity. Python is used to build a flexible backend infrastructure, while Solidity creates secure, immutable smart contracts for identity verification. This decentralized system enables seamless sharing and updating of patient information across providers while ensuring privacy and regulatory compliance. Smart contracts restrict access to authorized parties, improving interoperability, reducing identity fraud, and boosting healthcare efficiency. The scalable framework addresses persistent identity management challenges and can integrate with existing systems, promoting security, transparency, and trust in patient data handling, ultimately strengthening the healthcare ecosystem. Chapter 4 explores how IoT is revolutionizing healthcare by connecting medical devices, sensors, and systems to enable real-time communication and data exchange. This chapter explores IoT's transformative potential in enhancing patient care, resource utilization, and medical outcomes. IoT facilitates remote monitoring, proactive treatment, and early detection of health issues through wearable devices and smart sensors, reducing complications and hospital readmissions. It also optimizes healthcare operations with connected ambulances, RFID-enabled inventory management, and predictive maintenance of medical equipment, improving efficiency and reducing costs. By creating a patient-centric, intelligent healthcare ecosystem, IoT offers a safer, more efficient approach to healthcare delivery.

Chapter 5 introduces a gradient-based optimization technique for solving fuzzy-valued unconstrained optimization problems, where the objective function is fuzzy. By leveraging fuzzy centers and fuzzy arithmetic, the approach addresses uncertain systems, converting fuzzy systems of linear equations into optimization problems. It examines both partially fuzzy (with either fuzzy coefficients or right-hand side vectors) and fully fuzzy systems (with both fuzzy). Convergence analysis confirms solution existence, and various example problems are solved to validate the method. The results, compared with existing techniques, demonstrate strong agreement and highlight the effectiveness of the proposed approach. Chapter 6 utilizes convolutional neural networks (CNNs) for early detection and prediction of Alzheimer's disease progression, aiming to enable timely interventions and improve patient outcomes. Using data from the Open Access Series of Imaging Studies, the model undergoes preprocessing, including standardization and augmentation, to train on functional and structural biomarkers linked to Alzheimer's. Metrics like accuracy and receiver operating characteristic curves evaluate performance, with cross-validation ensuring broader applicability. The CNN achieved 99.95% accuracy, effectively distinguishing between Alzheimer's, mild cognitive impairment (MCI), and healthy controls by identifying key biomarkers such as hippocampal shrinkage and cortical thickness changes. The model's predictive capabilities were validated by accurately tracking MCI patients who progressed to Alzheimer's. This study highlights the potential of CNNs in clinical settings for early detection and disease monitoring, with future research focusing on integrating multimodal data and testing on larger, diverse populations. Chapter 7 addresses epilepsy, a brain disorder diagnosed after two seizures unrelated to medical conditions, by developing a computerized seizure detection technique to protect patients and alert caregivers promptly. Using EEG data, the proposed method employs a pre-processing unit for band-limiting, amplification, and signal rejection, while an adaptive algorithm detects seizure onset by identifying hypersynchronous pulses and filtering unwanted signals. Upon detection, the system notifies medical staff for immediate intervention. Demonstrating superior accuracy, sensitivity, and specificity, the approach is well suited for wearable devices within the Internet of Healthcare framework, enabling rapid detection and potentially saving lives.

Advancements in AI and IoT have revolutionized industries, particularly healthcare, ushering in the era of Health 4.0. Chapter 8 includes innovative technologies such as IoT, IoS, medical cyber-physical systems, health cloud, health fog, big data analytics, mobile networks, and blockchain. With healthcare handling vast patient data, clinical trial information, and research findings, secure data exchange remains a critical challenge. Protecting sensitive medical data on cloud platforms and internet-connected devices from breaches is a major concern. Blockchain, with its decentralized, secure, and immutable architecture, offers a promising solution for managing healthcare data. This chapter explores the benefits and challenges of blockchain in Health 4.0, its applications, and ongoing research, highlighting its potential to enhance data security and healthcare efficiency. Chapter 9 provides the development of AI in healthcare, particularly in cataract management, the leading cause of visual impairment worldwide, especially among aging populations. AI enhances efficiency and quality by addressing gaps in early detection, staging, and treatment of cataracts, which are challenging due to limited resources and increasing demand for surgeries. AI systems can automatically detect age-related eye diseases, extract high-level features, and improve diagnostic accuracy, making them invaluable for public medical systems and global advancements in cataract care.

In Chapter 10, combining machine learning with blockchain ensures secure and reliable image forgery detection. Blockchain stores image hashes and timestamps on a tamper-proof ledger, while machine learning models analyze features to distinguish authentic images from forgeries. This decentralized approach, reinforced by consensus mechanisms, prevents tampering and guarantees data integrity. Chapter 11 presents an efficient rectenna designed to harvest RF power at 3.5 GHz to charge wearable health monitoring (WHM) devices. The antenna, created using CST, achieves a 2.49 dB gain and a −40 dB reflection coefficient, while a serial harvester in ADS with a Schottky diode attains 63.5% power conversion efficiency at 0 dBm RF input. Optimization in CST and ADS improves impedance matching, gain, and RF-to-DC conversion, delivering a 2.5 V DC output sufficient for low-power medical devices. This work highlights the potential of rectennas to enhance WHM reliability by reducing reliance on traditional batteries. Chapter 12 examines how AI transforms telemedicine and remote patient monitoring, enhancing patient outcomes, accessibility, and resource efficiency. It explores AI-driven applications such as smart teleconsultations, advanced monitoring, and predictive analytics, which improve diagnostic precision, personalized treatments, and healthcare efficiency. The chapter underscores AI's transformative potential in advancing patient care and well-being. Chapter 13 explores the analysis of comfortably available technology to be carried out for health monitoring and net connectivity and formulates a layout of an IoT-based gadget that may be used to reveal patient's health. This study plays a major role in designing and improvement of a fitness tracking system and the use of IoT strategies for sufferers to let them live at domestic platform although the medical doctor has almost actual time get admission to their critical scientific measurements.

Chapter 14 explores the transformative potential of ChatGPT, a natural language model developed by OpenAI, in the medical field, focusing on its applications in diagnosis, treatment, patient interaction, and mental health support. It examines ChatGPT's history, mechanisms, and contributions to healthcare, including tasks such as improving efficiency, aiding communication, and providing second opinions. The chapter also addresses ethical considerations, challenges, and limitations, proposing a framework for integrating ChatGPT into healthcare settings. Structured across multiple sections, it provides a comprehensive review of literature, research methods, proposed systems, outcomes, and future directions, highlighting the model's benefits and areas for improvement. Chapter 15 delves into the multifaceted integration of IoT into healthcare analytics systems, highlighting its transformative potential for patient outcomes, data-driven decision-making, and healthcare delivery itself. We explore the diverse applications of IoT technology in healthcare analytics, encompassing population health management, remote diagnostics, real-time patient monitoring, and clinical research. Furthermore, we investigate the role of IoT gadgets such as wearables, sensors, and smart medical instruments in data collection. These devices capture a comprehensive picture of a patient's health through information on behavior, environmental factors, and physiological parameters, providing healthcare professionals with a holistic and continuous view. Additionally, the chapter addresses critical challenges associated with IoT integration, including data interoperability, security, and scalability. We examine how technologies such as edge computing, blockchain, and cloud computing play a vital role in safeguarding patient privacy and ensuring data integrity.

Chapter 16 explores lung cancer biology, traditional and emerging diagnostic methods, and key innovations such as noise reduction and feature extraction in imaging. Lung cancer remains one of the deadliest cancers worldwide, with early and accurate detection being crucial to improving survival rates. Traditional methods such as imaging and biopsies face limitations in sensitivity, accessibility, and cost. Advances in AI, machine learning, and radiomics have transformed lung cancer detection, enabling more precise and cost-effective approaches. It also addresses challenges, ethical implications, and the need for global collaboration to maximize the impact of these technologies, emphasizing their vital role in improving patient outcomes and combating this global health crisis. Chapter 17 provides a comprehensive survey on various research projects on prediction models and related marker discovery for retina-related disorders.

Moreover, from AI-driven diagnostic tools and optimized treatment paths to IoT-enabled smart hospitals and blockchain-based patient data management systems, this book offers a comprehensive roadmap for the future of hospital and healthcare services. This book is helpful for new mathematicians or AI or biomedical researchers to obtain new dimensions in their research career.

Chapter 1

Advanced WLDvG for medical image forgery detection

M. Arun Anoop[1], P. Karthikeyan[2], Ben Othman Soufiane[3] and S. Poonkuntran[4]

Abstract

Medical images of breast cancer may be transmitted through internet; also, clues hiding that are embedded in those images can turn out to be a serious threat to the transmission of medical images. Physicians and medical professionals do not recognize such type of forgery and in some cases this can happen through digital image transmission. A few middle-aged or less experienced doctors may not recognize such minute changes in digital images, which might lead to wrong prediction. Little modification may affect image quality and the identification of image falsification that was carried out. Proving the authenticity of medical images is a task performed using stacked versions of machine learning (ML) and deep learning models. In a way, there is only a slight change in medical X-ray images that can seriously affect patient's health, and this may lead to wrong patient diagnosis and seriously threaten patient's life; it might also have a mental impact on few patients. Early detection of such type of forgery is necessary to help patients overcome such dangerous situations. Features are extracted based on variants of Weber local descriptors and TensorFlow hub's feature vector-based CNN pretrained models. Subsequently, the most intelligent algorithm like ELM with pipeline and linear model options are used to check the authenticity of images. The proposed algorithm is used for breast cancer digital images, accuracy detection is performed to show the effectiveness of methods, and finally the results are compared with other methods. This chapter mainly utilizes 17 supervised ML techniques along with Alexnet, RESNET, RESNET50, VGG CNN pretrained stacked

[1]Department of Computer Science and Engineering, AJ Institute of Engineering and Technology, India
[2]Department of Electronics and Communication Engineering, Velammal College of Engineering and Technology, India
[3]Department of Computer Science, Higher Institute of Computer Sciences, Gouvernorat de Médenine, Tunisia
[4]School of Computer Science and Engineering, VIT Bhopal University, India

models for the detection of medical image forgery. Elephant herding optimization and genetic algorithm optimization are the best methods used with high accuracy.

Keywords: Breast cancer mammogram images; genetic algorithm; EHO algorithm; machine learning and deep learning hybrid genetic algorithm; supervised learning and pretrained feature vector methods

1.1 Introduction

Among the many difficult tasks that come up with digital photos globally is guaranteeing their validity. Though there is a lot of research being done on the subject, there are currently no permanent approaches for accurately and truthfully detect image modifications. Establishing the type of manipulation has been a key difficulty for academics. In the section on the proposed system, some of the creative methods for medical image forgery detection were discussed. Key-point-based approaches include SURF (speeded-up robust features) and SIFT (scale-invariant feature transform). The main plan uses only block-based techniques. So to extract features, use these methods normally. Use block-based methods (often quantitatively based feature extraction) to conduct research as they provide high accuracy based on the dataset(s). Hospital (server) and patients (client(s)) are the two communication entities in the healthcare system unit. A medium for communication is an interface for information exchanged between a server and its clients. Digital imaging is necessary in the modern era to enhance viewer's experience, showcase important information in the image that is not visible to the naked eye, and balance an image using geometric as well as photometric method. Many photo-editing applications, including the freely available GIMP and Adobe Photoshop, can be used to forge images.

Resize, brightness, splicing, metadata removal, GPS tag removal, and copy-move forgery are some of the major assaults. The biggest problem is the easy accessibility of tools for manipulating photos and the lack of original images for evaluation. Within the healthcare industry, photo fraud cannot be tolerated. In the rare circumstance that a radiography is compromised, and an attacker uses copy-move forgery to expand the tumor's region, the patient will be placed in a difficult position and the diagnosis will be wrong. The MIAS database [1] was utilized for medical image analysis.

If any patient data is altered, it could endanger his life. The biggest issue is that it could be fatal. Patients need to be aware of forgeries in order to prevent potentially fatal situations, and early identification can help. In certain situations, the abnormalities, if discovered, can treat the patients' life-threatening condition. Anybody can alter normal to abnormal and abnormal to normal to their own advantage. It is a major issue as well.

The study uses a dataset on breast cancer as a starting point to detect forgeries. In the preliminary research, typical nature photographs are examined (from the CASIA dataset and other datasets). Forgeries can occur in any digital image, whether it depicts an illness of a plant, a human organ, or a dental root canal. A smart healthcare system should include a mechanism to detect if hackers or other outside intruders have altered patient data while it is being transmitted.

The histogram of oriented gradients (HOG) algorithm before and after post-processing techniques was discussed by Lee *et al.* [2]. Their work focused on copy-move forensics that required the use of sophisticated methods in the future. Soni and colleagues [3] discussed the local binary pattern (LBP) histogram. The notion of focusing characteristics is used to detect image manipulation. In the future, they will focus on identifying the many fakes that are present in the picture and identifying precise fakes in high-risk areas. In the section on image forgery, Wang *et al.* [4] introduced an accurate detection method based on LBP and singular value decomposition (SVD). Additionally, they contrasted their findings using SVD alone and SVD + DCT combined.

To identify image forgeries, Yang *et al.* [5] suggested hybrid features based on SIFT and KAZE. Using their suggested hybrid technique, the authors assessed precision and recall based on post-processing changes in SIFT, SURF, Zernike Moments, and Bravo. A technique for identifying video frame forgeries based on LBP and cellular automata was presented by Tralic *et al.* [6,7]. They wanted to develop a new technique in the future to identify video copy-move forgeries (CMFs) in which a portion of a frame is copied and pasted to another frame within the same sequence. In the field of radiographic image forgery detection, Díaz-Flores-García *et al.* [8] developed a binomial distribution of the probability of right replies with the aid of relative frequency and random probability.

In 2008, Calberson and colleagues talked about technologies for detecting image tampering. Future research will determine whether more stringent guidelines or other precautions are indeed necessary to recover, bar, or prevent the fraudulent use of digital radiography in dentistry. A method to identify image falsification for a smart healthcare system was proposed by Ghoneim *et al.* [9]. They computed the accuracy of natural and mammography pictures in their experimental results. A technique for extracting palm features based on contactless capture mode was presented by Li and Yuan [10]. The authors created feature vectors using the relative radius of the centroids of H feature segments.

A novel texture recognition technique using multi-kernel LBP with pyramid structure and an evaluation of noisy image accuracy was proposed by Tuncer and Dogan [11]. For precise diagnosis and interpretation of medical imaging, particularly breast cancer X-ray images, it is imperative to rely on reliable medical facilities and staff. Several techniques can be used to extract features from breast cancer X-ray images like patterns, abnormality presence, or lung texture.

A few popular algorithms are discussed below:

a) One texture descriptor that captures the local fluctuations in pixel intensities is called LBP. It works well at collecting texture patterns, which is helpful for identifying abnormalities in lung texture.
b) Gray-level co-occurrence matrix (GLCM): This tool determines how frequently varied spatial distances and orientations result in the same pixel intensity co-occurrence. Textural characteristics such as homogeneity, contrast, energy, and entropy can be used to identify patterns or anomalies in X-ray pictures.

c) Histogram of oriented gradients (HOG): HOG calculates the gradient orientation distribution inside an image. It is frequently employed in object detection and can identify specific structural patterns or anomalies in X-ray images.
d) Deep learning models such as convolutional neural networks (CNNs) automatically extract pertinent information from photos. In many computer vision tasks, such as medical picture processing, they have demonstrated remarkable success. Features pertaining to anomalies or lung texture can be extracted by fine-tuning pretrained CNN models such as ResNet, VGG16, and Inception. Typical CNN layers that are able to extract information from breast cancer X-ray images are discussed below.
    (a) **Convolutional layers**: These layers extract different low-level and high-level characteristics, including edges, lines, forms, and textures, from the input images by applying convolutional operations to them. Features in a CNN model are automatically learned during training. The features are extracted by the network and are represented hierarchically. Nonetheless, the following general traits can be found if you are searching for a list of easily learned features or aspects that a CNN model can capture:
        (i) **Edges and lines**: CNNs are useful for identifying borders and contours in images because they can capture edges and lines in a variety of orientations and thicknesses.
        (ii) **Textures**: CNNs are able to recognize patterns in texture such as roughness and smoothness that are unique to lung tissue or anomalies shown in breast cancer X-ray images.
        (iii) **Forms and structures**: CNNs can identify a variety of forms and structures, including nodules that may be a sign of breast cancer infection, consolidation regions, and round or oval opacities.
        (iv) **Patterns**: CNNs are able to identify intricate patterns unique to breast cancer X-ray images such as erratic pavement patterns, patchy infiltrates, and ground-glass opacities.
        (v) **Spatial relationships**: CNNs can be trained to recognize the spatial relationships among various X-ray image regions or structures, which can reveal whether abnormalities or alterations linked to breast cancer are present.
    (b) **Pooling layers**: These layers reduce the spatial dimensions of the feature maps by downsampling them while maintaining the dominating features. In CNNs, max pooling is a frequently employed approach.
    (c) **Activation layers**: By introducing nonlinearity, activation functions such as rectified linear units (ReLU) aid in the network's ability to recognize intricate linkages and patterns in the input.
    (d) **Fully connected layers**: These layers allow the network to acquire higher-level representations and generate predictions by connecting all of the neurons from one layer to the next.

A healthcare framework that has an image forgery detection system in place can identify the fake before the diagnostic procedure even begins. In this chapter, copy-move forgery detection algorithm was utilized. Arun Anoop and Poonkuntran [12,13] discussed many forgeries and offered a novel approach that is hybrid intelligence fusion of feature extraction and can be evaluated using multiple classifiers. LPG is a combination of GLCM and LBP. Using LPG, a machine learning technique called ELM is employed as a classifier to identify image forgeries. Together, these increase accuracy rate and create a more precise detecting system, which is significant. Classifying and detecting remind of the increasing level of accurate detection as the features rise. Hence, this proposed system will find its application in medical image processing.

This chapter is organized as follows: Section 1.2 provides related works. Section 1.3 deals with problem utilized equations and proposed methodology. Section 1.4 discusses the experimentation and results, while Section 1.5 provides conclusion and future works.

1.2 Literature survey

Literature survey shown below is based on attacks and forgery detection methods (Figures 1.1–1.7) [23–29].

Figure 1.1 Copy-move forgery attack [23]

Figure 1.2 Photomontage attack [24]

Figure 1.3 Resizing attack [25]

(a) (b) (c)

Figure 1.4 Image splicing attack [26]

Figure 1.5 Colorized image attack [27]

Figure 1.6 Camera based image attack [28]

Figure 1.7 Format-based images [29]

Table 1.1 Image forgery attacks

Copy-move attack	**Photomontage or splicing or image compositing**	**Resize**
Copying a part of an image and pasting it into some other region of the image.	Combining two or more images and form one image.	Resizing the part of the image and recording.

The proposed approach is used to predict the attacks depicted in Table 1.1.

1.3 Proposed system

Types of WLD-variants method: Table 1.2 shows the different inventions of Weber local descriptors (WLDs).

a) Assign random values to the input weights (w_input) that connect the hidden layer of neurons with its input features.
b) Hidden layer biases are denoted as (b_hidden). Define the activation functions such as sigmoid and tanh.
c) Calculate the hidden layer output matrix denoted as H, by applying activation function to the weighted sum of inputs for each instance in the training dataset. H=g(z)=g(weighted sum of inputs)=g(x_train * w_input + b_hidden).
d) Find the output weights. The output weight matrix is denote as (w_output) using the Moore–Penrose pseudoinverse of the hidden layer output matrix and target values.
e) Regularization measure is C, identity matrix is I.
f) Output weight matrix is w_output=(H^T * H + C*I) ^-1 * H^T * y_train. Predict class label (y_pred) or classification output for new data.
g) That computing step can be denoted as H_test.
    H_test=g(x_test * W_input + b_hidden).
h) Predict class label (y_pred) = by multiplying hidden layer output with the output weight=H_test * w_output (Figure 1.8; Tables 1.3 and 1.4).

Table 1.2 WLD variants and GL variants

1	IWLD	$\gamma_{mx}(x_c)$= arctan($\propto X, J_x$), $\gamma_{mx}(x_c)$ = arctan($\propto X, J_x$) X = $\left(\frac{x_0 - x_c}{x_c}, \frac{x_1 - x_c}{x_c}, \ldots, \frac{x_{p-1} - x_c}{x_c}\right)$^T and $\langle , \rangle$ is the inner product operator. $J_x = (\cos\theta_0, \cos\theta_1, \ldots.., \cos\theta_{p-1})$^T and $J_y = (\sin\theta_0, \sin\theta_1, \ldots.., \sin\theta_{p-1})$^T denote the angle between X direction and $x_i - x_c$. Magnitude and orientation are defined, $\gamma_{mx}(x_c) = \sqrt{(\gamma_{mx}(x_c))^2 + (\gamma_{mx}(x_c))^2}$, $\gamma_0(x_0)$ = arctan $\frac{\gamma_{my}(x_c)}{\gamma_{mx}(x_c)}$
2	WLD	The WLD is a type of feature descriptor used in computer vision for image recognition and object detection. It was introduced by Jianxin Wu and Yunde Jia in 2011 as an extension of the local binary pattern (LBP) descriptor. The WBL is based on the Weber Law, which states that the perceived change in a stimulus is proportional to the magnitude of the stimulus. The Weber ratio measures the contrast of an image, while the orientation measures the direction of the gradient. Overall, the role of WLD in image forgery detection is to extract robust and discriminative features that can help to identify the presence of image manipulations and detect image forgeries.
3	GWLD	Gabor wavelet local descriptor (Gabor WLD) is a popular feature extraction method in image forgery detection. Gabor WLD is a combination of Gabor wavelet transformation and LBP that can be used to extract features from images. Gabor WLD can be used to extract features from the image that can be used to detect the boundaries between the different images. Overall, Gabor WLD is a powerful feature extraction method that can be used to detect various types of image forgeries. However, it is important to note that no single feature extraction method is perfect and researchers often use a combination of methods to improve the accuracy of image forgery detection.
4	SWLD	These statistical features are used to construct a local descriptor for each pixel in the image, which captures the texture and structural information of the surrounding region. In forgery detection, the sliding window-based leader detection (SWLD) algorithm can be used to compare local descriptors of different regions in an image to identify regions that are similar or have been copied from each other. By analyzing the statistical properties of the wavelet coefficients, the SWLD algorithm is able to detect regions that have been altered or manipulated. Overall, the SWLD algorithm is a powerful tool for image forgery detection, as it is able to capture both the texture and structural information of an image and can be used to identify a wide range of image manipulations.
5	WLD-ORI	Weber local descriptor-orientation (WLD-ORI) is used to extract features from an image by calculating the difference between pairs of pixels at different scales and orientations.
6	WLDV	WLDV stands for weighted local descriptor variance, and it is a method used in image forgery detection. In WLDV, the image is first divided into small blocks, and for each block, a feature vector is computed based on the variance of the local texture descriptors.
7	GLCM	Liu *et al.* [30] proposed an ELM with the help of the Cholesky decomposition method. GLCM stands for gray-level co-occurrence matrix, which is a statistical

Table 1.2 (*Continued*)

		method used to analyze the spatial relationships between pixels in an image. The GLCM is a matrix that represents how often different combinations of gray-level values occur at a given offset or distance within an image [14].
8	GLRLM	Wu *et al.* [31] proposed convolution feature extraction and feature coding approach and it is called autoencoder receptive fields ELM, which consist of local and global receptive fields. Feature coding consists of feature dimension reduction and two hidden layers for ELM processing. GLRLM stands for gray-level run length matrix, which is a texture analysis method used in image processing. GLRLM extracts features from an image by analyzing the distribution of gray-level runs in the image.
	GLDM	Zhao *et al.* [32] proposed random local weights-based ELM. GLDM stands for gray-level dependence matrix. It is a texture analysis method used to quantify the relationship between gray-level values of adjacent pixels in an image. The GLDM matrix is a 2D matrix that represents the frequency of occurrence of different pairs of gray-level values in an image.
	GLLV	The gray-level local variance (GLLV) histogram is a type of feature extraction method that can be used in image forgery detection. This method calculates the local variance of the gray-level values in an image and generates a histogram that represents the distribution of variance values across the image.
	GLSC	Gray-level spatial correlation (GLSC) involves analyzing the spatial distribution of gray levels in an image to identify areas that may have been manipulated or altered.
	NGLDM	Neighborhood gray-level difference matrix (NGLDM) is a texture feature extraction method that can be used in image forgery detection. This method involves computing the co-occurrence matrix of gray-level pairs in a small neighborhood around each pixel in an image. The NGLDM matrix contains information about the distribution of gray-level pairs in the neighborhood, which can be used to quantify the texture properties of the image.

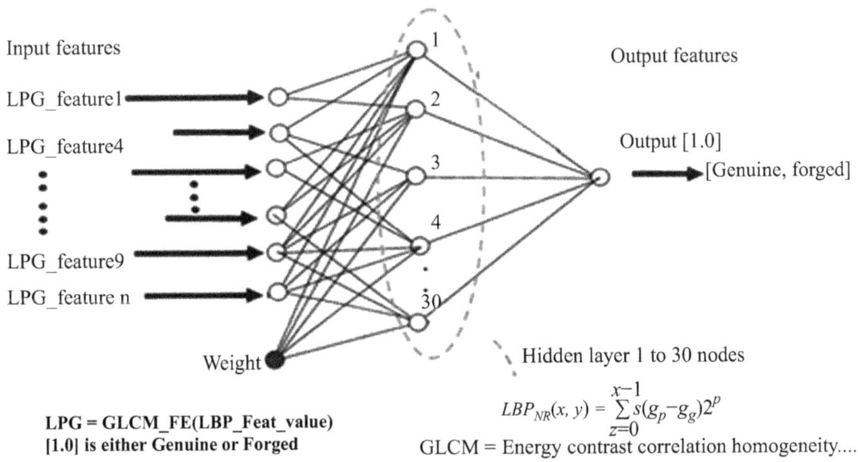

Figure 1.8 Modified ELM working mechanism [21,22,35,36]

Table 1.3 Entropy, information gain, Gini index, split information and gain ratio (process 1)

GLCM	GLRLM	GLDM	GLLV	GLSC	NGLDM	IWLD WLD GWLD SWLD	WLDVWOD	P	L
				Features				G F F G G G F	Y N N Y Y N N

Table 1.4 Entropy, information gain, Gini index, split information and gain ratio (process 2)

GLCM to WOD	Prediction	Labels
Features	Genuine	Yes
	Forged	No
	Forged	No
	Genuine	Yes
	Genuine	Yes
	Genuine	Yes
	Forged	No
	Forged	No

Information gain: Information gain (IG) is a concept used in the field of machine learning and decision trees to measure the importance of a feature in predicting or classifying a target variable (Table 1.5).

"Begin" -> "G=1" -> "Random Population" -> "Max_Gen" -> "G<MaxGen" -> "Sort elephants based on fitness" -> "CUO,SO" -> "Update position"

"Begin" -> "Initialization" -> "G=0" -> "Fitness evaluation" -> "Sort elephants/population from best to worst based on fitness" -> "CUO,SO" -> "Evaluate elephant based on its position" -> "G=G+1" -> "End condition(If yes, output best features and stop; else Go back to G=0)" (Figures 1.9 and 1.10).

The mammogram image can be processed based on GLCM and LBP variants. In LBPVG, different machine learning methods have been processed. Also feature scaling results are used to remove noises and redundant features. Supervised learning algorithms processed are CATBoosting, Naïve Bayes, Decision Trees, XGBoost, AdaBoost, Bagging Classifier, Extra Tree Classifier, Bernoulli Naïve

Table 1.5 Feature selection mathematical formations

Equation's name		Equation's details (*F* is sample, *A* is labels)
IG (*F*, Forged)	=	Entropy, $H(F) - \sum_{f \in FV} \frac{\lvert S_f \rvert}{\lvert S \rvert} H(S_f)$ Entropy, $H(F) - \frac{\lvert S_{\text{forged}} \rvert}{\lvert S \rvert} H(S_{\text{forged}}) - \frac{\lvert S_{\text{genuine}} \rvert}{\lvert S \rvert} H(S_{\text{genuine}})$
Entropy *H*(*F*)	=	$-\sum_{\text{all classes} \in C} P_c \log_2 P_c$ $-P_{\text{forged}} \log_2 (P_{\text{forged}})$ $-P_{\text{genuine}} \log_2 (P_{\text{genuine}})$
Naïve Bayes, P(Y|X)	=	$\frac{P(X \mid Y)P(Y)}{P(X)}$ Therefore, $\frac{P(\text{Forged}, F \mid \text{Genuine}, G)P(\text{Forged})}{P(\text{Genuine})}$ $= \frac{P(G \mid F)P(F)}{P(G)}$
Gini index, Gini (*F*)	=	$1 - \sum_{i=1}^{r} P_{i^2}$ $P_{1(g)} = \frac{\text{Number of data samples with class label Genuine}}{\text{Total number of data samples}}$ $P_{2(f)} = \frac{\text{Number of data samples with class label Forged}}{\text{Total number of data samples}}$
Split information (*F*, *A*)	=	$-\sum_{i=1}^{\text{all classes}} \frac{\lvert S_f \rvert}{\lvert S \rvert} \log_2 \frac{\lvert S_g \rvert}{\lvert S \rvert}$ $\frac{\lvert S_{\text{yes}} \rvert}{\lvert S \rvert} \log_2 \frac{\lvert S_{\text{yes}} \rvert}{\lvert S \rvert} - \frac{\lvert S_{\text{no}} \rvert}{\lvert S \rvert} \log_2 \frac{\lvert S_{\text{no}} \rvert}{\lvert S \rvert}$
Gain ratio (*F*, *A*)	=	$\frac{\text{Gain }(F, A)}{\text{Split information}(F, A)}$

Bayes, Passive Aggressive Classifier, Support Vector Machine, K-Neighbors Classifier, Logistic Regression, Random Forest Classifier, Gradient Boosting Classifier, and Linear Discriminant Analysis (Figure 1.11).

In future, the advanced efficient local LBP variants of the gray-level (ALBPVG) method will help to remove overfitting issues, with the help of some bioinspired optimization techniques before moving to classifiers.

The elephant herding optimization (EHO) algorithm is a new type of swarm-based metaheuristic search method that is implemented to solve optimization problems.

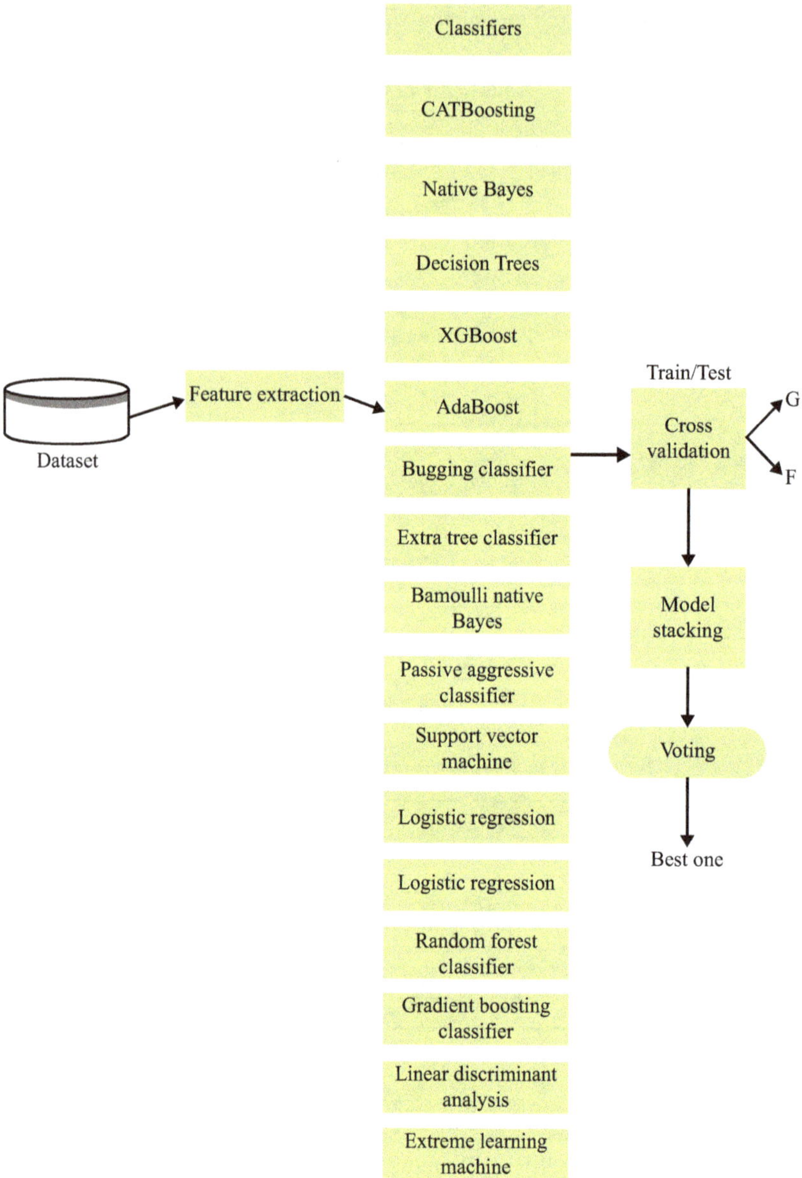

Figure 1.9 Proposed advanced WLDvG processed through EHO for image forgery detection

Based on scenario 1, the EHO algorithm has been designed. Scenario 1 is about once male elephant becomes adults, it leaves the group. Based on these two elephant behaviors, two subsequent operators have evolved using an EHO algorithm, that is the clan updating operator (CUO), and separating operator (SO).

```
Data: N_dimensional Feature Vector of AWLDvG
Result: Final Prediction either Genuine or Forged
Initialize data_file ← [featurevalues..];
data_file ← f1, f2, ....f31;
Base_Estimators ← [
if Classifier Accuracy is Higher then
    CATBoosting()
    RandomForest()
    NaiveBayes()
    DecisionTree()
    XGBoosting()
    AdaBoosting()
    BaggingClassifier()
    Extratree()
    BernoullisNB()
    PassiveAggressC()
    SupportVectorM()
    KNNClassifier()
    LogisticRegression()
    GaussianNaiveBayes()
    LinearDiscrimnantC()
    ExtremeLearningM()
    ];
    Choose Best Classifer
end
if FV is processed for Matching then
    Total Features: f1, f2, ....f31
    X ← f1, f2, ....f30
    Y ← f31
    Example:ELM only mentioned here to show Classification
     Accuracy
    from sklearn.model_selection import train_test_split
    from sklearn.preprocessing import StandardScaler
    from sklearn.base import BaseEstimator
    from sklearn.base import TransformerMixin
    from sklearn.pipeline import make_pipeline
    from sklearn.linear_model import LogisticRegression
    class MyOwnTransformer(BaseEstimator, TransformerMixin)
    elm ←
     make_pipeline(MyOwnTransformer(), LogisticRegression(random_tate =
     10, solver =' lbfgs', multi_lass =' auto'))
    elm.fit(X_test, y_test)
    elm_pred ← elm.predict(X_test)
    elm_score ← elm_score(X_test, y_test)
    elm_score
end
if Final_Prediction: then
    scores ← [model_A, model_B, .....model_n]
    algorithms ← [" Algorithm_1, Algorithm_2, ....Algorithm_n]
    for i ← range(len(Algorithms)) :
     print(algorithms_i + str(scores_i) + "%")
     sns.barplot(algorithms, scores)
end
```

Algorithm 1: AWLDvG-A NOVEL APPROACH FOR MEDICAL FORGERY DETECTION IN IMAGE TRANSMISSION

Figure 1.10 *Proposed flow of advanced WLDvG for image forgery detection using ML*

	Model	Accuracy
0	CAT Boosting	96.50
1	Random Forest	94.41
2	Naive bayes	86.01
3	DTC	90.21
4	XGB	97.20
5	Ada Boosting	95.80
6	Bagging C	99.06
7	ET	93.01
8	Bernoullis	92.02
9	Passive Aggress	97.18
10	SVM	96.71
11	KNN	96.48
12	LR	95.77
13	GB	97.20
14	LDA	94.41
15	ELM	96.50

Figure 1.11 Performance evaluation of proposed approach using ML methods

Based on scenario 2, matriarch, the female elephant, leads the group that has different elephant clans. The matriarch is the oldest female elephant in each elephant family. The matriarch is selected as the most suitable elephant in the family to model and solve optimization issues.

Clan updating operator:

- Updates current position and matriarch.
- Updates the distances of the elephant–individual in each clan with respect to the position of a matriarch.

Separating operator:

- Enhances population diversity at the leader search phase.
- Manly elephants leave their family group.
- Initiates the life features of male elephants.

Divide the whole elephant into some clans. Elitism strategy explains "save the best elephants in a temporary array." or "to protect best elephant persons from

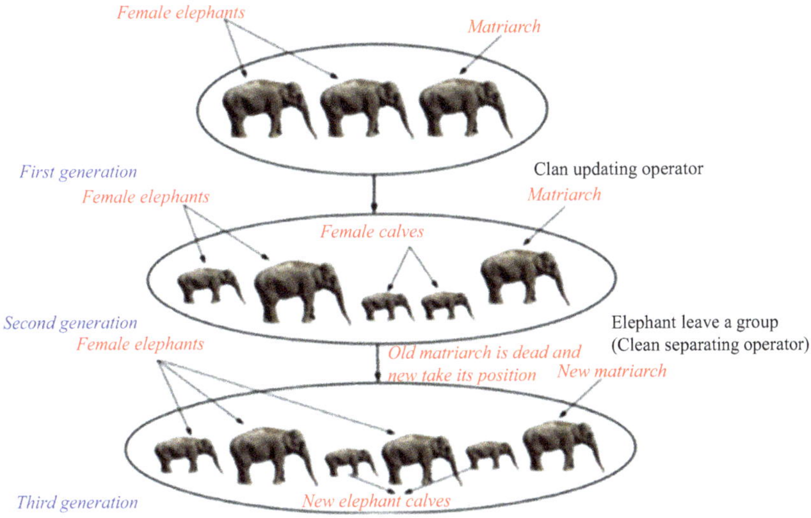

Figure 1.12 EHO process

being ruined." EHO is nothing but an optimization algorithm which is based on herding behavior of elephants. After an iteration is completed, the fitness value of the worst *m* elephants is completed with the best elephant entities that were saved (Figure 1.12).

EHO principles: EHO sets up CUO and SO to model its behavior.

Divide the entire population into some clans.

Each clan is led by a female individual, called a matriarch, denoting best selected individual in each iteration.

Worst individual in each of the iteration denotes male elephant who reached adulthood, leave its clan, and used to live alone.

Clan updating operator: Matriarch is the leader who leads the elephant in each clan. For the search individual (*j*) in clan C_i, its position must be altered according to their relationship with the clan leader.

$$x_{new,ci,j} = x_{ci,j} + \propto X \left(x_{best,ci} - x_{ci,j}\right) X \ r \text{ and}$$

$x_{new,ci,j}$ and $x_{ci,j}$ is the new individual or new position for search elephant individual j in clan C_i and the old position of elephant individual *j* in clan C_i.

$\propto$ denotes number between 0 and 1 that is $\in$ [0,1].

$x_{best,ci}$ is the position of the best elephant in the clan.

$$r \in [0, 1]$$

$$a \in [0, 1]$$

Clan leader's position is adjusted according to the present position of all individuals in the clan.

$x_{new,ci,j} = \beta X x_{center,C_i}$ where x_{center,C_i} is the central position of all individuals in clan C_i.

$$x_{center,C_i} = \frac{1}{n_{C_i}} \sum_{j=1}^{n_{C_i}} x_{ci,j}$$

$$\beta \in [0, 1]$$

n_{C_i} is the number of individuals in clan C_i.

Separating operator [34]: Initiates life characteristics of male elephant. *Male elephants leave their family group can be shown into separating operator* when solving optimization problems. This is implemented by elephants with worst fitness in each generation.

$$x_{\text{worst},ci} = x_{\min} + (x_{\max} - x_{\min} + 1)X\text{r}$$

$x_{\text{worst},ci}$ is the worst individual elephant in clan C_i.
$x_{\min}\&x_{\max}$ is the lower and upper bound of everyone.
r is the stochastic distribution between 0 and 1 that is $\in [0,1]$.

$$x_{center,ci,d} = \frac{1}{n_{ci}} X \sum_{j=1}^{n_{ci}} x_{ci,j,d}$$

$x_{center,ci,d}$ is the center of clan, C_i.
n_{ci} is the number of elephants in clan, C_i.
$\sum_{j=1}^{n_{ci}} x_{ci,j,d}$ is the sum of dth dimension of elephant $x_{ci,j}$.
$1 \leq \text{d} \leq \text{D}$

Main steps:

- Clan, C_i and elephant, j in clan, C_i used to generate $x_{new,ci,j}$ and update $x_{ci,j}$
- $x_{ci,j} = x_{best,ci}$ and generate $x_{new,ci,j}$ and update $x_{ci,j}$
- Clan, C_i then replace worst individual C_i.
- Calculate each elephant individual based on position (Figure 1.13).

Algorithm of elephant herding optimization

1. Begin
2. Iteration, G = 1
3. Population, P is random
4. Maximum Generation, MaxGen
5. Sort all elephants based on fitness.
6. Clan updating operator: when the elephants are moving together, in clan each elephant's next position is accessed by the matriarch. And determine the fitness of elephants in the clan.

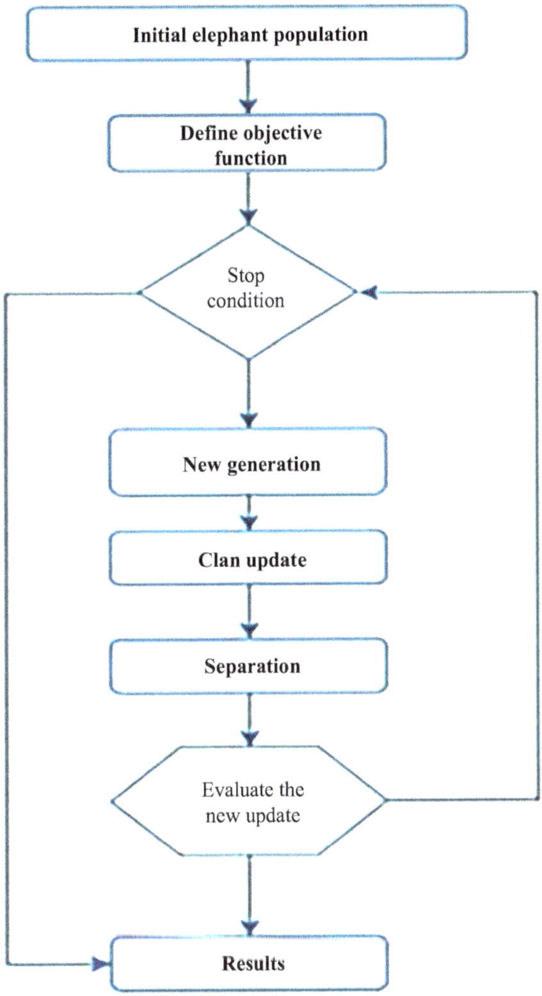

Figure 1.13 EHO flowchart

7. Separating operator: Male elephant will leave and live alone. Matriarch used to identify the worst elephant individual and replace it with new individual.
8. Evaluate each position based on position
9. Increment G by 1.
10. G < MaxGen

a) If yes, output the best solution and stop.
b) If no, then go to fitness evaluation step.

1.4 Experiments and results

The MIASDBv1 dataset was taken into consideration, and the suggested system's performance was evaluated using the following metrics. A variety of performance metrics can be employed to assess the effectiveness of an image forgery detection system. These are a few often used performance metrics and the formulae behind them.

a) Accuracy (ACC): Accuracy is the ratio of correctly categorized photos to the total number of images, and it indicates the overall correctness of the forgery detection system.

TP is the number of true positives (correctly detected forgeries), TN is the number of true negatives (correctly detected authentic images), FP is the number of false positives (authentic images wrongly classified as forgeries), and FN is the number of false negatives (forged images wrongly classified as authentic). The formula for ACC is TP + TN / (TP + TN + FP + FN).

b) Precision: Out of all photos categorized as forgeries, precision represents the percentage of accurately identified forgeries. Determine each classifier's accuracy score, which expresses the percentage of accurately predicted forged images relative to all predicted forged images.

$$\mathrm{TP}/(\mathrm{TP}+\mathrm{FP}) \text{ equals precision.}$$

c) **Recall (sensitivity):** Recall, which is the percentage of accurately identified forgeries out of all genuine forgeries, is also referred to as sensitivity or true positive rate (TPR). Determine each classifier's recall score, which expresses the percentage of accurately predicted forged photos relative to the total number of forged images.

$$\mathrm{Recall} = \mathrm{TP}/(\mathrm{TP}+\mathrm{FN})$$

d) Specificity: Out of all real authentic photographs, specificity quantifies the percentage of authentic images that are correctly identified.

$$\mathrm{TN}/(\mathrm{TN}+\mathrm{FP}) \text{ equals specificity.}$$

e) F1 score: This balanced indicator of the effectiveness of the forgery detection system is the harmonic mean of precision and recall. Determine each classifier's F-measure, which yields a balanced evaluation metric and is the harmonic mean of precision and recall.

$$\text{F1 score is equal to} 2 * (\text{recall} * \text{precision})/(\text{recall} + \text{precision}).$$

f) Receiver operating characteristic (ROC) curve: At different categorization thresholds, the ROC curve graphically illustrates the tradeoff between the TPR and the false positive rate (FPR). Better performance is indicated by a bigger area under the curve of the ROC curve, which is frequently employed as a performance metric (Tables 1.6 and 1.7).

Information gain: It is a concept used in decision trees and machine learning to quantify how significant a feature is in categorizing or predicting a target variable.

A variety of performance metrics can be employed to assess the effectiveness of an image forgery detection system. The following performance metrics are frequently used, along with the formulas for each (see Figures 1.14 and 1.15).

A variety of performance metrics can be employed to assess an image forgery detection system's effectiveness. The following performance metrics are frequently used, along with the formulas for each (Tables 1.8 and 1.9).

Fit function error: Keras Application VGG16. Additionally, I have employed "keras callback early stopping" to prevent overfitting. But it resulted in error.

TensorFlow hub (RESNET): Use of the "Resnet 50 V2 feature vector" Model_url and num_classes are among the arguments used and analyzed. The pretrained model is downloaded and saved using the hub Keras layer technique. The model's accuracy and loss were processed using epoch 5, the Adam optimizer, and the RESNET URL categorical cross-entropy (Figure 1.16; Tables 1.10–1.19).

Table 1.6 Entropy, information gain, Gini index, split information, and gain ratio (process 1)

GLCM	LBP	ILBP	MBP	LTP	ILTP	RLBP	SLBP	P	L
Features								G	Y
								F	N
								F	N
								G	Y
								G	Y
								G	N
								F	N

Table 1.7 Entropy, information gain, Gini index, split information, and gain ratio (process 2)

GLCM to SLBP	Prediction	Labels
Features	Genuine	Yes
	Forged	No
	Forged	No
	Genuine	Yes
	Genuine	Yes
	Genuine	Yes
	Forged	No
	Forged	No

```
(569, 33)
         id class      Feat1      Feat2     Feat3     Feat4     Feat5  \
0    842302     G  41.682006  41.806871  5.076269  4.814575  0.416853
1    842517     G  41.697157  41.725051  4.938046  4.765969  0.426069
2  84300903     G  41.781071  41.946037  4.999659  4.691446  0.412303
3  84348301     G  41.581866  41.700034  5.328744  5.084915  0.397539
4  84358402     G  37.825638  38.313333  7.512558  6.519371  0.363549

      Feat6     Feat7     Feat8  ...    Feat22    Feat23    Feat24     Feat25  \
0  0.445406  0.416853  0.445406  ...  0.567671  0.305453  0.308827  44.012333
1  0.447664  0.426069  0.447664  ...  0.569194  0.299584  0.299830  43.872187
2  0.445071  0.412303  0.445071  ...  0.573325  0.294960  0.296752  44.083043
3  0.424820  0.397539  0.424820  ...  0.557795  0.307961  0.306514  43.995015
4  0.445063  0.363549  0.445063  ...  0.556601  0.323710  0.330102  41.338459

      Feat26     Feat27     Feat28      Feat29      Feat30    Feat31
0  44.011932  12.628174  12.629099  119.318987  119.585443  2.284796
1  43.879259  12.627679  12.616446  119.145889  119.091534  2.289522
2  44.088136  12.653426  12.659337  119.462058  119.837002  2.287747
3  44.060497  12.621229  12.620969  119.224149  119.377414  2.283755
4  41.285737  11.946560  11.949779  108.331468  109.198167  2.342568
```

Figure 1.14 Various performance metrics (part 1)

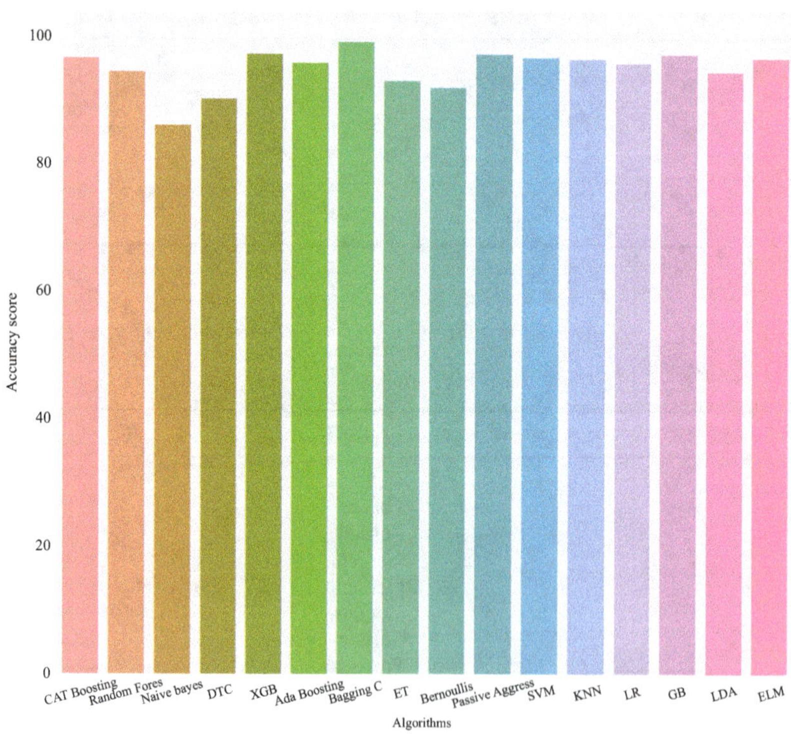

Figure 1.15 Various performance metrics (part 2)

Table 1.8 Performance metrics

Accuracy (ACC)	$\text{Accuracy} = \text{Overall correctness of the forgery detection system}$ $\text{Accuracy} = \frac{\text{Correctly classified images}}{\text{Total number of images}}$ $\text{ACC} = \frac{(\text{Number of correctly detected forgeries} + \text{Correctly detected authentic images})}{\text{Total number of images}}$ *or* (Correctly detected forgeries + correctly detected authentic images + authentic images incorrectly classified as forgeries + forged images incorrectly classified as authentic)
Precision	$\text{Precision} = \frac{\text{Correctly predicted forged images}}{\text{Total predicted forged images}}$ $\text{Precision} = \frac{\text{Number of true positives or (correctly detected forgeries)}}{\{\text{Number of true positives or (correctly detected forgeries)}\} + \{\text{number of false positives or (forged images incorrectly classified as forgeries)}\}}$
Recall (sensitivity)	$\text{Recall} = \frac{\text{Correctly predicted forged images}}{\text{Total actual forged images}}$ $\text{Recall} = \frac{\text{Number of true positives or (correctly detected forgeries)}}{\{\text{Number of true positives or (correctly detected forgeries)}\} + \{\text{number of false negatives or (forged images incorrectly classified as authentic)}\}}$

(Continues)

Table 1.8 (Continued)

Specificity	$$\text{Specificity} = \frac{\text{Correctly detected authentic images}}{\text{All actual authentic images}}$$ $$\text{Specificity} = \frac{\text{number of true negatives or (correctly detected authentic images)}}{\{\text{number of true negatives (or correctly detected authentic images)}\} + \{\text{number of false positives (or authentic images incorrectly classified as forgeries)}\}}$$
F1 score	a) The F1 score offers a fair assessment of the effectiveness of the forgery detection system. It is calculated as the harmonic mean of precision and recall. b) Determine each classifier's F-measure, which is a balanced evaluation metric and is the harmonic mean of precision and recall. c) F1 score is equal to 2 * (recall * precision) / (recall + precision).
Receiver operating characteristic (ROC) curve:	a) The tradeoff between the true positive rate (TPR) and the false positive rate (FPR) at different categorization thresholds is shown graphically by the ROC curve. b) A ROC curve's area under the curve (AUC) is frequently used to gauge performance; a greater AUC denotes superior performance.

Table 1.9 Performance evaluation of different classifiers

Sl. no.	Classifier	Precision (%)	Recall (%)	F1 score (%)
1.	Random Forest	97	94	96
2.	Naïve Bayes	83	98	90
3.	Decision Tree	95	88	91
4.	ADA Booster	97	97	97
5.	CAT Booster	100	94	97
6.	XG Booster	99	97	98
7.	Bagging Classifier	94	98	96
8.	Extra Tree	95	92	94
9.	Bernoulli's NB	94	94	94
10.	Passive Aggressive	97	92	94
11.	Support Vector Machine	96	96	96

(Continues)

Table 1.9 (*Continued*)

Sl. no.	Classifier	Precision (%)	Recall (%)	F1 score (%)
12.	K-nearest Neighbor	94	97	95
13.	Logistic Regression	97	97	97
14.	Gradient Boosting	98	98	98
15.	LDA	94	98	96
16.	Extreme Learning Machine with pipeline & linear model	97	98	97
	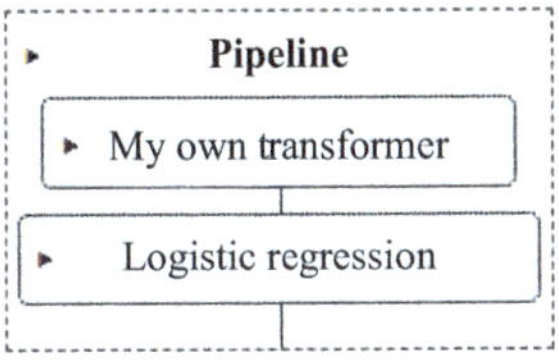			

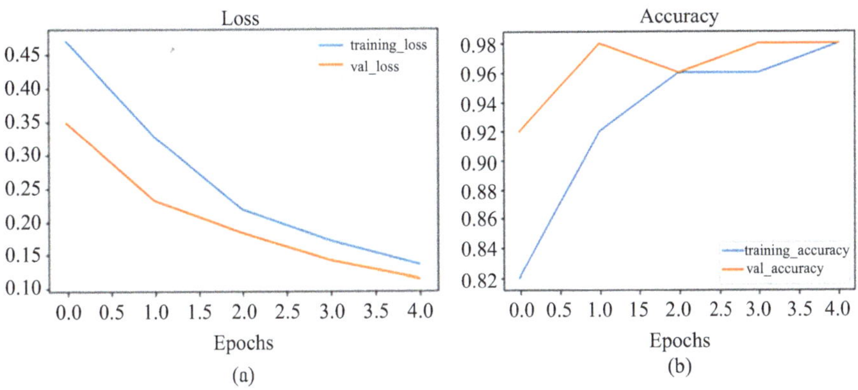

Figure 1.16 *RESNET: (a) loss vs epochs and (b) accuracy vs epochs*

Table 1.10 *Pretrained CNN models based on TensorFlow hub URLs for breast cancer detection based on small datasets*

Sl. no.	Models	Epochs	Val accuracy
1	RESNET	5	**98**
2	ALEXNET	5	96
3	RESNET 50	5	96
4	VGG16	5	82

Table 1.11 Pretrained CNN models based on TensorFlow hub URLs for COVERAGE, CoMoFoD datasets detection based on small and large datasets [37,38]

Sl. no.	Models	Epochs	Dataset	Val accuracy
1	RESNET	5	Small	58%
2	ALEXNET	5	Large	**62%**
3	RESNET 50	5	Small	60%
4	VGG16	5	Small	61%

Table 1.12 TensorFlow hub URLs of pretrained CNN models for breast cancer forgery (copy-move) detection based on small and large datasets [37,38]

Sl. no.	Models	Epochs	Val accuracy
1	RESNET	5	60%
2	ALEXNET	5	**67%**
3	RESNET 50	5	66
4	VGG16	5	61

Table 1.13 Performance evaluation based on genetic algorithm iterations 1 and 2 [33]

Iteration 1	Iteration 2
For chromosome 1: Accuracy: 0.9202127659574468 For chromosome 2: Accuracy: 0.9042553191489362 For chromosome 3: Accuracy: 0.9414893617021277 For chromosome 4: Accuracy: 0.9574468085106383 For chromosome 5: Accuracy: 0.8351063829787234 For chromosome 6: Accuracy: 0.9468085106382979 For chromosome 7: Accuracy: 0.9574468085106383 For chromosome 8: Accuracy: 0.9361702127659575 For chromosome 9: Accuracy: 0.9521276595744681 For chromosome 10: Accuracy: 0.9627659574468085 Maximum accuracy chromosome is 10 Second maximum accuracy chromosome is 4 Applying crossover and mutation between the selected top chromosomes. Replacing the selected chromosomes with the new chromosomes. Iteration 1 ended.	For chromosome 1: Accuracy: 0.9202127659574468 For chromosome 2: Accuracy: 0.9042553191489362 For chromosome 3: Accuracy: 0.9414893617021277 For chromosome 4: Accuracy: 0.9521276595744681 For chromosome 5: Accuracy: 0.8351063829787234 For chromosome 6: Accuracy: 0.9468085106382979 For chromosome 7: Accuracy: 0.9574468085106383 For chromosome 8: Accuracy: 0.9361702127659575 For chromosome 9: Accuracy: 0.9521276595744681 For chromosome 10: Accuracy: 0.9308510638297872 Maximum accuracy chromosome is 7 Second maximum accuracy chromosome is 4 Applying crossover and mutation between the selected top chromosomes. Replacing the selected chromosomes with the new chromosomes. Iteration 2 ended.

Table 1.14 *Performance evaluation based on genetic algorithm iterations 3 and 4*

Iteration 3	**Iteration 4**
For chromosome 1:	For chromosome 1:
Accuracy: 0.9202127659574468	Accuracy: 0.9202127659574468
For chromosome 2:	For chromosome 2:
Accuracy: 0.9042553191489362	Accuracy: 0.9042553191489362
For chromosome 3:	For chromosome 3:
Accuracy: 0.9414893617021277	Accuracy: 0.9414893617021277
For chromosome 4:	For chromosome 4:
Accuracy: 0.9202127659574468	Accuracy: 0.9202127659574468
For chromosome 5:	For chromosome 5:
Accuracy: 0.8351063829787234	Accuracy: 0.8351063829787234
For chromosome 6:	For chromosome 6:
Accuracy: 0.9468085106382979	Accuracy: 0.9468085106382979
For chromosome 7:	For chromosome 7:
Accuracy: 0.973404255319149	Accuracy: 0.9680851063829787
For chromosome 8:	For chromosome 8:
Accuracy: 0.9361702127659575	Accuracy: 0.9361702127659575
For chromosome 9:	For chromosome 9:
Accuracy: 0.9521276595744681	Accuracy: 0.9361702127659575
For chromosome 10:	For chromosome 10:
Accuracy: 0.9308510638297872	Accuracy: 0.9308510638297872
Maximum accuracy chromosome is 7	Maximum accuracy chromosome is 7
Second maximum accuracy chromosome is 9	Second maximum accuracy chromosome is 6
Applying crossover and mutation between the selected top chromosomes.	Applying crossover and mutation between the selected top chromosomes.
Replacing the selected chromosomes with the new chromosomes.	Replacing the selected chromosomes with the new chromosomes.
Iteration 3 ended.	Iteration 4 ended.

Table 1.15 *Performance evaluation based on genetic algorithm iterations 5 and 6*

Iteration 5	**Iteration 6**
For chromosome 1:	For chromosome 1:
Accuracy: 0.9202127659574468	Accuracy: 0.9202127659574468
For chromosome 2:	For chromosome 2:
Accuracy: 0.9042553191489362	Accuracy: 0.9042553191489362
For chromosome 3:	For chromosome 3:
Accuracy: 0.9414893617021277	Accuracy: 0.925531914893617
For chromosome 4:	For chromosome 4:
Accuracy: 0.925531914893617	Accuracy: 0.9202127659574468
For chromosome 5:	For chromosome 5:
Accuracy: 0.8351063829787234	Accuracy: 0.8351063829787234
For chromosome 6:	For chromosome 6:
Accuracy: 0.9680851063829787	Accuracy: 0.9414893617021277
For chromosome 7:	For chromosome 7:

(Continues)

Table 1.15 (Continued)

Iteration 5	Iteration 6
Accuracy: 0.9680851063829787 For chromosome 8: Accuracy: 0.9361702127659575 For chromosome 9: Accuracy: 0.9361702127659575 For chromosome 10: Accuracy: 0.9308510638297872 Maximum accuracy chromosome is 6 Second maximum accuracy chromosome is 3 Applying crossover and mutation between the selected top chromosomes. Replacing the selected chromosomes with the new chromosomes. Iteration 5 ended.	Accuracy: 0.9574468085106383 For chromosome 8: Accuracy: 0.9361702127659575 For chromosome 9: Accuracy: 0.9361702127659575 For chromosome 10: Accuracy: 0.9308510638297872 Maximum accuracy chromosome is 7 Second maximum accuracy chromosome is 6 Applying crossover and mutation between the selected top chromosomes. Replacing the selected chromosomes with the new chromosomes. Iteration 6 ended.

Table 1.16 Performance evaluation based on genetic algorithm iterations 7 and 8

Iteration 7	Iteration 8
For chromosome 1: Accuracy: 0.9202127659574468 For chromosome 2: Accuracy: 0.9042553191489362 For chromosome 3: Accuracy: 0.925531914893617 For chromosome 4: Accuracy: 0.9202127659574468 For chromosome 5: Accuracy: 0.8351063829787234 For chromosome 6: Accuracy: 0.9468085106382979 For chromosome 7: Accuracy: 0.9521276595744681 For chromosome 8: Accuracy: 0.9361702127659575 For chromosome 9: Accuracy: 0.9361702127659575 For chromosome 10: Accuracy: 0.9308510638297872 Maximum accuracy chromosome is 7 Second maximum accuracy chromosome is 6 Applying crossover and mutation between the selected top chromosomes. Replacing the selected chromosomes with the new chromosomes. Iteration 7 ended.	For chromosome 1: Accuracy: 0.9202127659574468 For chromosome 2: Accuracy: 0.9042553191489362 For chromosome 3: Accuracy: 0.925531914893617 For chromosome 4: Accuracy: 0.9202127659574468 For chromosome 5: Accuracy: 0.8351063829787234 For chromosome 6: Accuracy: 0.9202127659574468 For chromosome 7: Accuracy: 0.9414893617021277 For chromosome 8: Accuracy: 0.9361702127659575 For chromosome 9: Accuracy: 0.9361702127659575 For chromosome 10: Accuracy: 0.9308510638297872 Maximum accuracy chromosome is 7 Second maximum accuracy chromosome is 8 Applying crossover and mutation between the selected top chromosomes. Replacing the selected chromosomes with the new chromosomes. Iteration 8 ended.

Table 1.17 Performance evaluation based on genetic algorithm iterations 9 and 10

Iteration 9	Iteration 10
For chromosome 1: Accuracy: 0.9202127659574468 For chromosome 2: Accuracy: 0.9042553191489362 For chromosome 3: Accuracy: 0.925531914893617 For chromosome 4: Accuracy: 0.9202127659574468 For chromosome 5: Accuracy: 0.8351063829787234 For chromosome 6: Accuracy: 0.9202127659574468 For chromosome 7: Accuracy: 0.9468085106382979 For chromosome 8: Accuracy: 0.925531914893617 For chromosome 9: Accuracy: 0.9361702127659575 For chromosome 10: Accuracy: 0.9308510638297872 Maximum accuracy chromosome is 7 Second maximum accuracy chromosome is 9 Applying crossover and mutation between the selected top chromosomes. Replacing the selected chromosomes with the new chromosomes. Iteration 9 ended.	For chromosome 1: Accuracy: 0.9202127659574468 For chromosome 2: Accuracy: 0.9042553191489362 For chromosome 3: Accuracy: 0.925531914893617 For chromosome 4: Accuracy: 0.925531914893617 For chromosome 5: Accuracy: 0.8351063829787234 For chromosome 6: Accuracy: 0.9202127659574468 For chromosome 7: Accuracy: 0.9414893617021277 For chromosome 8: Accuracy: 0.925531914893617 For chromosome 9: Accuracy: 0.9361702127659575 For chromosome 10: Accuracy: 0.9308510638297872 Maximum accuracy chromosome is 7 Second maximum accuracy chromosome is 9 Applying crossover and mutation between the selected top chromosomes. Replacing the selected chromosomes with the new chromosomes. Iteration 10 ended.

Table 1.18 Performance evaluation based on genetic algorithm accuracy

Maximum accuracy is: 0.9414893617021277 Best columns for maximum accuracy are: ["Feat26," "Feat10," "Feat28," "Feat18," "Feat28"] Accuracy after performing extreme learning machine with pipeline and linear model is 98.03%.

1.5 Conclusion and future work

Authenticity prediction of an image has turned out to be essential because of a boundless utilization of images in different media to make genuine or counterfeit messages. Experimental results have shown that this method gives high accuracy and less false negative results. In the experiments, the proposed method outperforms some existing methods of image forgery detection with the help of proposed

Table 1.19 Performance evaluation based on author's published work [16]

Sl. no.	Classifier	Precision	Recall	F1 score
1	LORA [13]	84%	85%	84%
2	LPG+ELM [17]	89%	88%	88%
3	LPG+BAT+ELM [17]	96%	94%	95%
4	DBELM [15]	98%	96%	97%
5	LBPVG+ELM [18,19]	97.8%	98.8%	97.2%
6	LBPSOSA+ELM [19,20] ALEXNET+ELM Ca=97.47% ALEXNET+BAT+ELM Ca=94.24%	96.01%	95%	95%
7	LBFGS/GWF [20] AdaBoost classifier Ca=95.21–97.34%	Manual forged images prediction accuracy = 62–67% COMOFOD, CASIA dataset CNN results = 80–83% accuracy.		
8	Proposed system	With genetic algorithm plus ELM with pipeline and linear model, results 98.03% of accuracy.		

approach and rank optimizer. The proposed approach also proved authenticity and robustness based on different transformations. Among the four classifiers, ELM proved as efficient in work with a high accuracy rate. With this hybrid algorithm, it achieved 96.01%. As discussed, other hybrid combinations also require us to check and compare the results which have been achieved now. Therefore, future, work will focus on performance evaluation based on LPG with other optimization algorithms GA, ACO, PSO, and hybrids to improve the accuracy ratio and the feature selection criteria before optimization process. Evaluated accuracy based on ELM with testing and training samples will be extracted from the dataset mentioned. Future work will be based on *n*-fold cross-validation method(s).

References

[1] Suckling J, Parker J, Dance D, *et al.* (2015). *Mammographic Image Analysis Society (MIAS) database v1.21*. Apollo - University of Cambridge Repository. https://doi.org/10.17863/CAM.105113.

[2] Lee JC, Chang CP, and Chen WK (2015). Detection of copy–move image forgery using histogram of orientated gradients. *Inf Sci* 321:250–262. https://doi.org/10.1016/j.ins.2015.03.009.

[3] Soni B, Das PK, and Thounaojam DM (2017). Copy-move tampering detection based on local binary pattern histogram Fourier feature. In: *Proceedings of the 7th International Conference on Computer and Communication Technology (ICCCT-2017), November 2017*, pp 78–83.

[4] Wang Y, Tian L, and Li C (2017). LBP-SVD based copy move forgery detection algorithm. In: *IEEE International Symposium on Multimedia.* https://doi.org/10.1109/ISM.2017.108.

[5] Yang F, Li J, Lu W, and Weng J (2017). Copy-move forgery detection based on hybrid features. *Eng Appl Artif Intell* 59:73–83. https://doi.org/10.1016/j.engappai.2016.12.022.

[6] Tralic D, Grgic S, and Zovko-Cihlar B (2014a). Video frame copy-move forgery detection based on cellular automata and local binary patterns. In: *2014 X International Symposium on Telecommunications (BIHTEL).* Sarajevo, pp 1–4.

[7] Tralic D, Rosin PL, Sun X, and Grgic S (2014b). Copy-move forgery detection using cellular automata. In: Rosin P, Adamatzky A, and Sun X (eds), *Cellular Automata in Image Processing and Geometry* (Emergence, complexity and computation), vol 10, pp. 105–125. Springer International Publishing, Amsterdam. https://doi.org/10.1007/978-3-319-06431-4_6.

[8] Díaz-Flores-García V, Labajo-González E, Santiago-Sáez A, and Perea-Pérez B (2017). Detecting the manipulation of digital clinical records in dental practice. *Coll Radiogr* 23(4):e103–e107. https://doi.org/10.1016/j.radi.2017.05.003.

[9] Ghoneim A, Muhammad G, Amin SU, and Gupta B (2018). Medical image forgery detection for smart healthcare. *Adv Next Gener Netw Technol Smart Healthc IEEE Commun Mag* 56(4):33–37. https://doi.org/10.1109/MCOM.2018.1700817.

[10] Li W, and Yuan W (2024). Multiple palm features extraction method based on vein and palmprint. *J Ambient Intell Human Comput* 15:1465–1479. https://doi.org/10.1007/s12652-018-0699-1.

[11] Tuncer T, and Dogan S (2020). Pyramid and multi kernel based local binary pattern for texture recognition. *J Ambient Intell Human Comput* 11:1241–1252. https://doi.org/10.1007/s12652-019-01306-1.

[12] Arun Anoop M, and Poonkuntran S (2018). DFE approach for CMFD on digital images—a review and performance evaluation. *Int J Comput Technol* 5(11):148–159.

[13] Arun Anoop M, and Poonkuntran S (2019). Lora approach for image forgery detection and localization in digital images. *CnR's Int J Soc Sci Res* 4(3):1–14.

[14] Azemin MZC, Tamrin MIM, Hilmi MR, and Kamal KM (2015). GLCM texture analysis on different color space for pterygium grading. *ARPN J Eng Appl Sci* 10(15):6410–6413.

[15] Arun Anoop M, and Poonkuntran S (2020). Dbelm for image forgery detection. *Int J Comput Technol* 7(10):151–160.

[16] Arun Anoop M (2021). A novel approach for image forgery detection. PhD Thesis, Information and Communication Engineering. Anna University, Chennai.

[17] Arun Anoop M (2021). Local patterns of grey level (LPG) based image forgery detection. Indian Patent, Application Number: 202141002197.

[18] Arun Anoop M, Karthikeyan P, and Poonkuntran S (2023). Medical image detection and recognition using machine learning and deep learning. In Soufiane BO and Chakraborty C (eds.), *Machine Learning and Deep Learning Techniques for Medical Image Recognition*. Boca Raton, FL: CRC Press/Taylor & Francis, https://doi.org/10.1201/9781003366249.
[19] Arun Anoop M, Karthikeyan P and Poonkuntran S (2024). Unsupervised/supervised feature extraction and feature selection for multimedia data (Feature extraction with feature selection for Image Forgery Detection). In Swarnkar SK, Patra JP, Kshatri SS, Rathore YK, and Tran TA (eds.), *Supervised and Unsupervised Data Engineering for Multimedia Data*, pp. 27–61. Beverly, MA: Scrivener Publishing LLC.
[20] Arun Anoop M, and Karthikeyan P. Artificial intelligence and machine learning approaches for healthcare. In *GWF Algorithm for Image Forgery Detection*. River Publishers, Status: Review.
[21] Arun Anoop, M, and Poonkuntran, S (2021). RETRACTED ARTICLE: LPG: a novel approach for medical forgery detection in image transmission. *J Ambient Intell Human Comput* 12:4925–4941. https://doi.org/10.1007/s12652-020-01932-0.
[22] Calberson FL, Hommez GM, and De Moor RJ (2008). Fraudulent use of digital radiography: methods to detect and protect digital radio-graphs. *Am Assoc Endod JOE* 34(5):5. https://doi.org/10.1016/j.joen.2008.01.019.
[23] Dixit R, and Naskar R (2017) Review, analysis and parameterisation of techniques for copy–move forgery detection in digital images. *IET Image Process* 11(9):746–759.
[24] Agarwala A, Dontcheva M, Agrawala M, *et al.* (2004). Interactive digital photomontage, *ACM SIGGRAPH Conference Proceedings*.
[25] Dong WM, Bao GB, Zhang XP, and Paul J-C (2012) Fast multi-operator image resizing and evaluation. *J Comput Sci Technol* 27(1):121–134.
[26] Shi YQ, Chen C, and Chen W (2007). A natural image model approach to splicing detection. In: *Proceedings of the 9th Workshop on Multimedia & Security (MM&Sec'07)*, pp 51–62.
[27] Guo Y, Cao X, Zhang W and Wang R (2018). Fake colorized image detection. *IEEE Trans Inf Forensics Secur* 13(8):1932–1944.
[28] Wu L, and Du X (2014). Security threats to mobile multimedia applications: camera-based attacks on mobile phones. *IEEE Commun Mag* 52:80–87.
[29] Tan LK (2006). Image file formats. *Biomed Imag Intervent J* 2(1):e6.
[30] Liu Q, Zhao X, Hou Z, and Liu H (2017). Epileptic seizure detection based on the kernel extreme learning machine. *Technol Health Care* 25(S1):399–409. https://doi.org/10.3233/THC-171343.
[31] Wu C, Li Y, Zhao Z, and Liu B (2020). Extreme learning machine with multi-structure and auto encoding receptive fields for image classification. *Multidimensional Systems and Signal Processing* 31(4):1277–1298. https://doi.org/10.1007/s11045-020-00708-1.
[32] Zhao SX, Wang XZ, Wang LY, Hu JM, and Li WP (2017). Analysis on fast training speed of extreme learning machine and replacement policy.

International Journal of Wireless and Mobile Computing 13(4). https://doi.org/10.1504/IJWMC.2017.089327.

[33] McCall J (2005). Genetic algorithms for modelling and optimisation. *J Comput Appl Math* 184(1):205–222. https://doi.org/10.1016/j.cam.2004.07.034.

[34] Wang GG, Deb S, and Coelho LDS (2016). Elephant Herding Optimization. *Proceedings – 2015 3rd International Symposium on Computational and Business Intelligence, ISCBI 2015*. https://doi.org/10.1109/ISCBI.2015.8.

[35] Yaseen ZM, Deo RC, Hilal A, *et al.* (2018). Predicting compressive strength of lightweight foamed concrete using extreme learning machine model. *Adv Eng Softw* 115:112–125. https://doi.org/10.1016/j.advengsoft.2017.09.004.

[36] Li W, Chen C, Su H, and Du Q (2015). Local Binary Patterns and Extreme Learning Machine for Hyperspectral Imagery Classification. *IEEE Trans Geosci Remote Sens* 53(7):3681–3693. https://doi.org/10.1109/TGRS.2014.2381602.

[37] Hu F, Xia G. S, Hu J, and Zhang L (2015). Transferring deep convolutional neural networks for the scene classification of high-resolution remote sensing imagery. *Remote Sens* 7(11):14680–14707. https://doi.org/10.3390/rs71114680.

[38] Nagiub Abdelsalam EM, Hussain KF, Omar NM, and Ali QT (2019). Acute Myeloid Leukemia Diagnosis Using Deep Learning. *Clin Lymphoma Myeloma Leuk* 19(S1):S206. https://doi.org/10.1016/j.clml.2019.07.067.

Chapter 2

Blockchain in healthcare: transformative cases and emerging use cases

Himadri Nath Saha[1], Reek Roy[2], Shreea Bose[3], Abhishek Sengupta[4], Snehashis Kayal[5], Shuvam Roy[6], Arup Halder[6] and Pritam Manna[6]

Abstract

Blockchain technology has become an agent of change in the healthcare sector, resolving longstanding problems with patient-centered care, interoperability, and data security. This chapter investigates the different uses of blockchain in healthcare, presenting compelling and emerging use cases that highlight the revolutionary power of this decentralized and tamper-resistant technology. Initially, the importance of blockchain in securing patient data is discussed, with a focus on its ability to construct a tamper-proof ledger of health records, ensuring data integrity and improving patient privacy. Several real-world examples demonstrate the successful use of blockchain to protect sensitive information, restrict unwanted access and streamline data sharing among healthcare providers. Blockchain's decentralized architecture addresses interoperability, a longstanding challenge in healthcare. Smart contracts enable frictionless data transmission between heterogeneous systems, increasing collaboration and decreasing inefficiencies. Case studies demonstrate how blockchain has enabled secure and standardized data sharing, resulting in enhanced care coordination and patient outcomes. As the healthcare business digitizes and faces new difficulties, blockchain technology is proving to be a driver of positive change. This chapter emphasizes blockchain's varied influence in healthcare by giving compelling case studies that demonstrate its ability to disrupt the industrial landscape and redefine stakeholder relationships.

[1]Department of Computer Science, Surendranath Evening College, Calcutta University, India
[2]Department of Computer Science, Belda College, Vidyasagar University, India
[3]Department of Computer Science, St. Xavier's College, Calcutta University, India
[4]Department of Computer Science and Engineering, Indian Institute of Information Technology, India
[5]Department of Information Technology, Techno International New Town, Maulana Abul Kalam Azad University of Technology, India
[6]Department of Computer Science and Engineering, The Neotia University, India

Keywords: Smart healthcare; blockchain; blockchain healthcare; health monitoring; future trends; pervasive healthcare; smart contracts

2.1 Introduction

Blockchain technology has grown increasingly recognized as an innovative force in the healthcare industry, opening up new opportunities in patient care, data management, and supply chain logistics while providing creative answers to persistent problems. At its core, blockchain is a decentralized, transparent open ledger system [1] that records transactions across a computer network. In the healthcare sector, where data security, privacy, and integrity are crucial, blockchain can disrupt the way that information is currently managed, shared, and provided. When blockchain increased the security of the cryptocurrency Bitcoin, it became the catchphrase [2].

Blockchain provides a high degree of data security by utilizing cryptographic techniques to make sure that once data is stored, it cannot be changed or tampered with. This feature is essential in the healthcare sector, where protecting patient data's safety and reliability is paramount. Further, patients now have more control over their health data due to blockchain technology. Blockchain-based platforms allow patients to securely collaborate on their medical records with researchers, medical providers, and other authorized parties. Patients are given the ability to actively participate in their healthcare decisions, which improves not only treatment outcomes but also care coordination.

One of the biggest advantages of blockchain technology for the clinical sector is its ability to reduce expenses and simplify administrative processes. By eliminating intermediaries and automating transactions through smart contracts, blockchain can reduce administrative overheads and improve efficiency in areas such as billing, claims processing, and supply chain management. Moreover, blockchain has the potential to revolutionize clinical research and development. By providing an easily accessible and secure platform for transferring clinical trial data, blockchain can expedite the development of new drugs, improve patient outcomes, and encourage researcher collaboration. By providing an easily accessible and secure platform for transferring clinical trial data, blockchain can expedite the development of new drugs, improve patient outcomes, and encourage researcher collaboration [3].

Blockchain can revolutionize traditional banking and financial services by enabling secure and transparent transactions without the need for intermediaries like banks [4]. Cryptocurrencies such as Bitcoin and Ethereum utilize blockchain technology to facilitate peer-to-peer transactions, cross-border payments, and smart contracts for automated financial agreements. By documenting the path taken by goods from producer to customer, blockchain can also assist us in managing supply chains and improving their transparency and traceability. Identity management systems built on blockchain technology can verify people's identities in a safe and unchangeable manner, lowering the possibility of fraud and identity theft. Blockchain technology can be utilized to build transparent, safe, and distant voting systems that allow votes to be verified. Each vote is recorded on the blockchain, ensuring tamper-proof and auditable elections, and increasing trust in the democratic process. Patents, copyrights, and trademarks are examples of proprietary rights that can be securely recorded and managed using blockchain technology. This can help protect creators' rights, prevent infringement, and facilitate licensing and royalty payments. Iansiti and Lakhani [5] point that it might take a long time for blockchain to produce the anticipated levels of corporate transformation because of societal, organizational, and implementation constraints like security or governance.

Among the transformative cases of blockchain in healthcare is secure patient data management. Healthcare systems throughout the world are grappling with the problems of managing massive volumes of sensitive patient data while maintaining its confidentiality and accessibility. Conventional electronic health record (EHR) systems are often divided, compartmentalized, and vulnerable to unauthorized access or data breaches. Patients and medical professionals can access and update patient health records securely in real time while preserving the confidentiality and integrity of information thanks to blockchain technology, which offers a decentralized, immutable platform for doing so. Patients are granted greater authority over their health information thanks to blockchain-based EHR systems. They guarantee that their data is securely stored and shared across numerous providers and systems, and they give patients the ability to grant or withdraw access as needed. The application of blockchain technology to the logistics of the healthcare supply chain is another new application of blockchain technology. Healthcare supply chains are intricate and multifaceted, with various parties, processes, and regulations. Conventional supply chain management systems can occasionally experience interruptions, inefficiencies, and a shortage of information. A decentralized and transparent platform for tracking and tracing medical supplies, equipment, and products across the supply chain is made possible by blockchain technology. This enables stakeholders to lower the risk of expired or counterfeit goods, monitor inventory levels, and expedite procurement procedures. Healthcare firms can improve operational efficiency, cut costs, and ensure timely delivery of key supplies by utilizing blockchain-based supply chain solutions, resulting in better patient care and outcomes.

2.1.1 Distributed ledger and decentralization

Blockchain technology offers an impermeable, translucent, and secure platform for data management and storage. It works on the distributed ledger and decentralization principles. Since data in a traditional centralized system is usually managed and maintained by a single person, it is susceptible to manipulation, hacking, and illegal access. However, a distributed ledger system, like blockchain, disperses copies of the ledger throughout a network of computers, or nodes, in which every member keeps an immutable and synchronized record of transactions. When there is no central authority or middleman in charge of the network, consensus processes are used to guarantee that all participants agree on the legitimacy of transactions. This is known as decentralization.

When healthcare data is kept across several network nodes thanks to the distributed ledger, it is extremely resistant to manipulation or tampering. Cryptographically linking every transaction to its predecessor creates an open and immutable record of the data history. By doing this, the danger of fraud, data breaches, and unauthorized access is decreased, and data integrity and security are improved. By keeping medical records in a decentralized, safe manner, blockchain technology gives patients more control over their health information. To maintain privacy and consent in healthcare transactions, patients can grant or cancel access to their data as needed. Blockchain enables people to take control of their health information, which advances patient-centric treatment and builds patient–provider confidence. Interoperability problems strike healthcare systems frequently, impeding the smooth transfer of patient data between various providers and systems. By standardizing data formats and access protocols, blockchain's decentralized design makes secure and interoperable data exchange possible. To improve care coordination and patient outcomes, healthcare organizations can safely share patient health records, diagnostic results, and treatment plans across various platforms. By tracking the movement of medical supplies, equipment, and products on a distributed ledger, blockchain improves supply chain transparency and traceability in the healthcare industry. Participants in the network timestamped and authenticated each transaction, allowing stakeholders to follow the origin and legitimacy of goods from production to distribution. This reduces the risk of supply chain disruptions, ensures regulatory criteria are met, and helps prevent counterfeit medications. Secure and transparent collaboration between academics, clinicians, and other stakeholders in the healthcare ecosystem is made possible by blockchain technology. Blockchain-based platforms allow researchers to securely share data, work together on studies, and trace the origin of research findings. As a result, medical research proceeds more quickly, data exchange is encouraged, and healthcare innovation is encouraged.

2.1.2 Smart contract

A digital contract that is self-executing and encoded with predetermined rules and conditions is known as a smart contract. Operating on a blockchain network, these contracts automatically carry out and uphold their terms in response to certain triggers. Smart contracts eliminate the need for middlemen or other parties to oversee transactions since once they are placed on the blockchain, they become irrevocable and automated. Smart contracts have a lot of promises to increase data interoperability, expedite administrative procedures, and improve patient care in the healthcare industry. By carrying out predetermined rules and conditions based on the patient's medical data, smart contracts can automate the processing of insurance claims. For instance, the smart contract can minimize errors and administrative costs by automatically determining coverage, confirming eligibility, and triggering payment to the healthcare provider when a patient has a medical procedure. By automating tasks such as patient recruiting, permission administration, and data collecting, smart contracts can make managing clinical trials easier. Smart contracts can track patient involvement in trials, guarantee adherence to trial protocols, and automatically compensate researchers or trial participants based on predetermined milestones or results. Healthcare supply chain logistics can be improved by smart contracts by automating tasks such as product monitoring, inventory management, and procurement procedures. To ensure transparency, authenticity, and regulatory compliance, pharmaceutical companies can utilize smart contracts to track the transportation of medications from manufacturing facilities to distribution centers and pharmacies. Secure and interoperable data sharing across various healthcare systems and stakeholders can be facilitated via smart contracts. Healthcare organizations may guarantee the smooth integration and sharing of patient health records, diagnostic results, and treatment plans across diverse systems while ensuring data privacy and security by incorporating defined data exchange protocols and access controls into smart contracts.

Blockchain technology is being applied in the healthcare sector by combining automation and artificial intelligence (AI) to enhance data interoperability and streamline administrative processes. Self-executing contracts, or smart contracts, allow predefined actions to be automatically executed when certain conditions are satisfied. They do this by explicitly writing the contents of the agreement into code. Smart contracts can save paperwork, minimize errors, and increase efficiency in the healthcare industry by automating a variety of administrative operations like patient billing, insurance claim processing, and clinical trial management. The integration of diverse data sources is made possible by blockchain-based smart contracts, which enhance the accuracy and accessibility of patient information by facilitating safe and interoperable data transmission between various healthcare systems and stakeholders. Healthcare companies can decrease administrative expenses and streamline operations by utilizing automation and smart contracts.

Blockchain technology is altering traditional approaches to patient data management, medication traceability, supply chain logistics, and administrative procedures. It is also driving innovative and new use cases in the healthcare industry. Blockchain offers an autonomous, readily apparent and safe platform for data sharing, management, and archiving that could improve patient outcomes, business efficiency, and the healthcare industry. As blockchain grows and matures, it is expected to have a greater impact on healthcare, ushering in an unprecedented period of patient-centered care, collaboration, and innovation.

Section 2.2 consists of a literature survey done on various existing systems mentioning the results, pros, and cons of these systems. In Section 2.3, the system design of the proposed model has been elaborated. Sections 2.4–2.6 have the methodology of the proposed model implemented in various cases. In Section 2.7, the future research direction of the proposed model is explained. Section 2.8 has a short conclusion of this book chapter.

2.2 Background and related works

Through secure and efficient data storage, blockchain technology upgrades traditional healthcare practices into a more all-encompassing approach to optimal treatment and recuperation. Blockchain is a platform to enable the most recent and present advancements in the healthcare industry. Sharma *et al.* [3] propose Healthify, a safe distributed application that encrypts medical data to create a safer environment. Healthify is a comprehensive approach to healthcare data protection centered on shared ledger technology. Providing a useful application with easy accessibility to the devices and a persistent database is the aim of this technique. This architecture consists of three layers: layer of data collection, layer of data processing, and layer of storage. The key elements of the architecture, including distributed applications, are interplanetary file system (IPFS) storage and a smart contract. With the help of this application, patients, doctors, diagnostic facilities, and healthcare analyzers can all effortlessly and safely upload and access patient data. To achieve safe and adaptable healthcare data management, a smart contract that handles file sharing, access restriction, authentication, and token management has been created. To guarantee confidentiality and privacy, users can also verify the documents' integrity at any moment. The plan demonstrates compliance with safety and storage standards, as evidenced by results, security analysis, comparison study, and performance evaluation. Adding extra services for users in the healthcare industry would be an easy way to expand the suggested application.

The supply chain of the pharmaceutical business benefits from blockchain technology by offering visibility, traceability, and privacy, as noted by Kumari *et al.* [6]. The consequences on patients, the usefulness of medications, data kept in a blockchain database, and the authorized blockchain that records transactions in the pharmaceutical sector have all been covered. We also talked about the permitted blockchain, which was used to record the transactions for a later investigation, the medication's usability, its effects on the patient, and the data added to the blockchain database. The public's access to high-quality healthcare, the safety of highly trusted pharmaceuticals, and the use of real, modern digital devices in recipes have all been discussed [7]. Employing blockchain technology in health sciences research raises several risks, including the possibility of misconduct and data fabrication or manipulation. This risk arises from data's immutability once registered on the blockchain. Deleting or rectifying erroneous or fraudulent data from the blockchain is more difficult, which could result in unethical behavior or erroneous study conclusions. Interoperability problems between various data platforms and silos within the ecosystem supporting bio-sciences research present another difficulty. The many formats, standards, and protocols that different stakeholders employ to store and handle data give rise to these interoperability issues. Blockchain technology offers a standardized, decentralized framework for safely exchanging data, which can help address these issues. Life sciences research uses blockchain-based technology for several purposes, including data protection, obtaining consent from individuals for the use of their data, and streamlining the analysis of large datasets. Blockchain protects the confidentiality and integrity of sensitive data while promoting cooperation and data analysis among researchers by offering a transparent and safe platform for data capture and exchange. Charles [8] explores several applications of blockchain technology in life sciences research and provides guidelines for conducting such applications morally and legally. It provides insightful information for researchers, blockchain developers, and life sciences organizations who are thinking about using blockchain technology in their studies.

Blockchain technology has been effectively used to handle distributed data. Blockchain databases are designed to be only-ever-created and not edited or deleted. By using blockchain, we can reduce the burden on the healthcare ecosystem. The development of next-generation health data-sharing systems can be built upon its decentralization and inviolability features. Blockchain technology in healthcare is constantly collecting patient data, information, and reports. The doctor takes the patient's details and verifies them before starting the treatment. The mobile device helps efficient healthcare data collection, real-time communication with stakeholders, and improved workflow. Medical equipment provides patients with managing device fleet efficiently while enhancing patient safety and building a data foundation [9]. XDS includes a registry for querying which patient information is stored in an EHR repository, as well as methods for obtaining them. On a blockchain, it is cheap to verify the integrity of an individual transaction. Any network participant can access the integrity of a single piece of information and audit it in real time. Therefore, it is possible to execute costless verification economically. For instance, an audit of healthcare accounting data, which may be assembled with integrity from the most basic transactional units, used to be time-consuming and expensive. Blockchains now allow this procedure to operate continuously in the background while adhering to laws.

Blockchain technology has the power to fundamentally transform a variety of industries, including healthcare. Blockchain is perfect for safely keeping sensitive patient data because it provides a decentralized, tamper-resistant ledger. Blockchain technology can facilitate the safe and easy exchange of patient data between various healthcare providers, guaranteeing the accuracy, timeliness, and accessibility of the data when required. Patients can have more control over their health data because of blockchain technology [10]. They can let researchers, medical professionals, and other outside parties safely and openly access their data. This model proposes a patient-centric healthcare management system that leverages blockchain technology to modernize and optimize healthcare services. Integrating blockchain into the healthcare ecosystem is aimed to enhance data security, streamline administrative processes, improve interoperability, and empower patients to take control of their health data. Blockchain helps to securely store patient's health records on a decentralized blockchain ledger, ensuring data integrity, security, and privacy. It reduces paperwork and streamline healthcare procedures by using smart contracts to automate administrative operations such as billing, insurance claim processing, appointment scheduling, and drug refills [11]. It enables seamless sharing of patient health data among healthcare providers using blockchain-based interoperability standards, ensuring continuity of care and avoiding duplicate tests or treatments. Using blockchain-based identity management and permission systems, blockchain provides patients with the ability to decide who has access to their health information. It can grant or revoke permissions to researchers, healthcare professionals, or other parties as needed.

Blockchain also provides patients with personalized health management tools and features such as reminders for appointments, medication adherence, wellness tips, and insights based on their health data analysis.

Blockchain is an emerging technology that offers safe transaction processing, open environment trust-building, and data storage. Put in place safeguards to guarantee the blockchain ledger's immutability, prohibiting illegal changes to data recorded and improving data integrity. It prevents unwanted access to sensitive healthcare data stored on the blockchain and implements access control and robust encryption methods. Blockchain uses methods such as homomorphism encryption and proofs of no knowledge to reduce the quantity of private information that is exposed on the blockchain while maintaining the ability to conduct safe transactions and share data. It assures that the blockchain network's transactions are transparent and auditable so that users may confirm the accuracy of the data and develop faith in the system. Sophisticated encryption methods are employed to protect private medical information kept on the blockchain. Blockchain utilizes the access control mechanisms to restrict unauthorized access to healthcare records and transactions and conduct regular audits of smart contracts to identify and mitigate potential security vulnerabilities before deployment. Blockchain uses privacy-enhancing technology to reduce the amount of sensitive healthcare data that is exposed on the blockchain, including homomorphism encryption and zero-knowledge proofs. It considers deploying permission or private blockchain networks to restrict access to sensitive data to authorized participants only. Blockchain implements reputation systems or trust scores to evaluate the reliability and trustworthiness of participants, facilitating informed decision-making. Blockchain ensures transparency and audibility of transactions on the blockchain network to build trust among participants. It explores scalability solutions such as sharding, sidechains, or off-chain protocols to increase the throughput and scalability of blockchain networks. Blockchain continuously optimizes consensus algorithms to reduce computational overhead and latency associated with transaction validation [12].

Blockchain technology has enormous potential for enhancing EHR sharing systems' effectiveness, security, and privacy. Patients who use EHRs have complete control over their records and can authorize or deny hospital access to them. IPFS, which offers the benefit of distribution and guarantees record immutability, is used to store records. The suggested system makes use of IPFS/multi-cloud and blockchain immutability to guarantee the availability, confidentiality, and integrity of the health record [13]. Blockchain technology can be efficiently used to update healthcare systems, improve data security and privacy, build participant confidence, and optimize performance for real-world applications by solving security, privacy, trust management, and performance optimization issues. Continued research, innovation, and collaboration are essential to overcome these challenges and unlock the full potential of blockchain in healthcare and other industries [14]. More significantly, the owners of the record are the patient processing signup who provide the unique address and public and private key combination. The doctor opens the file containing the hospital's public and private keys, validates every detail about the patient, and performs a fingerprint test. After that, the hospital has the option to review its current records or make new ones. To submit the record, the healthcare provider fills it out (deciphers the file containing the hospital's public and private keys). The hospital can then choose to inspect existing records or create new ones. The physician fills out the record details form and decrypts the file containing the public and private keys of the hospital to submit a record. A user creates eHealth records, diagnoses kits, and decrypts files including public or private keys for clinicians. They also refer to the patient's continued care. They also refer to the patient's continued care. For insurance claims and statistical analysis, the agent consults the eHealth records. The hospital can review the records they have access to by inputting the decryption password, which decodes the public-private key combination file stored in the browser's local storage. Ultimately, IPFS decrypts files containing public or private key combinations kept in the browser's local storage.

While most of the systems in use now cannot be interconnected with such a wide range of devices, the Internet of Things (IoT) has this potential. It is possible that the data that is kept will not be safeguarded. There are a lot of disconnects between the different actors, phases, and processes in complex systems. Although gaps are rarely the cause of accidents, an examination of the events leading up to an accident will usually uncover multiple gaps. Enhancing general safety involves being able to comprehend and reaffirm practitioners' typical capacity to close gaps. The smart contracts accept all processes and a user registration form to visit the doctor. The doctor handles the patient's address, and details checks the Aadhar card, and consults with other doctors to carry on the treatment process. After the doctor gives a prescription to the patient to buy medicine through the pharmacy. The agent uses IoT-based healthcare reports for statistical analysis and insurance claims [15].

2.3 Proposed model

Blockchain technology in the healthcare sector enables revolutionary situations, including secure patient data management, transparent medicine traceability, and efficient supply chain logistics. Blockchain technology improves data security, interoperability, and patient empowerment, which completely changes the way healthcare is delivered. Better patient outcomes, trust, and transparency are ensured by doing this (Figure 2.1).

1. Identification of use case: Determine which healthcare domains, including patient data management, medication traceability, supply chain management, or data exchange for medical research, could benefit greatly from the application of blockchain technology.
2. Requirement analysis: Perform a comprehensive investigation and collect requirements particular to the selected use case. Understanding the relevant parties, data flow, legal requirements, and security considerations is necessary for this.
3. Blockchain platform selection: Based on features such as consensus process, scalability, security, and interoperability, select a suitable blockchain platform [16,17]. Requirements for the use case may dictate the adoption of consortium, private, or public blockchains as options.

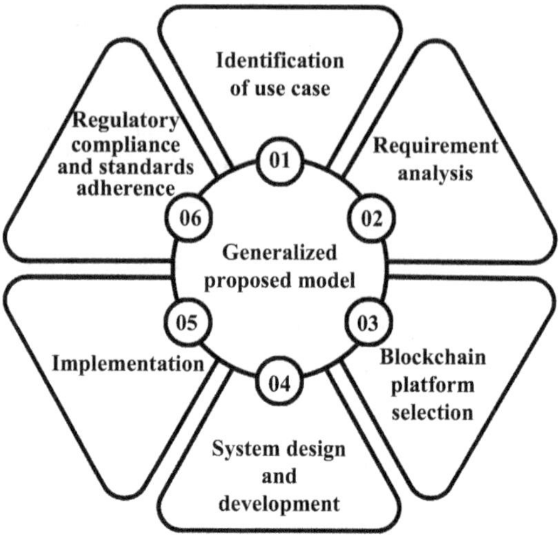

Figure 2.1 Diagrammatic representation of the proposed model

4. System design and development: Data structures, smart contracts, consensus techniques, and interaction with current healthcare IT systems should all be taken into account while designing the blockchain system architecture. Create smart contracts and the required data exchange protocols.
5. Implementation: Conduct a pilot project to evaluate the blockchain solution's viability and efficacy in an actual healthcare setting. This entails working with stakeholders, putting the solution into practice, and gathering input for future enhancements.
6. Regulatory compliance and standards adherence: Ensuring compliance with healthcare regulations such as the Health Insurance Portability and Accountability Act (HIPAA) and the General Data Protection Regulation (GDPR). Adhere to industry standards for data security, privacy, and interoperability to maintain credibility and legality.

The goal of this proposed model flowchart is to meet growing use cases in the healthcare business and achieve disruptive effects by providing a high-level overview of the stages needed in integrating blockchain technology.

In the following sections, this chapter illustrates how the proposed model can be implemented in various sections of society, discussing in detail about the methodology and the results that can be achieved using blockchain.

2.4 Clinical trial management using blockchain

Clinical trial management with blockchain technology offers a unique way to improve clinical trial transparency, integrity, and security. Clinical trial data, from patient consent to study findings, is securely stored and easily verifiable owing to the blockchain's decentralized and immutable ledger. This solution addresses common problems such as data tampering, privacy breaches, and inefficient data administration [18]. Smart contracts automate participant permission, data verification, and regulatory compliance, dramatically decreasing the risk of human mistakes and streamlining trial management.

2.4.1 Methodology

1. **Project initialization**: The clinical study's objectives are established from the start, including what kind of data will be collected, how long the experiment will last, and what expected results. Researchers, healthcare providers, patients, and regulatory authorities have all been onboarded. This foundational step is crucial for setting the trial's scope and governance, as well as ensuring that all participants are on the same page and understand their roles within the blockchain architecture.
2. **Blockchain setup**: Choosing the right type of blockchain (public, private, or consortium) is critical. A private or consortium blockchain is commonly used in clinical trials due to its mix of transparency and privacy, allowing only authorized participants' access. Smart contracts are then created and deployed on the blockchain. Without requiring human intervention, self-executing contracts automate and enforce trial rules such as data validation and consent management. The terms of the agreements are specifically codified in these contracts.
3. **Patient enrollment**: Blockchain technology is essential for managing consent and participant identity verification during patient enrollment. Every consent form and patient's identification are safely stored on the blockchain, providing an unchangeable record that improves compliance and confidence. This is an essential step in maintaining patient privacy and making sure that all data gathering complies with legal and ethical norms.
4. **Data collection and entry**: Encrypted patient data, medical device data, and researcher data are directly added to the blockchain. This guarantees that once recorded, data is not only safe but also unchangeable, avoiding any post-hoc changes that would jeopardize the integrity of the study. Smart contracts automatically validate incoming data against pre-established criteria, ensuring the completeness and quality of the data acquired.
5. **Data access and sharing**: Trial data can be accessed with controlled access because of blockchain technology. Permissioned access and cryptographic keys ensure that the data is only seen or interacted with by authorized stakeholders. Without jeopardizing patient confidentiality or data integrity, this safe environment facilitates real-time monitoring and transparency, enabling stakeholders to follow the course of the experiment and obtain relevant data.
6. **Reporting and analysis of data**: After the data collection process is over, the data is combined and anonymized, and conclusions regarding the clinical trial's results are made through analysis. The blockchain's immutable ledger provides a transparent and verifiable record of all data analysis, guaranteeing

that the results published are accurate and unaltered. The final reports benefit from the traceability of the blockchain, which creates an auditable record of the trial's findings and makes them eligible for publishing and regulatory submission.

7. **Audit and compliance**: Regulatory audits are greatly streamlined by blockchain's transparent and immutable nature. On the blockchain, auditors can independently confirm trial procedures and data, guaranteeing adherence to all moral and legal requirements. To further improve the trial's integrity and regulatory compliance, smart contracts can also automate some compliance checks.
8. **Post-trial activities**: Blockchain continues to store trial data securely after it has been concluded, making it easier for long-term research or future reference. The blockchain is used to handle ongoing patient involvement, enabling safe and effective follow-ups and outcome reporting (Figure 2.2).

With its decentralized ledger, blockchain technology powers every stage of the clinical trial process. Once data is submitted, it remains unchangeable and

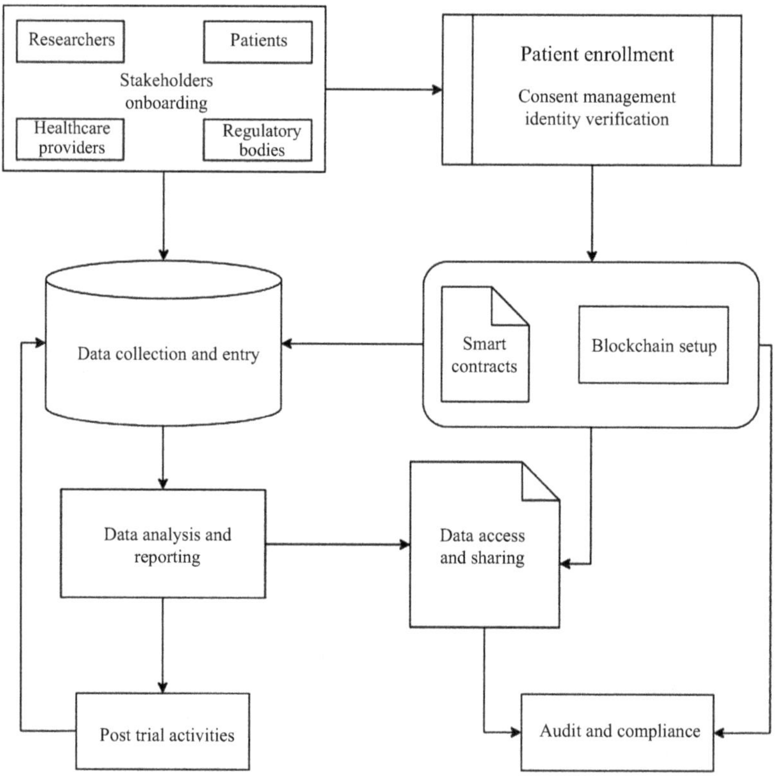

Figure 2.2 Flowchart representing the workflow of blockchain in clinical trial management

time-stamped, offering an unquestionable record of all actions and data entries [19]. Cryptographic signatures and encryption guarantee the confidentiality and validity of data, and smart contracts automate and enforce trial standards to minimize mistakes and inconsistencies. This all-encompassing use of blockchain technology not only improves the effectiveness and integrity of clinical trials but also fosters confidence among all parties involved by offering a transparent, safe, and unchangeable record of the whole trial procedure.

2.4.2 Result

Clinical trial administration is transformed by the application of blockchain technology, which enhances data security, quality, and transparency through the study. Blockchain builds confidence among stakeholders, including academics, participants, and regulatory organizations, by ensuring that data that has been recorded is immutable and verifiable. Smart contracts automate critical processes, reducing errors and streamlining operations, while safe and selective data sharing protects patient privacy and meets regulatory criteria. Finally, these developments make clinical trials more efficient, trustworthy, and ethical, potentially quickening the discovery and approval of new medical treatments and cures. The transparency and accountability inherent in the blockchain-supported method also pave the path for increased public trust in clinical research, creating a climate conducive to innovation based on data integrity and participant safety.

2.5 Detecting fake drugs and managing the supply chain

The pharmaceutical industry plays a critical role in global healthcare, providing life-saving medications and treatments to millions of people worldwide. However, amid its noble pursuits, the industry faces a persistent challenge: the proliferation of counterfeit drugs. Counterfeit medications not only endanger patient health but also undermine the integrity of the entire supply chain. The pharmaceutical industry's current supply chain is antiquated and has unclear visibility throughout the whole system. Furthermore, there has been a rise in the market's circulation of fake medications. The WHO research states that around 10.5% of pharmaceuticals in lower- and middle-income nations are counterfeit and that these drugs can be fatal or seriously dangerous to the public's health.

Detecting and combating fake drugs require a multifaceted approach that addresses vulnerabilities throughout the whole supply chain, from production to delivery and everything in between. Traditional methods of authentication often fall short in the face of increasingly sophisticated counterfeiters. To overcome these challenges, innovative solutions leveraging blockchain technology are emerging as a promising means to enhance drug traceability, transparency, and security [20]. By integrating blockchain into the pharmaceutical supply chain, stakeholders can establish an immutable and transparent ledger of drug transactions, ensuring that each product's journey from manufacturer to end-user is verifiable and tamper-proof. This revolutionizes the way we manage the supply chain, offering real-time

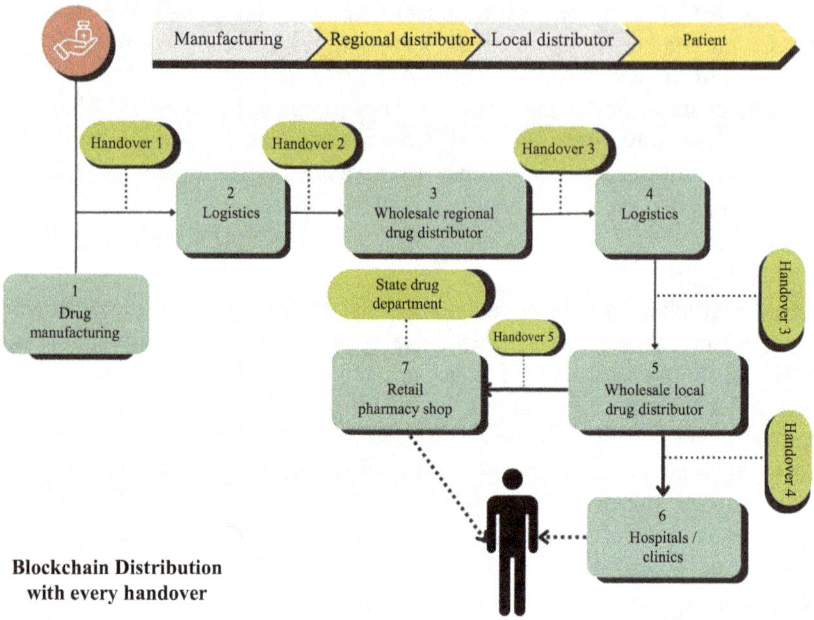

Figure 2.3 Workflow model of detecting fake drugs and managing the supply chain

visibility into the movement of medications and enabling swift detection of any anomalies or counterfeit products.

In this context, Figure 2.3 presents a comprehensive overview of how blockchain technology is leveraged to detect fake drugs and manage the pharmaceutical supply chain effectively. It illustrates the key components and processes involved in ensuring the authenticity, safety, and integrity of medications, ultimately safeguarding public health and bolstering trust in the pharmaceutical industry.

2.5.1 Methodology

1. Manufacturing phase: Pharmaceutical companies produce drugs. Each drug batch is assigned to a unique identifier or serial number.
2. Blockchain integration: Each step of the supply chain is recorded on a blockchain ledger. Information includes manufacturer details, batch numbers, and production dates.
3. Distribution phase: Drugs move through various stages: wholesalers, distributors, and pharmacies. Each transfer of ownership or location change is recorded on the blockchain.
4. Authentication and verification: By entering a serial number or scanning a QR code, pharmacies or end-users can confirm the legitimacy of drugs. By accessing the blockchain, they can see the entire journey of the drug, ensuring it is not counterfeit.

5. Smart contracts: Smart contracts automate certain processes such as payment upon delivery confirmation. They can also trigger alerts for any discrepancies or irregularities in the supply chain.
6. Data integrity: Blockchain ensures data immutability, preventing tampering with records. Timestamps and cryptographic hashes verify the authenticity of information.
7. Traceability and transparency: A clear and auditable record of the drug's travel is available to all parties involved. This openness makes it easier to track out the source of problems or fake medications.
8. Regulatory compliance: Regulatory bodies can access the blockchain for compliance checks and audits. It ensures adherence to quality standards and regulations throughout the supply chain.
9. Feedback loop: End-users can provide feedback or report any issues, which can be recorded on the blockchain. This feedback loop helps in continuous improvement and problem-solving.
10. Continuous monitoring and updates: The medication is constantly being updated on the blockchain ledger as it travels through the supply chain. Because changes and updates are recorded in real time, all parties involved have access to the most recent data.

2.5.2 Result

The use of blockchain-based technology in drug supply chain management and fraudulent medication identification produces observable benefits such as increased consumer safety, regulatory compliance, greater transparency, and efficient anomaly detection. By leveraging the inherent properties of blockchain technology – transparency, immutability, and decentralization – all relevant stakeholders may collaborate to combat the issue of counterfeit medications, protecting the integrity of the pharmaceutical supply chain and promoting public health.

2.6 Pharmaceutical medicine supply chain

Supply chain management is the process of planning, implementing, and supervising supply chain operations to enhance customer satisfaction through efficiency. Transportation and storage of completed items, inventories for work-in-progress, and raw materials from the point of origin to the location of consumption are all included in supply chain management. The supervision of materials, data, and finances as they flow from supplier to manufacturer to wholesaler to retailer to customer is known as supply chain management (SCM) [21]. SCM is the purposeful and intentional management, integration, and control of corporate functions. SCM influences and adds to the business's supply chain to improve efficiency, lower costs, increase flexibility, and other factors that ultimately benefit customers. The supply chain function is composed of numerous sub-areas, including purchasing and procurement, operations, logistics, transportation, warehousing, distribution, customer support, and inventory management. It is difficult to find a conventional supply chain oversight model in the business world, particularly in the pharmaceutical sector.

The integration and balancing of various streams inside and between enterprises is known as supply chain management. Managing internal supply chains is a cross-sectional business function, and managing supply chains on behalf of clients is a vertical industry sector. A business may function as a provider of supply chain services inside the vertical sector. Nonetheless, a company's supply chain employees work horizontally throughout their companies for every client they service. For every organization, forecasting and planning are essential to predicting the number of resources and materials needed to deliver goods and services to clients on schedule. SCM includes tasks such as capacity planning, inventory control, demand forecasting, and more in this domain [22]. Purchases or procurements are crucial to the supply chain's commercial domain. This entails finding vendors to provide the required goods and services, haggling over prices and conditions, and creating contracts. Consequently, this domain is used to manage supplier performance and ongoing contractual agreements. This field is also known as buying, sourcing, purchasing, or procurement. Purchasing, when strictly defined, refers to the flow of materials or things, whether they are going out, through, or arriving.

In certain manufacturing organizations, the logistics department may handle planning and forecasting, whereas in others, logistics is just responsible for the movement and transportation of supplies and goods. A wide range of management tasks are included in operations to efficiently allocate resources to meet customer obligations. Typically, inventory management entails controlling storage and issuing procedures, keeping stock levels stable, and replenishing physical inventory. Stock may include completed goods that are waiting to be sold or shipped, work-in-progress, or materials obtained from suppliers. Managing a fleet of company-owned cars, arranging for the pickup, delivery, or exchange of materials and goods, or supervising outside transportation providers are all included in the field of transport management. Managing company-owned warehouse space or space supplied by outside sources is what is meant by warehousing. Physically delivering the company's goods to clients or sub-distributors is referred to as distribution.

Customer service is an important component of supply chain management, although being frequently disregarded. It entails making sure that clients' expectations are satisfied and doing what is required to keep promises and perform duties. The effectiveness of a company's supply chain management system is critical to its success. Many businesses credit their present success to having efficient supply chain management systems that increase satisfaction among both customers and suppliers. An organization can convey its requirements to suppliers and marketers clearly and concisely by using SCM. Since pharmaceutical items are unique, pharmaceutical supply chain management is very important.

2.6.1 Methodology

1. Planning: To accomplish desired results, planning entails examining the present situation, assessing requirements, setting goals, and defining quantifiable

targets in addition to creating strategies, allocating duties, and obtaining required resources.

2. Testing: Coordinating the steps required to get raw materials, produce goods, carry out quality checks, package goods for transportation, and set up delivery schedules are all included in testing.
3. Supplier: Suppliers offer commodity products in the chemical industry that can be obtained from a variety of sources. These raw materials are essential for the development of drugs.
4. Manufacturing: The large-scale manufacturing – referred to as manufacturing in this context – of pharmaceutical drugs is one of the most significant parts of the pharmaceutical industry.
5. Storehouse: Warehouses play a crucial role in the healthcare supply chain by acting as short-term storage facilities and enabling the organized transportation of goods to medical facilities such as pharmacies and hospitals.
6. Distribution: By maximizing resource use, distribution management seeks to guarantee a steady supply of supplies and drugs to the facilities where they are needed.
7. Hospital: Hospitals depend on a smooth supply chain to ensure that drugs are available when needed, improving patient care and reducing errors in all areas.
8. Patient: A well-planned supply chain assures the timely availability of the right pharmaceuticals to suit specific healthcare demands, which benefits patients (Figure 2.4).

2.6.2 Result

Increasing the number of reliable suppliers while lowering the risk of interruptions or inefficiencies are two ways to grow the network of suppliers and lower the risk in healthcare supply chains. This growth improves the adaptability and durability of locating essential medical equipment and supplies. Improving supply chain transparency is essential to building stakeholder trust and guaranteeing the integrity and authenticity of goods. Healthcare businesses can track the path of medical products from manufacturing facilities to end-consumers by utilizing blockchain technology or other technologies. This allows for real-time monitoring and verification of each transaction or transfer.

To find weaknesses and take preventive measures against possible supply chain threats like natural catastrophes, unstable geopolitical conditions, or cybersecurity breaches, risk assessments are crucial. Predictive risk modeling, inventory management, and forecasting accuracy can all be enhanced with proprietary technologies, encompassing predictive analytics, the IoT, and AI. By enabling healthcare companies to foresee and minimize possible disruptions, these technologies guarantee a steady and dependable supply of vital medicinal supplies. Overall, healthcare supply chains may become more resilient, responsive, and effective in meeting the changing demands of patients and healthcare providers by growing supplier networks, enhancing transparency, conducting risk assessments, and utilizing cutting-edge technology [23,24].

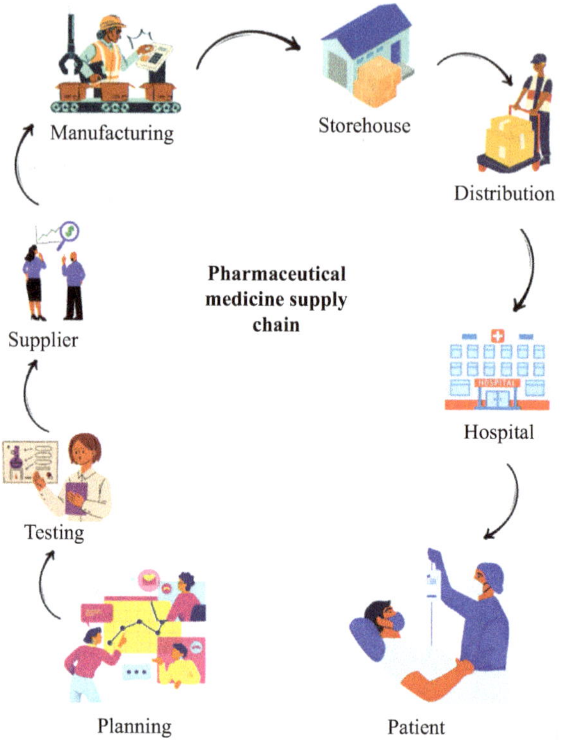

Figure 2.4 Diagrammatic representation of the medicine supply chain

2.7 Future research direction

To further progress the field of study, future research initiatives in blockchain technology for healthcare could concentrate on several important areas. Research on ways to make blockchain networks in healthcare settings more scalable and performant so they can manage the growing amount of data and transactions without sacrificing security or efficiency. Interoperability: Study ways to make existing healthcare systems and other blockchain platforms more interoperable so that data may be seamlessly exchanged and integrated across various healthcare networks and apps. To ensure patient confidentiality and safeguard against unauthorized access or data breaches, develop stronger security measures and improved privacy-preserving procedures specifically adapted to the particular requirements of healthcare data. Examine how regulatory frameworks may affect the application of blockchain technology in the healthcare industry and suggest ways to maintain compliance with pertinent laws and rules, like HIPAA and GDPR while taking advantage of blockchain's advantages. Patient empowerment: Investigate methods to enable patients to take charge of their health data by providing them with blockchain-based solutions that enable secure sharing, manage consent, and provide incentives for patients to participate in data-sharing programs. Automation and smart contracts: To improve efficiency and cut down on administrative burden, look at the possible uses of automation and smart contracts in healthcare procedures, including managing clinical trials, processing insurance claims, and organizing the supply chain for medical supplies. Measure results such as cost savings, efficiency gains, and improvements in patient outcomes. Conduct longitudinal studies and real-world implementation trials to assess the impact, effectiveness, and usability of blockchain solutions in various healthcare settings. Examine how blockchain technology may affect data ownership, consent, equity, and the potential to worsen already-existing inequities in healthcare outcomes and access. One should also look at the ethical and legal implications of blockchain technology in the healthcare industry. The healthcare sector may continue to leverage blockchain technology's revolutionary potential to improve patient care, strengthen data security and privacy, and streamline healthcare delivery procedures by concentrating on these research areas.

2.8 Conclusion

Blockchain-based technology can revolutionize the healthcare industry by providing solutions for critical problems related to data management, supply chain transparency, and patient empowerment. Blockchain provides special solutions to improve data quality, security, and interoperability in healthcare settings through its decentralized, immutable, and transparent ledger system. Healthcare companies may lower the risk of supply chain disruptions and increase the size of their supplier network by utilizing blockchain technology. A more transparent supply chain guarantees authenticity and safety by allowing stakeholders to track the path of medicinal supplies. Supply chains are further strengthened by using cutting-edge technologies and conducting risk assessments, which increase their resilience and ability to adapt to changing demands. Studies of actual deployment are also required to assess the impact and efficacy of blockchain solutions in various healthcare environments. Carefully weighing the ethical, legal, and social ramifications is necessary to reduce potential hazards and guarantee that healthcare services offered by blockchain are accessible to all. In summary, despite several challenges and uncertainties, blockchain technology has undeniable potential advantages for the healthcare sector. Blockchain has the potential to revolutionize the healthcare sector and advance patient-centered care, efficiency, and transparency for years to come with further study, cooperation, and innovation.

References

[1] Tandon, A., Dhir, A., Islam, A.N., and Mäntymäki, M. Blockchain in healthcare: A systematic literature review, synthesizing framework and future research agenda. *Computers in Industry*, 122, p. 103290, 2020.

[2] Verma, R., and Dhanda, N. Blockchain types: a characteristic view. In *Distributed Computing to Blockchain* (pp. 69–85). Academic Press; 2023.

[3] Sharma, P., Jindal, R., and Borah, M.D. Healthify: a blockchain-based distributed application for health care. In *Applications of Blockchain in Healthcare* (pp. 171–198). Singapore: Springer Nature; 2020.

[4] Benchoufi, M., and Ravaud, P. Blockchain technology for improving clinical research quality. *Trials*, 18(1), pp. 1–5, 2017.

[5] Iansiti, M., and Lakhani, K.R. The truth about blockchain. *Harvard Business Review*, 95(1), pp. 118–127, 2017.

[6] Kumari, M., Gupta, M., and Ved, C. Blockchain in pharmaceutical sector. In *Applications of Blockchain in Healthcare* (pp. 199–220). Singapore: Springer Nature; 2020.

[7] Srivastava, S., Singh, S.V., Singh, R.B., and Shukla, H.K. Digital transformation of healthcare: A blockchain study. *International Journal of Innovative Science, Engineering & Technology*, 8(5), 2021.

[8] Charles, W.M. Accelerating life sciences research with blockchain. In *Applications of Blockchain in Healthcare* (pp. 221–252). Singapore: Springer Nature; 2020.

[9] Gupta, M., Jain, R., Kumari, M., and Narula, G. Securing healthcare data by using blockchain. In *Applications of Blockchain in Healthcare* (pp. 93–114). Singapore: Springer Nature; 2020.

[10] Ciampi, M., Esposito, A., Marangio, F., Sicuranza, M., and Schmid, G. Modernizing healthcare by using blockchain. In *Applications of Blockchain in Healthcare* (pp. 29–67). Singapore: Springer Nature; 2020.

[11] Narikimilli, N.R.S., Kumar, A., Antu, A.D., and Xie, B. Blockchain applications in healthcare–a review and future perspective. In *International Conference on Blockchain* (pp. 198–218). Cham: Springer International Publishing; 2020.

[12] Swarnkar, M., Bhadoria, R.S., and Sharma, N. Security, privacy, trust management and performance optimization of blockchain technology. In *Applications of Blockchain in Healthcare* (pp. 69–92). Singapore: Springer Nature; 2020.

[13] Spano, R., Massaro, M., and Iacuzzi, S. Blockchain for value creation in the healthcare sector. *Technovation*, 120, p. 102440, 2023.

[14] Shuaib, K., Abdella, J., Sallabi, F., and Serhani, M.A. Secure decentralized electronic health records sharing system based on blockchains. *Journal of King Saud University-Computer and Information Sciences*, 34(8), pp. 5045–5058, 2022.

[15] Vahdati, M., Gholizadeh HamlAbadi, K., and Saghiri, A.M. IoT-Based healthcare monitoring using blockchain. In *Applications of Blockchain in Healthcare* (pp. 141–170). Singapore: Springer Nature; 2020.

[16] Yadav, A.K., Singh, K., Amin, A.H., Almutairi, L., Alsenani, T.R., and Ahmadian, A. A comparative study on consensus mechanism with security threats and future scopes: Blockchain. *Computer Communications*, 201, 102–115, 2023.

[17] Praveen, G., Anand, M., Singh, P.K., and Ranjan, P. An overview of blockchain consensus and vulnerability. *Information and Communication Technology for Intelligent Systems: Proceedings of ICTIS 2020*, 1, 459–468, 2021.

[18] Curbera, F., Dias, D.M., Simonyan, V., Yoon, W.A., and Casella, A. Blockchain: An enabler for healthcare and life sciences transformation. *IBM Journal of Research and Development*, 63(2/3), pp. 8:1–8:9 2019.

[19] Maslove, D.M., Klein, J., Brohman, K., and Martin, P. Using blockchain technology to manage clinical trials data: a proof-of-concept study. *JMIR Medical Informatics*, 6(4), p. e11949, 2018.

[20] Dutta, P., Choi, T.M., Somani, S., and Butala, R. Blockchain technology in supply chain operations: Applications, challenges and research opportunities. *Transportation Research Part E: Logistics and Transportation Review*, 142, p. 102067, 2020.

[21] Ghadge, A., Bourlakis, M., Kamble, S., and Seuring, S. Blockchain implementation in pharmaceutical supply chains: A review and conceptual framework. *International Journal of Production Research*, 61(19), pp. 6633–6651, 2023.

[22] Abbas, K., Afaq, M., Ahmed Khan, T., and Song, W.C. A blockchain and machine learning-based drug supply chain management and recommendation system for smart pharmaceutical industry. *Electronics*, 9(5), p. 852, 2020.

[23] Kumar, M., Raj, H., Chaurasia, N., and Gill, S.S. Blockchain inspired secure and reliable data exchange architecture for cyber-physical healthcare system 4.0. *Internet of Things and Cyber-Physical Systems*, 3, pp. 309–322, 2023.

[24] Shah, R., and Chircu, A. IoT and AI in healthcare: A systematic literature review. *Issues in Information Systems*, 19(3), 2018.

Chapter 3

Blockchain-enabled patient identity management

Shreea Bose[1], Snehashis Kayal[2], Reek Roy[3], Susanta Ghosh[4], Sayandeep Rana[4], Dipendu Ray[4], Arghadip Jana[4] and Himadri Nath Saha[5]

Abstract

In the continuously changing world of healthcare, protecting patient data is critical. This chapter presents a novel strategy to address the essential issue of patient identity management by integrating Python and Solidity computer languages into blockchain technology. The suggested approach uses blockchain technology's transparent, immutable, and decentralized characteristics to improve patient-identifying data security and integrity. Python is used to build the backend infrastructure because of its flexibility and efficiency. In contrast, Solidity, a language created for Ethereum blockchain smart contracts, is used to create a safe and unchangeable identity verification procedure. The synergy between these languages enables the creation of a robust, decentralized system that facilitates the seamless sharing and updating of patient information across healthcare providers while maintaining privacy and compliance with regulatory standards. Smart contracts are used by the system to manage patient identities, guaranteeing that only authorized parties can view and modify relevant data. The blockchain-enabled patient identification management system improves interoperability across healthcare providers, lowering the risk of identity fraud and increasing overall healthcare efficiency. This suggested framework addresses persistent issues with patient identification management and contributes to the growing corpus of knowledge regarding blockchain applications in healthcare. It offers a framework that is flexible and scalable which can be included in current healthcare systems to promote security, openness, and confidence in the handling of patient IDs. As the healthcare industry continues its digital revolution, this innovative approach has the

[1]Department of Computer Science, St. Xavier's College, Calcutta University, India
[2]Department of Information Technology, Techno International New Town, Maulana Abul Kalam Azad University of Technology, India
[3]Department of Computer Science, Belda College, Vidyasagar University, India
[4]Department of Computer Science and Engineering, The Neotia University, India
[5]Department of Computer Science, Surendranath Evening College, Calcutta University, India

potential to fundamentally alter patient data management, which would ultimately enhance patient outcomes and fortify the healthcare ecosystem.

Keywords: Smart healthcare; blockchain; cloud server; Internet of Things (IoT); health monitoring; pervasive healthcare; smart contracts

3.1 Introduction

In the modern digital age, the healthcare industry is undergoing an enormous shift, leveraging advanced technologies to ensure data security, streamline processes, and improve patient care. Blockchain is a technology that can change the game in patient identification management, better than any other. Blockchain, a decentralized and unchangeable ledger system, transforms the way healthcare organizations handle sensitive data by offering a transparent and safe platform for storing patient identities [1]. The issues associated with traditional patient identification management solutions include identity theft, data breaches, and disjointed healthcare systems. Since patient data is frequently fragmented across several systems, it is susceptible to manipulation and unwanted access. Moreover, the sharing of patient data is made more difficult by the incompatibility of various healthcare providers, which results in inefficiencies and lower-quality care.

Blockchain technology provides a decentralized, unbreakable patient identification management solution to get around these challenges. Blockchain uses cryptography and distributed consensus mechanisms to maintain the confidentiality and integrity of medical data, allowing individuals greater control regarding their medical information [2]. The creation of a single source of truth for patient data is one of the key elements of blockchain-enabled patient identification management. A decentralized network of nodes securely stores and updates patient IDs and medical data in real time, eliminating inefficiencies and ensuring data consistency for all parties involved. Healthcare providers can more easily share information thanks to this coordinated approach, which enhances medical results and care coordination. By encrypting patient data and distributing it among several nodes, blockchain improves data security by practically eliminating the ability for malevolent actors to alter or steal private information. Every transaction performed on a distributed ledger is by encryption linked to each other, creating an immutable verification record that anybody can view and review. Instilling confidence in the data's integrity through openness and traceability reduces the likelihood of data breaches and guarantees adherence to the Health Insurance Portability and Accountability Act (HIPAA), among other regulatory requirements.

Likewise, blockchain's self-sovereign identity management gives patients greater authority over their health data. Blockchain-based digital identities enable patients to securely access and share their medical records with researchers, healthcare providers, and other authorized parties, eliminating the need for middlemen and enhancing data privacy. Additionally, patients can choose which information they give while staying anonymous, protecting their privacy and self-determination.

Looking up, patient identity management enabled by blockchain signifies a revolutionary change in the healthcare industry, presenting hitherto unseen chances to improve data security, interoperability, and patient empowerment. By using the inherent properties of blockchain technology, medical institutions may create a transparent and secure environment for patient identification management, encouraging collaboration, creativity, and trust in hospital services.

3.1.1 Patient identity management

Patient identity management is the process of precisely and securely identifying individuals in the healthcare system and managing their personal and medical information during their contacts with healthcare providers. It entails gathering, storing, and retrieving patient information, including demographics, medical histories, treatment plans, and billing information. Effective patient identity management is critical for assuring high-quality care, improving care coordination, and preserving patient privacy and confidentiality. Patient identification management in conventional healthcare systems has frequently been fragmented, and prone to errors and security breaches [3]. Patients may have many medical records spread among various healthcare providers, resulting in duplication of effort, inefficiencies, and gaps in service. Added to that, antiquated identification systems, such as paper-based records, and reliance on personal identifiers like social security numbers, raise the danger of identity theft and fraud.

With the development of digital technology, patient identity management has expanded to incorporate electronic health records (EHRs), patient portals, and identity verification systems. These solutions are designed to improve patient involvement, expedite administrative tasks, and boost data accuracy by giving individuals the ability to access their health information and take an active role in their care. Interoperability difficulties, disparities in information, and cybersecurity concerns continue to exist, emphasizing the need for novel solutions to solve these complexities. Blockchain technology has shown promise as a substitute for patient identity management. It provides a safe, decentralized environment for transferring and storing patient data across healthcare networks. By providing a single source of truth for patient data and enabling safe data interchange, blockchain has the potential to completely alter patient identity management. This will improve care coordination, strengthen data security, and empower patients more.

3.1.2 Blockchain ledger

A distributed, decentralized database called a blockchain ledger keeps track of transactions over a network of computers in a way that guarantees security, immutability, and transparency. The term "blockchain" refers to the method by which all transactions are arranged into blocks and connected in a sequential chain. Each block is named after a cryptographic hash of the block before it. Because of this structure, any attempt to change a transaction would have to change every block that came after it, which makes it extremely impractical and computationally impossible.

A transaction cannot be changed or removed once it is registered on the blockchain. This feature safeguards patient data confidentiality by thwarting illicit modifications or manipulations. Medical records and patient identities saved on the blockchain are unchangeable, giving patients and healthcare professionals access to a trustworthy source of truth. All network users who have been granted permission can view and access blockchain ledgers. By allowing stakeholders to see the provenance and history of patient data, this transparency promotes confidence. Both patients and healthcare professionals may verify the correctness of their medical records and track the transfer of patient data between different systems and organizations. To safeguard sensitive patient data and ensure the security of transactions, blockchain uses cryptographic algorithms. Since each transaction is verified by network consensus and cryptographically hashed, it is nearly difficult for unauthorized parties to access or change patient data without the necessary authority. By doing this, data security and confidentiality are enhanced while the risk of fraudulent transactions and financial breaches is decreased. Since blockchain functions as a decentralized network of nodes, managing patient data no longer requires a central authority. This decentralized architecture improves resilience against cyberattacks and system breakdowns while lowering the possibility of single points of failure. Additionally, it encourages patient empowerment and data sovereignty by granting people more control over their medical records.

In the healthcare industry, blockchain ledger technology provides a stable and safe platform for managing patient identities and medical records. It is ideally equipped to handle the intricate problems related to patient identity management because of its immutability, transparency, security, and decentralization, which will ultimately improve data integrity, interoperability, and patient-centric care delivery.

Blockchain technology has numerous advantages for patient identification management in healthcare. For starters, the blockchain provides an independent, opaque framework for safely preserving patient identities and health records [4]. By distributing patient data over a network of nodes and encrypting it using cryptographic techniques, blockchain lowers the danger of data breaches and illegal access while maintaining the reliability and safety of the data. Subsequently, blockchain improves interoperability by establishing a single and standardized framework for exchanging patient data among various healthcare providers and systems. Blockchain allows for easy data transmission while preserving data integrity and consistency via smart contracts and consensus processes. Patients have more control over their health data due to identity management on the blockchain. By doing away with brokers and enhancing data protection, patients utilizing blockchain-based digital identities can safely browse and discuss their medical records with physicians, researchers, and other approved parties. In general, blockchain has enormous potential to transform patient identity management in healthcare, resulting in better care coordination, improved data security, and increased patient autonomy.

In Section 3.2, a review of the literature on several systems now in use is presented, along with an analysis of their benefits and drawbacks. The system design of the suggested model has been further upon in Section 3.3. The methodology of the suggested model, used in a variety of cases, is presented in Sections 3.4–3.6. The performance analysis and the suggested model's future research path are described in Sections 3.7 and 3.8. This book chapter concludes briefly in Section 3.9.

3.2 Background and related works

An inventive identity management (IdM) solution designed for remote healthcare has been presented by Javed *et al.* [5], which makes use of a cooperative Ethereum blockchain under regulatory oversight. Individual health IDs allow patients and healthcare providers to be identified from one another, allowing for more efficient access to healthcare services. Whether they are consumers looking for medical attention or medical experts providing specialized care, users in the remote healthcare IdM framework take on the role of identity owners. Organizations like the Department of Health, the Nursing Council, and the Pharmaceutical Council, which are in charge of overseeing provider registration and licensure, are given regulatory authority over healthcare providers.

The consortium approach used by the selected blockchain infrastructure as proposed by Mohammed *et al.* [6] supports Ethereum and Hyperledger, guaranteeing a stable and decentralized identity management system designed with the healthcare industry in mind. The Ethereum blockchain's Health SC and Registry SC smart contracts, in particular, are crucial for controlling identity-related activities. On the Ethereum blockchain, the health ID functions as the deployed smart contract's unique address. To create a JSON Web Token (JWT), the unique traits of individuals and medical professionals are formatted in JSON, saved in cloud-based storage, and verified by regulatory bodies. Identity owners can choose between distributed and centralized cloud storage for attribute management, and the identity attribute hash-enabled targeted attribute downloads and retrievals. Healthcare providers generate their digital health IDs by deploying a smart contract on the blockchain platform Ethereum using their externally owned account. Providers receive an encrypted challenge message from the regulatory authority after sending a registration request with their health ID. They successfully decrypt the message using their private key, proving ownership. In compliance with regulatory standards, providers validate their identification by displaying a current practice license in person or virtually. Following a successful verification process, the regulator issues a signed identification token (JWT) to the provider and registers the health ID and public key in the registry smart contract. Patients receive health IDs from providers; they then use the registry smart contracts, verify attestation function to confirm the information, guarantee registration with a particular regulatory agency. By acquiring hashes and utilizing retrieve_hash to obtain matching hashes from the provider's health smart contract, patients authenticate providers. Blockchain records every authentication occurrence, and providers safely exchange a symmetric key for the decryption of identity tokens. Using the Registry smart contracts verify_regulator method to get the regulator's public key, patients verify regulator signatures to confirm registration. By verifying regulator attestation, this validation procedure validates the identity token's legitimacy. With the use of smart contracts and a collaborative Ethereum blockchain, the remote healthcare IdM system enables users with distinct health IDs to easily manage their identities. It is supervised by healthcare authorities. This strong structure guarantees safe registration, deployment, and verification procedures, and regulatory attestation gives the system legitimacy. Blockchain-recorded procedures simplify communication between patients and providers, and the inclusion of cloud storage enhances attribute management capabilities. The straightforward architectural layout emphasizes user liberty and creates a safe, secure environment for the provision of remote healthcare services.

Harrell *et al.* [7] proposed MediLinker, which is a virtual wallet, that includes the following six forms of verified credentials: credit card, health ID, insurance, prescription drugs, consent for research, and MPOA. These credentials, informed by clinical necessities, offer essential patient information for encounters. While the insurance credential holds information about health insurance, the health ID credential contains demographic data based on an identification issued by the government. The COVID-19 vaccination status, medication, and prescription data are all covered by the drug credential. Patient billing information is stored on financial institution credentials, and study participation details are recorded on investigation authorization credentials. Supervisors and their dependents' information is included in the MPOA credential. MediLinker empowers users to receive and store verifiable credentials in their digital wallet, issued by various entities known as "Issuers" and held by the individual as the "Holder." Except for research consent, institutional representatives issue credentials after patient review. The workflow involves securely digitizing government-issued or third-party credentials through blockchain. Patients establish trust by scanning a clinic's QR Code, entering medical data, and verifying identity with physical cards. An iOS app streamlines this process, ensuring accuracy and incorporating a revocation feature. Then, approved credentials can be electronically distributed to collaborating institutions, doing away with the requirement for paper records. MediLinker is built on the Hyperledger Indy blockchain framework, chosen for its decentralized identity features, providing patients with maximum data protection and control. Supporting the idea of decentralized trusted identification, Hyperledger Indy makes it easier to store record structures and identity requirements for provider–patient connections. According to World Wide Web Consortium specifications, patients can confirm their identification without depending on an authoritative registry or identification provider by using decentralized identifiers. Furthermore, Hyperledger Aries functions as a gateway layer, facilitating the development, transfer, and archiving of verifiable digital credentials and establishing a connection between Hyperledger Indy and the customer end. To make MediLinker easier to use on Android and iOS devices, mobile applications were developed.

Using the React Native framework, Sharma *et al.* [8] developed native applications with a shared code base, ensuring consistency across platforms. By providing patients with more engaging smartphone experience, this method improves patient contact. User interface elements were implemented using the Material-UI framework, guaranteeing a modern and responsive design. MediLinker is designed to be compatible with any public cloud, but it was opted for by AWS due to its convenience and the University's existing contract. Hyperledger Aries agents and Hyperledger Indy servers were hosted via AWS, a HIPAA-compliant cloud service as mentioned by Abdul-Moheeth *et al.* [9]. For flexibility, each patient's agent needs a virtual machine (VM). Docker containers on a single VM are advised for larger patient volumes, even though they work well in our case. While the mobile app can be offered through stores like the Apple Store and Google Play Store for clinical adoption, the web application does not need to be installed. MediLinker is a cutting-edge blockchain solution leveraging Hyperledger Indy and Hyperledger Aries for patient identity management. Through a user-friendly web and mobile app, patients can securely log in, verify credentials, and share information across clinics using blockchain wallets. The architecture empowers patients with control over their identity data, ensuring privacy and personalized health data management. The proof-of-concept design showcases a scalable and operational patient-centric identity system for future integration in healthcare settings.

Blockchain technology combines with healthcare 4.0, which is characterized by digitization and data-driven decision-making, to transform patient identity management. Blockchain enhances patient privacy and confidence by ensuring safe, transparent, and interoperable health data sharing. Tanwar *et al.* [10] concluded that this combination should lead the healthcare industry into a new era of effectiveness, responsibility, and individualized treatment.

The transformational potential of blockchain technology in healthcare is examined in the study by Ouaguid *et al.* [11], with a focus on how it can manage EHR and handle various system constraints. It assesses widely used blockchain-based methods for managing healthcare data, including its benefits and drawbacks. Important obstacles are noted, including inadequate stakeholder involvement, a lack of regulatory supervision, and regulatory flexibility.

A patient's digital medical history, including diagnoses, prescriptions, test results, and treatment plans, is called an EHR. For more coordinated treatment, it enables safe access to and sharing of patient data by healthcare professionals. This architecture of the e-healthcare ecosystem addresses the difficulties in creating a synchronized and safe environment for exchanging EHR data while taking ethical, technical, and legal considerations into account. It emphasizes the significance of patient sovereignty over their data, compliance with laws like GDPR and HIPAA, and transparent data handling. Various EHR management systems are explored, including cloud-based solutions and those leveraging blockchain technology to address reliability, security, and interoperability concerns. The paper discusses four approaches to EHR management integrating blockchain technology, with a focus on their architecture and specificities compared to other methods. The first approach proposed by Alam *et al.* [12] combines IoT and blockchain for real-time patient health data collection. Patient sensors capture data such as glucose levels and blood pressure, which are stored in a distributed manner. Overall, it explores the structural and architectural considerations for blockchain-based e-health systems, highlighting their impact on regulatory compliance, security, and interoperability. It evaluates various blockchain approaches in healthcare, emphasizing the importance of key functionalities and collaboration among stakeholders for effective implementation. Challenges such as inadequate entity representation and regulatory compliance are identified, underscoring the need for close collaboration and regulatory flexibility to ensure global compliance and data interoperability.

Blockchain offers a secure and decentralized solution for record-keeping, ensuring data immutability and transparency. The pairing of blockchain based on artificial intelligence healthcare organizations is investigated in the study conducted by Kshetri *et al.* [13] to improve security, efficacy, and safety. It proposes a novel AI-based healthcare blockchain model, healthAIChain, aimed at improving patient data security. AI holds a significant promise for revolutionizing healthcare by improving various aspects of care, including diagnosis accuracy, treatment planning, and personalized care. It can streamline processes, reduce costs, and handle vast amounts of healthcare data efficiently. However, ethical and governance considerations are essential to ensure their responsible use and address concerns regarding algorithmic biases and ethical dilemmas [14]. Overall, while AI offers immense potential for healthcare advancement, careful oversight is necessary to maximize its benefits and mitigate risks. Blockchain technology offers promising solutions to address security, safety, transparency, and trust concerns in healthcare systems by decentralizing transactions and enhancing security features. However, the rapid growth of online systems and the increasing prevalence of cyber threats, exacerbated by events like the COVID-19 pandemic, highlight the urgent need for enhanced security measures as mentioned by Bittins *et al.* [15] Adapting to emerging technologies such as AI, augmented reality, and blockchain is crucial for businesses, including the healthcare sector, to maintain competitiveness and ensure customer safety. The healthAIChain concept proposes combining the use of blockchain with powered artificial intelligence healthcare solutions to enhance patient data security. The combination of AI and the technology of blockchain work together to offer a powerful solution for enhancing security and transparency in healthcare systems. It is suggested to use the HealthAIChain paradigm to protect patient data and enhance system functionality. Subsequent studies ought to examine supplementary advantages of blockchain technology beyond safeguarding data and examine obstacles inside the healthcare domain. Blockchain and AI enable automation that has the potential to dramatically alter medical treatment, even beyond data protection.

A blockchain-based method for protecting medical data in cyber-physical systems is suggested in the research held by Kumar *et al.* [16]. It makes use of decentralized, immutable, and transparent blockchain technology to guarantee data security and privacy. The patient-centric approach grants users' full control over their data, enhancing security. Experimental results demonstrate the system's robustness against security threats and its ability to recover data. Overall, as proposed by Tandon *et al.* [17], the model strengthens healthcare data security while empowering patients with control over their information. Key components are interplanetary file system (IPFS) which enables decentralized storage and retrieval of data. BigchainDB facilitates real-time transactions and asset management. Tendermint provides fault tolerance, ensuring system reliability even with potential failures. MongoDB stores additional local information efficiently. Symmetric encryption is used for data security with AES encryption. To solve issues with healthcare systems, the study suggests architecture for safe and dependable data sharing in Cyber-Physical Healthcare 4.0 that is supported by blockchain technology. It draws attention to the drawbacks of the current systems such as non-patient-centric methodologies, security lapses, and privacy issues. The proposed solution leverages blockchain technology to decentralize EHRs, ensuring transparency, integrity, and security. Utilizing technologies such as Tendermint and IPFS, it addresses issues such as data fragmentation and unauthorized access. Additionally, by using the blockchain-enabled AES-256 algorithm, privacy and security are improved. The proposed framework offers a cost-effective and robust platform for healthcare data management, with potential extensions to other applications and integration with intelligent technologies to improve patient care.

3.3 Proposed model

By utilizing blockchain technology, the suggested model for blockchain-enabled patient identity management transforms the handling of healthcare data. Ensuring data integrity, interoperability, and patient privacy creates a transparent and safe platform for maintaining patient identities and medical information. The approach empowers patients and promotes easy data exchange between healthcare providers by enabling digital identity formation, safe data storage, consent management, and compliance procedures. The concept establishes a new benchmark for effective, patient-centered healthcare delivery by connecting with already existing health information exchange (HIE) networks and abiding by legal requirements (Figure 3.1).

1. **Blockchain infrastructure**: Create a blockchain network with interoperability, security, and scalability in mind for healthcare applications. With the sensitive nature of patient data, permissioned blockchain frameworks like Ethereum Enterprise or Hyperledger Fabric can be utilized to manage compliance requirements and adhere to legal requirements.
2. **Digital identity creation**: Create a reliable blockchain-based digital identity management solution. Every patient receives a digital wallet, which is a unique cryptographic identification that holds their verified identifying traits, including medical history, consent preferences, and demographic data.
3. **Patient onboarding and verification**: To guarantee the security and accuracy of patient identities, provide a streamlined procedure for patient onboarding and verification that incorporates multi-factor authentication and biometric authentication. To automate identification verification procedures while protecting patient privacy and consent, use smart contracts.

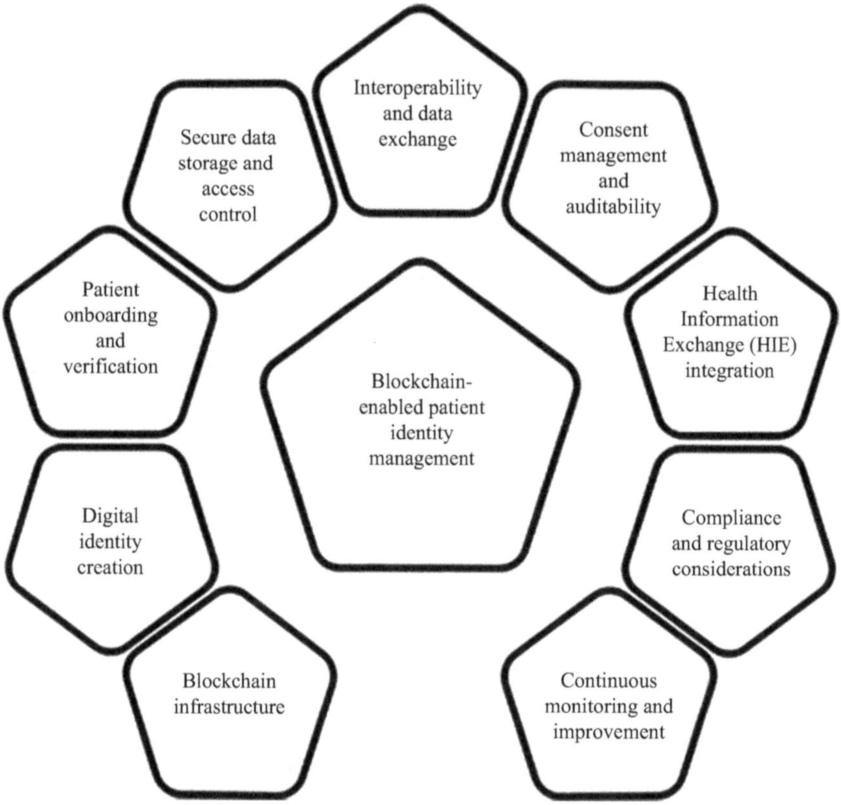

Figure 3.1 Diagrammatic representation of the proposed model

4. **Secure data storage and access control**: Make use of the distributed ledger feature of blockchain technology to safely handle and store patient data, making sure that sensitive data is protected by encryption and access controls. Employ cryptographic methods to facilitate safe data transfer while maintaining patient confidentiality such as homomorphic encryption and zero-knowledge proofs.
5. **Interoperability and data exchange**: Enable seamless data transmission by standardizing data formats and protocols to promote interoperability across healthcare systems and providers. Create APIs and interoperability layers to interface with the current healthcare IT infrastructure and provide real-time patient information access across various systems.
6. **Consent management and auditability**: Establish a system for consent management so that patients may regulate and oversee who has access to their health information. Employ the transparent and auditable characteristics of blockchain technology to monitor data access and utilization, giving patients insight into who has accessed their information and why.

7. **Health information exchange**: To promote safe and compatible patient data sharing across healthcare providers, insurers, and other stakeholders, integrate blockchain-enabled patient identity management with the current HIE networks. Utilize blockchain technology's decentralized design to boost HIE effectiveness and data integrity.
8. **Integration compliance and regulatory considerations**: Make sure that all applicable healthcare laws and regulations—including HIPAA, GDPR, and local data protection ordinances—are followed. Protect patient rights and reduce legal risk by implementing privacy-enhancing technologies and compliance measures while managing sensitive health information.
9. **Continuous monitoring and improvement**: Set up procedures for the patient identity management system powered by blockchain to be continuously monitored, assessed, and improved. Get input from interested parties, carry out frequent security assessments, and make system iterations to solve new issues and enhance efficiency.

Through the implementation of this proposed framework, healthcare establishments can leverage blockchain technology to revolutionize patient identity management, granting patients greater autonomy over their medical records and fostering trust, safety, and collaboration across the healthcare network.

In the following sections, this chapter illustrates how the proposed model can be implemented in various sections of society, discussing in detail the methodology and the results that can be achieved using blockchain in healthcare.

3.4 Blockchain and federated learning for privacy-preserving and collaborative health data analysis

The potential to use massive volumes of data for research and better patient outcomes in healthcare is enormous. This possibility, however, raises considerable privacy concerns, as health data is sensitive and personal. Traditional centralized data analysis approaches increase the danger of data breaches and misuse. Combining blockchain technology with federated learning (FL) to overcome these difficulties creates a new paradigm for privacy-preserving and collaborative health data analysis. Blockchain technology, an autonomous shared ledger, provides a solid foundation for safe data sharing and access management in healthcare applications by using data transaction integrity and accessibility [18]. It offers an environment that ensures data confidentiality and privacy by enabling participants to verify transactions for a central authority. The learning of the model is aggregated without disclosing the raw data, protecting the privacy of the individual data providers.

FL is the neural network-based approach that allows several people or devices to train a model at the same time without sharing raw data. Instead of storing data in a single location, FL spreads computational duties across multiple nodes (such as smartphones, hospitals, or IoT devices). Each node trains the model locally using its data before sending only model changes or gradients to a central server or aggregator. The server integrates these changes to improve the global model, which is subsequently returned to the nodes for more training. This procedure protects privacy and data protection, making FL especially useful in sensitive industries such as healthcare.

Combining blockchain and FL allows for the creation of a decentralized network of healthcare providers and researchers [6]. In this network, participants can collaborate to enhance machine learning models while keeping patient data localized and secure. This method not only improves privacy and security but also allows for the building of more comprehensive and accurate models by learning from a variety of diverse, distributed data sources. This cooperative, privacy-protecting strategy has promised to boost healthcare innovation, enhancing patient outcomes, and opening the door to a new wave of safe, data-driven healthcare solutions.

3.4.1 Methodology

1. Data collection
   The first step in the procedure is gathering health data from multiple sources such as clinics, wearable technology, hospitals, and research facilities. First, this data is preprocessed locally. It can include patient records or data from real-time health monitoring. Preprocessing entails deleting any personal information to preserve privacy while also cleaning, standardizing, and anonymizing the data to make sure it is in a state that can be used for analysis.
2. Federated learning setup
   - Initialization: By establishing a baseline model with starting parameters, a central authority or server starts the FL process. Subsequently, this model is disseminated to all nodes that are involved, which may be healthcare providers or research institutes that own their respective data subsets.
   - Local model training: Using its local dataset, each node trains the distributed model. This is an important step because it lets the data stay in its original location as the model learns from a variety of data sets.
   - Local update encryption: The nodes encrypt their gradients, or model updates, following training. To guarantee that any data sent over the network cannot be intercepted or used improperly, encryption is an essential step.
3. Secure aggregation via blockchain
   - Transmission and aggregation: Encrypted updates are transferred to the blockchain network and aggregated. A smart contract built on the blockchain oversees this aggregating procedure. It guarantees that all modifications are combined in a safe, unchangeable way to enhance the overall model while safeguarding personal information.
   - Model update: Following aggregation, the global model is updated with new parameters that reflect the combined learning from all nodes.

4. Model validation and feedback
   - Validation: The modified global model is transmitted back to the nodes for validation against their respective local datasets. This stage is critical for determining the model's performance and generalizability across multiple data sources.
   - Feedback loop: Based on performance, nodes can send feedback, which is encrypted and sent via the blockchain. This feedback helps to shape future iterations and enhancements to the model.
5. Iteration
   The process of local model training and feedback is iterative. The model refines and improves with each cycle, becoming more accurate and resilient in its predictions or assessments.
6. Deployment
   The model is put to use in real-world scenarios after it performs well enough. This can entail applying the model to research projects, patient monitoring apps, or clinical decision support systems. Crucially, the model retains its accuracy and relevance over time by continuously learning from fresh data updates.
7. Access control and data sharing
   To control who has access to the model and any shared findings, blockchain is essential. Ensuring that the model may only be accessed or contributed to by authorized organizations further enhances the system's security and privacy. In the healthcare industry, where data sensitivity cannot be emphasized, this degree of control is essential.

3.4.2 Result

A revolutionary method for handling and making use of sensitive data is presented by the combination of blockchain technology with FL for health data analysis. This novel approach permits decentralized, cooperative machine learning while protecting patient privacy and security. The solution tackles important issues related to data privacy and misuse by letting data stay on local devices and only exchanging model updates over a safe, encrypted blockchain network. Robust and highly accurate predictive models are developed by an iterative process that involves global model refining, secure update aggregation, and local model training. Without jeopardizing individual privacy, these models can greatly improve patient care and medical research. Additionally, the blockchain component ensures that only authorized participants can contribute to and profit from the collective learning process by facilitating visible and auditable access control in addition to securely securing data transfers. The result is a novel approach to health data analysis that strikes a compromise between the requirement for highly developed analytical skills and the requirement to protect patient privacy. This establishes a new benchmark for the safe, moral, and effective use of sensitive health data (Figure 3.2).

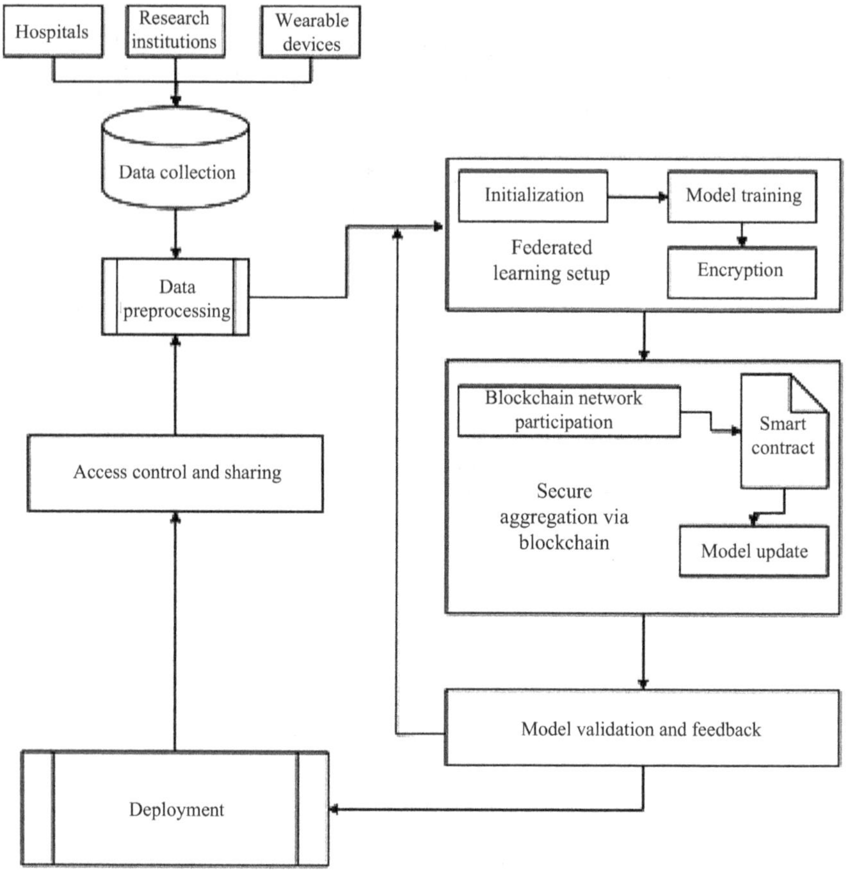

Figure 3.2 Flowchart to represent the working of the model

3.5 Blockchain and Internet of Things for remote patient monitoring and care coordination

Blockchain technology offers significant potential in remote patient monitoring (RPM) and care coordination by providing enhanced data security, interoperability, and transparency. Utilizing blockchain, patient health data can be securely stored and shared across disparate platforms, ensuring integrity and privacy through its decentralized and immutable nature. Smart contracts can automate processes such as alerts and notifications based on predefined conditions, while patients gain greater control over their data-sharing preferences. Additionally, blockchain facilitates transparent audit trails for regulatory compliance and streamline payment processes. Although there are implementation and regulatory compliance hurdles, using blockchain to RPM and care coordination can ultimately enhance patient outcomes and operational efficiency within the healthcare sector. The Internet of Things (IoT) and blockchain technology are merged to create a new paradigm for RPM and care coordination to overcome these obstacles. The IoT is a network of networked devices that collect and exchange data over the network to enable remote automation, control, and monitoring of processes and systems.

The IoT can significantly improve RPM and care coordination. Healthcare providers can access real-time health data collected by IoT devices. Examples of these devices are wearable monitoring and smart medical equipment. This enables timely intervention, continuous care, and personalized treatment plans. Additionally, IoT platforms facilitate communication between patients and healthcare professionals, enhancing care coordination and patient engagement. Combining blockchain and IoT can create a robust, secure, and efficient RPM and care coordination system. Blockchain ensures secure data storage and sharing, while IoT devices provide real-time health data. This integration enables transparent and tamper-proof data records, enhancing patient privacy and consent management. Moreover, smart contracts can automate processes, such as medication reminders and insurance claims, streamlining care coordination and improving patient outcomes (Figure 3.3).

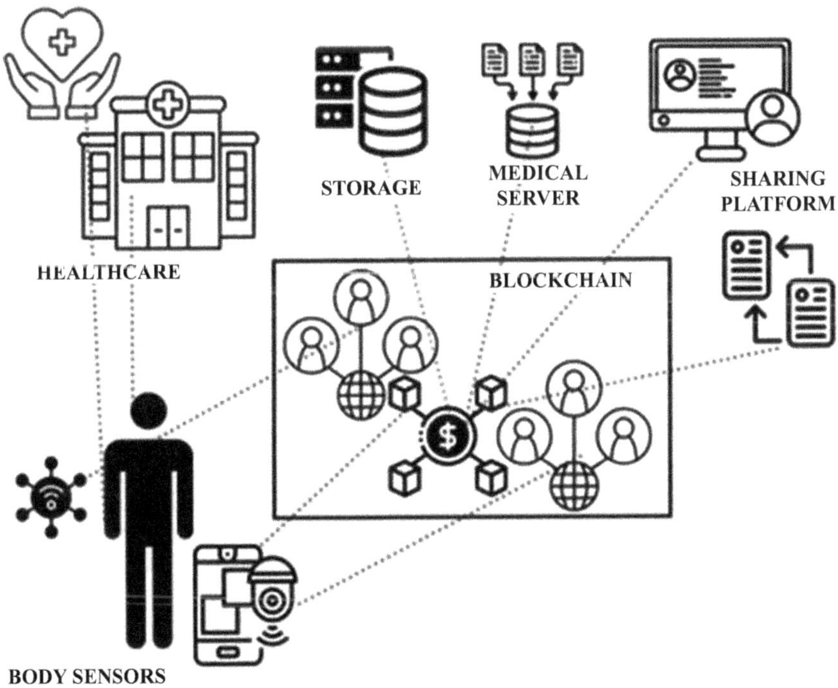

Figure 3.3 Workflow model of remote patient monitoring and care coordination

3.5.1 Methodology

1. Body sensors:
   A. Electrocardiogram: An IoT-based electrocardiogram (ECG) monitoring system measures heart rates and waveforms and sends the data to a database and web server for real-time monitoring.
   B. Temperature: A temperature monitoring system employing IoT sensors consists of sensor-equipped devices gathering temperature data, which is then wirelessly transmitted to a centralized platform. This platform, capable of real-time monitoring and analysis, facilitates proactive intervention to maintain optimal conditions across diverse environments, including industrial facilities, warehouses, healthcare settings, and smart homes.
   C. Blood glucose: Blood glucose prediction using IoT sensors involves deploying sensors to monitor glucose levels in diabetic patients, transmitting data wirelessly to a central platform, and applying predictive models to anticipate future glucose levels, enabling proactive management and personalized care.
   D. Heart rate: Heart rate monitoring via IoT involves wearable devices with heart rate sensors continuously tracking and wirelessly transmitting data to a centralized platform. This platform analyzes the data, offering real-time monitoring and alerts for abnormal readings. Users can access their heart rate trends, set personalized alerts, and share data with healthcare providers for remote monitoring. This approach enables proactive cardiovascular health management and personalized interventions based on individual heart rate patterns.
   E. Blood oxygen: Measuring blood oxygen using an IoT device involves sensors, like pulse oximeters, integrated into wearables or standalone devices. These sensors emit light to detect oxygen levels in the blood, transmitting data wirelessly to a platform. Users access real-time readings via an app, enabling proactive respiratory health management and personalized interventions, etc.
2. Medical data storage:
   Medical data storage facilitated by IoT sensors involves the collection of healthcare-related information from various sensors, devices, and systems, typically deployed in healthcare facilities or worn by patients. Vital signs, medication adherence, activity levels, and other metrics relating to health are all recorded by these sensors. IoT connection methods like Bluetooth, Wi-Fi, or cellular networks are used to securely transfer the collected data to a cloud-based platform or centralized storage system. To maintain patient privacy and data security, this platform stores and arranges the data by legal regulations like HIPAA. The saved data can be accessed and retrieved by authorized users, such as patients and healthcare professionals, as needed for analysis, decision-making, and monitoring. This approach enables efficient and centralized management of medical data, supporting RPM, care coordination, and personalized healthcare interventions facilitated by IoT technology.

3. Healthcare:
   In healthcare, patient data must be securely stored and maintained to ensure continuity of care, accurate diagnosis, and effective treatment. This data includes a wide range of information such as medical history, diagnoses, medications, lab results, imaging studies, and treatment plans. EHR systems, which act as centralized repositories for patient data accessible to authorized healthcare practitioners, are required to store this data. To protect patient privacy and data security, EHR systems are made to adhere to healthcare legislation such as HIPAA. Redundancies and backups are also frequently used to guard against data loss and guarantee data accessibility in the event of crises or system failures. Healthcare institutions make significant investments in cybersecurity and strong infrastructure to guard patient data against breaches, hacks, and unwanted access. The ability to efficiently store and retrieve patient data is crucial for delivering high-quality, coordinated care and facilitating informed decision-making by healthcare providers.
4. Data secure via blockchain:
   Blockchain-based patient data security has many benefits, including increased accessibility, privacy, and integrity. Since the blockchain is decentralized and unchangeable, patient records are kept across a dispersed network of nodes, making it extremely difficult for one institution to alter or tamper with the data. An audit trail that is clear and impenetrable is produced when every transaction or modification to the patient data is cryptographically linked and documented in a block. Additionally, by enabling patients to monitor who has viewed or modified their records and to grant access authorization, blockchain gives patients more control over the information about their health. By leveraging blockchain for patient data security, healthcare organizations can enhance trust, transparency, and interoperability while mitigating the risks of data breaches and unauthorized access.
5. Model validation and feedback:
   Model validation is a crucial step in predictive modeling where the model's performance is assessed using unseen data to ensure its accuracy and reliability. The model can be trained on one subset of the dataset and assessed on the other by dividing it into training and testing sets. This allows for the evaluation of the model's prediction performance. The model's capacity to accurately anticipate outcomes in real-world events and generalize effectively to new information is strengthened by this approach.

   Model feedback involves collecting input from users or systems based on the performance of a predictive model and using it to improve the model's accuracy and relevance over time. This iterative process allows for adjustments in parameters or algorithms, ensuring the model aligns better with desired outcomes or user preferences. Continuous monitoring and feedback help keep the model effective and up to date in addressing evolving needs and challenges.
6. Deployment:
   The model is put to use in real-world scenarios after it performs well enough. This can entail applying the model to research projects, patient monitoring

apps, or clinical decision support systems. Crucially, the model retains its accuracy and relevance over time by continuously learning from fresh data updates.

3.5.2 Result

Blockchain and IoT integration improves data security, interoperability, and patient empowerment, leading to promising outcomes in RPM and care coordination. Blockchain technology ensures the integrity and immutability of patient data, while IoT devices enable remote monitoring and data collection. By enabling smooth data transfer between healthcare systems, this combination gives people more control over their health information and increased involvement in their care. Streamlined administrative processes and automated tasks improve operational efficiency, while the traceability provided by blockchain ensures compliance with regulatory requirements. Overall, by enhancing data management, patient involvement, and care coordination, the combination of blockchain technology and IoT presents a revolutionary opportunity to transform the delivery of healthcare services.

3.6 Blockchain and smart contracts for secure and interoperable health data exchange

The dynamic fusion of blockchain technology and smart contracts is ushering in a new era of innovation in health data management. Blockchain, recognized for its decentralized and secure architecture, acts as an incorruptible ledger, while smart contracts, automated and self-executing, introduce efficiency and reliability. This complementary combination offers a cutting-edge foundation for the safe and seamless interchange of health data in the healthcare industry. Blockchain's decentralized structure, which distributes data throughout a network, guarantees immutability and transparency. It has become a haven for medical records in the healthcare industry, reducing the hazards connected with centralized storage solutions.

Smart contracts, serving as self-executing agreements, bring automation to critical aspects of healthcare, including consent management, data access, and interoperability [19]. By enforcing predetermined conditions, these contracts elevate the precision and reliability of health data exchanges. The integration of blockchain and smart contracts directly addresses security concerns in health data exchange. Streamlining patient consent and leveraging blockchain's decentralized structure fortify the system against potential breaches. Interoperability is further fortified by adopting standardized data formats, facilitating seamless communication across diverse healthcare systems. In essence, the marriage of blockchain and smart contracts in healthcare not only fortifies security but also promotes transparency and interoperability. This transformative synergy empowers patients, fosters collaborative healthcare practices, and positions the industry for a digitally advanced and interconnected future.

3.6.1 Methodology

1. *Patients:* In this context, patients take an active role in initiating and triggering the process of exchanging health data securely. It signifies that patients are actively engaging in actions or procedures that kickstart the secure health data exchange mechanism. This initiation could involve providing explicit consent, granting access to their health information, or initiating data-sharing requests within the established system. Patients Initiate Secure Health Data Exchange emphasizes the proactive role of individuals in the healthcare process, where patients play a crucial part in instigating the secure exchange of their health data through designated mechanisms, ensuring control and consent over the sharing of their personal health information.
2. *Healthcare providers:* Healthcare providers are actively involved in the process of securely exchanging health data. Their involvement highlights the use of innovative methods to preserve the confidentiality and integrity of health data and suggests a hands-on role in the application of smart contracts and blockchain systems. By taking part, healthcare providers most likely make use of the automation, immutability, and transparency provided by digital contracts and blockchain technology to guarantee safe, uniform, and compatible health data sharing.
3. *External entities:* Wearable devices, external to the core healthcare system, play a significant role by supplying health-related data. These devices contribute to the secure health data exchange process, likely sharing real-time health metrics, activity data, or other relevant information. The integration of wearable device data into the blockchain and smart contract framework ensures the security, traceability, and interoperability of this external data source.
4. *External system:* External systems, operating independently from the primary healthcare infrastructure, actively engage in and benefit from the secure health data exchange process. These external systems likely access or contribute data within the blockchain and smart contract framework, enhancing interoperability and ensuring security in the exchange of health-related information. The utilization of secure health data exchange by external systems emphasizes the collaborative nature of the healthcare ecosystem, where diverse entities contribute to and leverage a standardized, secure, and interoperable framework.
5. *Blockchain network:* The blockchain network serves as the overarching infrastructure responsible for managing and orchestrating the secure exchange of health data. It makes use of the decentralized and impenetrable characteristics of blockchain technology to guarantee the security, transparency, and immutability of medical information transactions. The management role includes overseeing the execution of smart contracts that automate various aspects of the health data exchange process and enforcing predefined rules and conditions. Through the blockchain network, interoperability is promoted by providing a standardized and reliable platform for diverse entities, such as patients, healthcare providers, wearable devices, and external systems, to engage in secure data exchange (Figure 3.4).

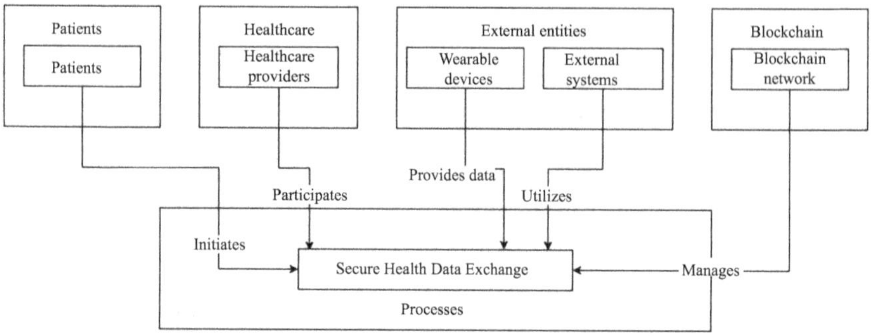

Figure 3.4 Diagrammatic representation of the smart contracts for secure and interoperable health data exchange

3.6.2 Result

The integration of blockchain and smart contracts in healthcare has yielded transformative outcomes, notably enhancing the security, interoperability, and efficiency of health data exchange. Blockchain's decentralized structure ensures robust security, reducing the risks of unauthorized access and tampering. Smart contracts automate processes such as consent management, fostering transparency and trust for both patients and healthcare providers.

The use of blockchain has greatly addressed healthcare's interoperability problems. Its decentralized architecture and standardized data formats remove previous barriers to smooth communication between various healthcare organizations.

Patients now experience greater autonomy in managing and exchanging their health data. Smart contracts streamline processes, leading to increased operational efficiency by automating tasks like data access approvals. The integration of blockchain and smart contracts has ushered in a more interconnected, secure, and patient-centric healthcare landscape, promising ongoing innovation in health data management.

3.7 Future research direction

The upcoming possibilities for blockchain in patient identification management are among the many innovative tactics aimed at enhancing security, privacy, and interoperability within healthcare systems. Investigating how blockchain technology can be used in conjunction with FL, and smart contracts with the IoT to take advantage of several opportunities and issues in patient identification management is one exciting line of inquiry. Ideally, FL and blockchain integration offer a viable path for cooperative and privacy-preserving health data analysis. By collaborating on model training without providing raw data, FL protects patient privacy across institutions. Researchers can make sure that FL processes are traceable and intact by using blockchain's tamper-proof ledger to handle access control and data provenance. In the future, studies may concentrate on enhancing the scalability and efficiency of blockchain-FL frameworks while tackling issues like model synchronization over decentralized networks and heterogeneous data.

Additionally, new possibilities for RPM and care coordination are presented by the convergence of blockchain technology and the IoT. Healthcare providers may guarantee data integrity, authenticity, and patient privacy by safely storing IoT-generated health data on a blockchain. Novel blockchain-based designs for IoT-enabled healthcare systems may be investigated in future studies, taking into account real-time data analytics, data aggregation, and compatibility with current healthcare infrastructure. Additionally, to support large-scale deployments and a variety of use cases, research efforts might concentrate on improving the scalability and resilience of blockchain IoT ecosystems. Blockchain technology and smart contracts enable safe and interoperable health data exchange among multiple parties. Self-executing contracts that are programmed on a blockchain are known as smart contracts. They can automate data-sharing agreements, access control regulations, and consent management, which can streamline patient information interchange while guaranteeing regulatory compliance. Standardized smart contract templates for healthcare transactions, improving the effectiveness of data validation and verification procedures, and investigating cutting-edge methods for cross-chain interoperability and data portability are some potential future study areas.

To handle the changing opportunities and difficulties in healthcare, future research in blockchain-based patient identity management should concentrate on combining cutting-edge technologies such as FL, IoT, and smart contracts. Researchers can create novel solutions that improve security, privacy, and interoperability while giving patients and healthcare providers more control over the management and sharing of health data by synergistically fusing various technologies. Moreover, to spur innovation and guarantee the useful applicability of blockchain technologies in actual healthcare settings, interdisciplinary cooperation between researchers, medical practitioners, legislators, and industry players will be crucial.

3.8 Conclusion

In summary, the application of blockchain technology to patient identity management has great potential to transform healthcare systems across the globe. Healthcare practitioners may improve the security, privacy, and interoperability of patient identity data by utilizing the immutability, decentralization, and transparency that are among the intrinsic features of blockchain technology.

Healthcare businesses may guarantee the accuracy of EHRs, expedite data exchange procedures, and give patients more control over their private health information by implementing blockchain technology. Blockchain builds confidence between patients, healthcare providers, and other stakeholders by enabling safe authentication, effective data sharing, and strong consent management.

Moreover, blockchain's potential goes beyond just managing patient identities. Its integration with smart contracts, FL, and the IoT creates new opportunities for safe health data interchange, RPM, and privacy-preserving data analysis. Even though blockchain has a lot of potential, there are still issues. Regulatory compliance, scalability, interoperability, and integration with the current healthcare infrastructure are some of the major obstacles that call for additional study and development. Moreover, the successful adoption and implementation of blockchain technologies in the healthcare industry depend on interdisciplinary collaboration and stakeholder engagement. Unlocking the full potential of blockchain in patient identity management will require sustained research, innovation, and cooperation in the upcoming years. Through problem-solving, optimizing current models, and investigating new uses, the healthcare sector may use blockchain technology to enhance patient outcomes, increase productivity, and revolutionize healthcare delivery worldwide.

References

[1] Jaoude, J.A. and Saade, R. Business applications of blockchain technology–a systematic review. *IEEE Access*. 7(1), pp.10–1109, 2019.

[2] Gupta, M., Jain, R., Kumari, M. and Narula, G. Securing healthcare data by using blockchain. In *Applications of Blockchain in Healthcare* (pp. 93–114). Singapore: Springer Nature; 2020.

[3] Swarnkar, M., Bhadoria, R.S. and Sharma, N. Security, privacy, trust management and performance optimization of blockchain technology. In *Applications of Blockchain in Healthcare* (pp. 69–92). Singapore: Springer Nature; 2020.

[4] Hylock, R.H. and Zeng, X. A blockchain framework for patient-centered health records and exchange (HealthChain): evaluation and proof-of-concept study. *Journal of Medical Internet Research*, 21(8), p. e13592. 2019.

[5] Javed, I.T., Alharbi, F., Bellaj, B., Margaria, T., Crespi, N. and Qureshi, K. N. Health-ID: A blockchain-based decentralized identity management for remote healthcare. *Healthcare*, 9(6), p. 712, 2021. doi:10.3390/healthcare9060712.

[6] Mohammed, A.H., Abdulateef, A.A., and Abdulateef, I.A. Hyperledger, Ethereum and blockchain technology: A short overview [Internet]. *Proceedings of the 3rd International Congress on Human-Computer Interaction, Optimization and Robotic Applications (HORA)* (pp. 1–6). IEEE; 2021.

[7] Harrell, D.T., Usman, M., Hanson, L., *et al.* Technical design and development of a self-sovereign identity management platform for patient-centric health care using blockchain technology. *Blockchain in Healthcare Today*, 5(S1), pp. 10-30953, 2022.

[8] Sharma, P., Jindal, R. and Borah, M.D. Healthify: A blockchain-based distributed application for health care. In *Applications of Blockchain in Healthcare* (pp. 171–198). Singapore: Springer Nature; 2020.

[9] Abdul-Moheeth, M., Usman, M., Harrell, D.T. and Khurshid, A. Improving transitions of care: designing a blockchain application for patient identity management. *Blockchain in Healthcare Today*, 5(S1), pp. 10-30953, 2022.

[10] Tanwar, S., Parekh, K. and Evans, R. Blockchain-based electronic healthcare record system for healthcare 4.0 applications. *Journal of Information Security and Applications*, 50, p. 102407, 2020.
[11] Ouaguid, A., Hanine, M., Chiba, Z., Abghour, N. and Ghazal, H. Analysis of Blockchain Integration in the e-Healthcare Ecosystem. In *2023 6th International Conference on Advanced Communication Technologies and Networking (CommNet)* (pp. 1–8). IEEE; 2023.
[12] Alam, S., Bhatia, S., Shuaib, M., *et al.* An overview of blockchain and IoT integration for secure and reliable health records monitoring. *Sustainability*. 15(7), p. 5660, 2023.
[13] Kshetri, N., Hutson, J. and Revathy, G. HealthAIChain: Improving security and safety using Blockchain Technology applications in AI-based healthcare systems. In *2023 3rd International Conference on Innovative Mechanisms for Industry Applications (ICIMIA)* (pp. 159–164). IEEE; 2023.
[14] Charles, W.M. Accelerating life sciences research with blockchain. In *Applications of Blockchain in Healthcare* (pp. 221–252). Singapore: Springer Nature; 2020.
[15] Bittins, S., Kober, G., Margheri, A., Masi, M., Miladi, A. and Sassone, V. Healthcare data management by using blockchain technology. In *Applications of Blockchain in Healthcare* (pp. 1–27). Singapore: Springer Nature; 2020.
[16] Kumar, M., Raj, H., Chaurasia, N. and Gill, S.S. Blockchain inspired secure and reliable data exchange architecture for cyber-physical healthcare system 4.0. *Internet of Things and Cyber-Physical Systems*, 3, pp. 309–322, 2023.
[17] Tandon, A., Dhir, A., Islam, A.K.M.N. and Mäntymäki, M. Blockchain in healthcare: A systematic literature review, synthesizing framework and future research agenda. *Computers in Industry*, 122, p. 103290, 2020. doi:10.1016/j.compind.2020.103290.
[18] Zhao, Y., Zhao, J., Jiang, L., *et al.* Privacy-preserving blockchain-based federated learning for IoT devices. *IEEE Internet of Things Journal*, 8(3), pp. 1817–1829. 2020.
[19] Venkatesan, S., Sahai, S., Shukla, S.K. and Singh, J. Secure and decentralized management of health records. In *Applications of Blockchain in Healthcare* (pp. 115–139). Singapore: Springer Nature; 2020.
[20] Qi, Y., Hossain, M.S., Nie, J. and Li, X. Privacy-preserving blockchain-based federated learning for traffic flow prediction. *Future Generation Computer Systems*, 117, pp. 328–337. 2021.
[21] Yadav, A.K., Singh, K., Amin, A.H., Almutairi, L., Alsenani, T.R. and Ahmadian, A. A comparative study on consensus mechanism with security threats and future scopes: Blockchain. *Computer Communications*, 201, pp. 102–115, 2023.
[22] Praveen, G., Anand, M., Singh, P.K. and Ranjan, P. An overview of blockchain consensus and vulnerability. *Proceedings of the Information and Communication Technology for Intelligent Systems – ICTIS 2020*, 1, pp. 459–468, 2021.

Chapter 4

Integration of Internet of Things (IoT) in healthcare: a paradigm shift toward smart and efficient patient care

Reek Roy[1], Shreea Bose[2], Snehashis Kayal[3], Partha Pratim Mandal[4], Sandip Patra[4], Sarmin Ahmed[4] and Himadri Nath Saha[5]

Abstract

The Internet of Things (IoT) has become a disruptive technology across many industries, and its application in healthcare is revolutionizing the way medical services are delivered. This chapter provides an overview in details of the impact and potential of IoT in medical field, emphasizing its role in enhancing patient particular attention, enhancing the utilization with resources as well as increasing standard medical outcomes. IoT in healthcare involves connecting medical devices, sensors, and systems through the Internet, enabling seamless communication and data exchange. This connectivity facilitates real-time monitoring, remote patient management, and intelligent decision-making by healthcare professionals. Wearable devices, smart sensors, and connected medical equipment contribute to the creation of a comprehensive healthcare ecosystem that is both efficient and patient-centric. One of the main benefits of IoT in healthcare is its capability to supply remote, constant surveillance of patients. By utilizing wearable technology with sensors that record health-related data and vital signs in real -time, medical practitioners may keep an eye on their patients even while they are not in traditional hospital settings. This promotes proactive treatment and early health problem identification, decreasing complications and lowering hospital readmission rates. Furthermore, IoT applications contribute to the optimization of healthcare operations and resource management. Smart healthcare infrastructure, including connected ambulances, RFID-enabled inventory management, and predictive medical

[1]Department of Computer Science, Belda College, Vidyasagar University, India
[2]Department of Computer Science, St. Xavier's College, Calcutta University, India
[3]Department of Information Technology, Techno International New Town, Maulana Abul Kalam Azad University of Technology, India
[4]Department of Computer Science and Engineering, The Neotia University, India
[5]Department of Computer Science, Surendranath Evening College, Calcutta University, India

equipment maintenance, improves healthcare delivery efficiency. This leads to cost savings, better utilization of resources, and an overall improvement in the quality of healthcare services. In comparison to other systems currently in use, our suggested approach looks for to be safer and more efficient.

Keywords: Smart healthcare; wireless body sensors; cloud server; Internet of Things (IoT); health monitoring; pervasive healthcare

4.1 Introduction

The Internet of Things (IoT) is a technology that has revolutionized an array of networked objects with sensors and distinct identities. These devices, ranging from smartphones and wearable gadgets to industrial machinery and medical equipment, are capable of collecting, exchanging, and analyzing data in real time. The main objective of the IoT is to create a network of self-reporting devices that can interact with humans and one another. This will allow for intelligent decision-making, automation, and seamless integration across multiple domains. The IoT is a network of physical objects, or "things," that are combined with sensors, software, and other technologies to communicate and share data with other systems and devices over the Internet. There is much more to the IoT than just computers and other technology. Anything having a sensor and a unique identifier (UID) is included. The primary goal of the IoT is to develop self-reporting devices that can communicate with people in real time.

The idea of the IoT has been around for a long, but only recently has it become practical because of certain technological developments. The availability of low-cost, energy-efficient sensor technologies has helped IoT technology to now be accessible to more firms because of affordable, reliable sensors. Interaction: Many internet network protocols have made it easy to connect sensors to the cloud and other "things" for efficient data transfer across cloud computing platforms. With the increasing availability of cloud platforms, individuals and businesses may now access the infrastructure they require to expand without having to manage it all.

4.1.1 Sensors and Arduino Uno

Sensors are parts or apparatus that measure and identify external factors or ambient circumstances, then transform the data into signals that electronic instruments or systems can understand. Sensors come in a variety of forms, each intended to identify particular kinds of physical events. The following are some typical types of sensors: light, humidity, motion, pressure, temperature, accelerometer, gyroscope, biometric, proximity, light, and so forth. These sensors can detect environmental changes, monitor equipment performance, track movement, and assess physiological indicators, among other uses. Businesses and organizations that integrate sensors into IoT devices can obtain important insights into their operations, streamline processes, and improve decision-making. Sensors are essential for many applications, including consumer electronics, healthcare, and industrial automation.

In order to enable automation, monitoring, and control in a wide range of sectors and applications, sensors are crucial parts of contemporary technology [1].

Arduino sensors are sensors that are compatible with Arduino microcontrollers. Computer programs and hardware that are easy to use form the foundation of Arduino, an open-source electronics platform. It comprises a development environment for creating, compiling, and uploading code to the board and a programmable microcontroller board. Arduino sensors are designed to interface with Arduino boards and provide input data to the microcontroller. Arduino sensors come in various types and can measure a wide range of physical properties or environmental conditions such as temperature, humidity, light intensity, motion, proximity, and more. These sensors typically interface with Arduino boards using standard communication protocols such as analog voltage, digital signals, or serial communication. Arduino provides libraries and code examples for interfacing with various sensors, making it easier for users to integrate sensor functionality into their projects. In addition, a large variety of Arduino-compatible sensors are available from numerous third-party manufacturers and providers, letting users select the sensors that best suit the needs of their projects. All things considered, Arduino sensors are essential to extending the functionality of Arduino-based projects by allowing them to communicate with and react to the real environment. They make it possible to create a wide variety of innovative and interactive electronic devices and systems.

Sensors and Arduino Uno play a key role in many projects, ranging from basic home-based experiments to intricate IoT systems. The designs' eyes and ears are sensors that gather data from the real world. The Arduino Uno functions as the brain, interpreting the data and directing actuators according to preprogrammed logic. The Arduino Uno is a great platform for integrating with different kinds of sensors because of its adaptability and simplicity of usage. It may be connected to a variety of sensors, such as motion, gas, humidity, temperature, and more, thanks to its digital and analog input/output ports. Professionals, students, and hobbyists choose Arduino Uno because of its extensive library and shield compatibility, which further enhances its sensor integration capabilities. In most sensor-Arduino Uno projects, the workflow entails attaching the sensor to the Uno, utilizing analog or digital pins to read the sensor's data, processing the data with the Arduino microcontroller, and then taking action based on the information obtained. Additionally, because Arduino Uno is compatible with many different IoT platforms and communication protocols, including as Bluetooth, Wi-Fi, and LoRa, it is a fantastic option for designing IoT applications that require sensor integration. Developers may design IoT solutions that enable real-time communication and decision-making, gather data from field-deployed sensors, and remotely monitor and control equipment by connecting Arduino Uno to the Internet.

All things considered, sensors, IoT, and Arduino Uno work well together to create intelligent, networked systems that can detect, comprehend, and react to environmental changes. As the IoT advances, the combination of sensors and Arduino Uno will be critical to promoting creativity and creating new opportunities across a range of industries.

4.1.2 IoT in healthcare

IoT technology adoption in the healthcare sector indicates a mentality shift in favor of a more clever and successful patient care approach. Healthcare providers may transform the delivery, monitoring, and management of healthcare services by utilizing IoT-enabled devices, sensors, and data analytics [16]. In the end, this will result in better patient outcomes, more enjoyable interactions, and efficient use of resources. The IoT in the healthcare industry includes a variety of devices in addition to standard PCs and equipment. It encompasses everything that has a sensor that has been given a unique identity (UID), including ambient environmental sensors, wearable health monitors, and smart medical equipment. Numerous real-time data points, such as vital signs, patient activity levels, medication adherence, environmental variables, and more, can be gathered by these devices. Creating self-reporting gadgets that can communicate with consumers and with one another in real time is the primary goal of the IoT in the healthcare sector. This makes it possible to continuously monitor the health state of the patient, identify health concerns or irregularities early on, and take prompt action to stop unfavorable outcomes. IoT-enabled devices also make it easier for data to be integrated and interoperable, which gives healthcare professionals access to full patient data from many sources and the capacity to make well-informed decisions about diagnosis, treatment, and care management.

IoT is revolutionizing healthcare by creating a more connected and data-driven approach to patient care [2]. Telemedicine and remote patient monitoring (RPM) are two important uses of IoT in healthcare. IoT-enabled wearable health monitors can remotely monitor patients recovering from surgery or those with chronic diseases by continuously monitoring vital signs and sending data to healthcare providers in real time. Other uses of IoT are as follows:

- Remote monitoring: IOT devices allow doctors to monitor patients from far away, especially those with chronic conditions needing close attention, it tracks heart rate, blood sugar, and blood pressure.
- Better medication management: IOT-enabled pill dispensers can prompt patients to take their medication and even track if they have done so, this improves medication adherence.
- Early disease detection: Wearable technology with integrated sensors can monitor a person's activity levels, sleep habits, and other medical information. This information can be a goldmine for the identification of early signs of disease like heart trouble or diabetes, allowing for preventive measures.
- Hospital asset tracking: Within healthcare institutions, IOT technology can monitor the whereabouts and condition of medical supplies, equipment, and pharmaceuticals. It helps in optimizing inventory management, prevents loss or theft, and ensures availability of essential resources.
- Chronic disease management: By giving patients tools for self-monitoring and adherence to treatment plans, IoT solutions help with the management of chronic illnesses such as diabetes, hypertension, and asthma.

Smart medical equipment with IoT capabilities, such as insulin pumps, pacemakers, and medicine dispensers, improve patient safety and treatment compliance.

These gadgets can lower the chance of side effects and improve patient outcomes by automatically adjusting pharmaceutical dosages, reminding users to take their medications and warning medical professionals of possible problems or difficulties. Hospital administration and operations are another area in which IoT is revolutionizing healthcare. The integration of IoT sensors and automation technology into smart hospital infrastructure can optimize resource allocation, expedite operations, and improve patient flow and experience. IoT-enabled smart beds, for instance, can reduce the effort of healthcare personnel by automatically adjusting settings based on patient preferences and providing real-time alarms for falls or pressure ulcers. This enhances patient comfort and safety.

Section 4.2 consists of a literature survey done on various existing systems mentioning the results, pros, and cons of these systems. In Section 4.3, the system design of the proposed model has been elaborated. Sections 4.4–4.6 have the methodology of the proposed model implemented in various cases. In Sections 4.7 and 4.8, the performance analysis and the future research direction of the proposed model are explained, respectively. Section 4.9 has a short conclusion of this book chapter.

4.2 Background and related works

A linked environment made possible by wearable and IoT-enabled personalized healthcare technologies allows for the provision of a range of healthcare services, including fitness tracking, remote health monitoring, and dietary plans. The potential of an IoT-based virtual rehabilitation system was demonstrated. These included the capacity to offer real-time remote health monitoring, enhance the quality of life for the elderly and disabled, forecast disease, prescribe preventive medication, and oversee medical services in hospitals and emergency rooms. It emphasizes how the Internet of Medical Things (IoMT) is intentionally being used to stop the spread of COVID-19, improve worker safety, and increase the effectiveness of pandemic management. Strong security measures are essential because of problems such as replay assaults, man-in-the-middle attacks, malware, and password guessing that have been brought up by challenges like the rapid adoption of IoMT internationally. It also mentioned the future of IoT in healthcare, focusing on RPM for real-time health tracking, telemedicine, drug adherence monitoring, enhanced data analytics for predictive modeling, and the transformative potential of machine learning and artificial intelligence (AI) in healthcare applications [3].

Mohammed and Hasan [4] explain how IoT technology is being used to create a smart healthcare monitoring system. It highlights the value of IoT in healthcare and how it makes it possible to track vital health indicators such as body temperature, heart rate, and SPO2 in real time. In order to configure the hardware, sensors monitoring body temperature, heart rate, and SPO2 levels are connected to the Raspberry Pi 4B, which serves as a microcontroller. The patient's real-time location can be tracked and efficient communication made possible by an attached 4G GSM, GPRS, and GNSS HAT module. The data gathered from the hardware and sensors is stored in the cloud using the MySQL database. This allows for the recording of patient information, abnormal data, and past patient status for reference. The system also shares real-time measured health parameters to both doctors and patients through a mobile application. The measured data is compared to the normal range of data; if the measured data deviates from the normal range of data, an emergency alert is sent by SMS notification, together with the patient's current location, to physicians and the patient's family.

Kokkonis *et al.* [5] illustrate the development of an e-health jacket that continuously collects data from medical sensors and sends it to the cloud, as well as the wearable health monitoring system (WHMS). It focuses on collecting large-scale medical data in real time from the e-health jacket worn by patients with chronic or rare health conditions. When the unique distress button on the jacket is depressed, the clinic will receive an instant alarm and dispatch an ambulance to the patient's residence. Additionally, a GPS positioning capability will be available to transmit the patient's precise location whenever he leaves his house or hospital. To overcome challenges such as managing and analyzing the increasing data traffic in health care, emphasizing the importance of data security, and wireless data transmission, the sensors used for monitoring patient health are the exceptional levels of system design of the WHMS and the hybrid transport protocol used for wireless data transfer. The hardware components used in the architecture are reasonably priced and readily available in the market. The recommended communication protocol accounts for the wearable system's battery life, the patient's proximity to the connection, and the kind of health information being transmitted. Among the information they were able to gather are mobility and gesture detection, signal auditing, customized bio-signal evaluation, and condition sampling.

A mobile application focusing on chronic health conditions such as hypertension and diabetes to allow users to communicate with doctors through the application was developed. The system involves users such as patients, doctors, and home caregivers, enabling remote health monitoring and communication. It also includes a system architecture involving home appliances, sensors, and actuators for data collection and transmission. The system focuses on monitoring physiological parameters such as blood pressure, sugar levels, and temperature with wireless communication between the user and home appliances. The users can register on the mobile application, upload health data, chat with doctors, and book appointments through the mobile app. The system also supports remote monitoring of COVID-19 symptoms, analyzing user input, and facilitating communication between patients, doctors, and family members for efficient healthcare management. The system aims to improve convenience for patients with health challenges enabling real-time monitoring, remote consultation, and cost-effective healthcare. This intelligent health assistance platform allows for distant patient observation, particularly in times of quarantine such as the COVID-19 outbreak. It implements IoT-based solutions to reduce hospital burden, provides essential comforts, and extracts health data for medical diagnoses and prescriptions [6].

An IoT-driven intelligent medical system for efficient diagnosis in emergency care collects linked devices and the latest technologies to track and evaluate patients' health indicators in real time. This, therefore, implies that the current system envisioned is driven by wearable sensors, a wireless channel, cloud storage and processing, machine learning algorithms, and mobile applications driving the overall enhanced emergency healthcare experience. The sensors are helpful for keeping an eye on a number of critical health indicators such as high blood pressure, saturation level of oxygen, respiratory rate, heart rate, and body temperature. These gadgets send real-time data to a centralized cloud-based platform via wireless communication. This is ensured through the use of low-powered communication protocols that further extend the life of the batteries of the wearable devices. In the cloud, the accumulated health data is securely stored and later processed using advanced analytics and machine learning algorithms. Predictive diagnostics was possible as these algorithms take historical data about a patient to identify patterns, trends, and possible anomalies. The system design will give immediate alerts to healthcare providers in case of abnormal readings or a critical health condition that demands immediate attention on the part of healthcare providers. Mobile applications form a very integral part of this system, providing easy-to-use interfaces to both the patient and the healthcare provider. Patients can leverage individual health insights, get health recommendations, and track their health status in more proactive health management. However, healthcare providers can have real-time access to the data of each patient [7].

A cutting-edge idea for individualized skincare and early identification of dermatological issues is displayed by smart skin health monitoring through the use of an AI-enabled cloud-based IoT system. The smart wearable gadgets including the sensors involved in this system record the vital parameters of skin constantly such as moisture, temperature, and UV exposure. The collected data as a whole is transmitted over a secure data network of IoT onto a cloud-based platform so that the data is centrally stored and processed in real-time. AI algorithms play a pivotal role in data analysis associated with skin health. Machine learning models can identify patterns, anomalies, and early warning signs of dermatological conditions, constantly improving accuracy through learning. Another vital area is that of monitoring the exposure of UV, by sensors being applied to monitor in real time with the feedback of the user if the danger of the sun occurs. The app also provides educational resources where information about skin care, prevention, and the proper products according to the condition of the individual's skin is provided. Overall, the outlook with AI-based cloud IoT revolutionizes the skincare sector in providing a proactive, data-driven, and personalized approach toward preserving and improving skin health [8].

An IoT platform for a wearable device for alerting seizures makes use of sensors that monitor physiological signs related to epileptic seizures. It connects through IoT protocols with a cloud-based infrastructure, where in effect, it could do its analysis in real time, making it relatively simple. The alert level could be customized by the user, and the app was user-friendly with machine-learning algorithms. Geo-location tracking will help get timely assistance by notifying the nearby contacts or the emergency services of the location of the user during the seizure. The optimized battery ensures prolonged use of the device, while the security and privacy features do not give access to the health information of the user. Integration with healthcare providers helps in the monitoring from remote locations and better management of the user's condition, The platform's user-centric design, coupled with continuous improvements, ensures an effective and supportive solution for individuals at risk of seizures [9].

IoT-based applications in healthcare devices have revolutionized patient care and management integration of sensors, connectivity, and data analytics has brought immense change to patient care and management in medical devices. It has transformed the way healthcare providers monitor, diagnose, and treat patients. This means it has improved patient outcomes through the enhancement of efficiency in providing healthcare. For example, other applications that can be enabled by IoT are RPM, where IoT-enabled devices, for example, wearable sensors or home-based monitoring equipment, gather patient data in real time and send it to medical professionals. This will support in continuous tracking of vital signs, compliance in taking medications, and health status in getting early intervention and personalized treatment plans. It is particularly useful in monitoring chronic diseases such as diabetes, hypertension, and heart diseases, where it prevents recurring visits to the hospital and lowers the cost of healthcare that utilizes IoT devices such as video conferencing tools and connected medical peripherals to make it possible for remote consultations by patients with their care providers. IoT allows the safe sending of health data, incorporating pictures, test outcomes, and other knowledge among the devices and people involved. IoT-based applications in healthcare devices empower both patients and healthcare providers by delivering personalized, accessible, and efficient care while driving innovation in the healthcare industry [10].

The IoT enables various devices, sensors, and objects to exchange data and interact with each other by being connected to the Internet. This connectivity makes it easier to automate procedures, monitor them remotely, and increase overall productivity. Utilizing identification technologies to gain secure access to patient data and accomplish clear information sharing is one of the most important aspects of developing a healthcare IoT (HIoT) system. A UID is assigned to each entity in the system to guarantee correct identification and data integrity. Connection inside an HIoT network is made possible in large part by communication technology. Short-range communication protocols such as Wi-Fi, Bluetooth, and Radio-Frequency Identification (RFID) are frequently included with these technologies. RFID technology is used to identify and track assets or products in real time, while wireless networking and Bluetooth are commonly used to send data between devices that are next to each other. HIoT systems can provide effective data sharing and communication by utilizing these communication technologies to ensure smooth connectivity across diverse devices, sensors, and medical equipment [11]. The potential of HIoT systems is further enhanced by the incorporation of location technology, especially in healthcare settings. Medical equipment may be tracked and monitored thanks to location technology, which guarantees its availability when needed and speeds up emergency responses. Healthcare institutions can increase patient care quality, expedite processes, and increase safety by adding location-tracking technologies. For example, location technology can assist in the rapid location of particular medical supplies or devices in a hospital setting, hence decreasing reaction times and increasing overall efficiency [12].

4.3 Proposed model

The healthcare sector has undergone significant transformation since the introduction of IoT technology, which has changed patient care and management. Improving healthcare outcomes, using available resources most, and improving patient experiences are potential benefits of this paradigm shift toward intelligent and effective patient care. We explore the common generalized framework for implementing IoT in healthcare with this proposed model, concentrating on important elements and their connections.

1. Sensor technology integration:
   The foundation of the IoT in healthcare is sensor technology. A range of sensors, such as temperature, heart rate, blood pressure, glucose level, and motion sensors, are incorporated into medical equipment, wearable technology, and even infrastructure. These sensors gather data in real time from medical devices, patients, and environmental factors. The collected data is securely moved to centralized servers or cloud platforms for analysis and management.
2. Data transmission and communication:
   Effective communication routes are essential for sending sensor-collected data. Bluetooth, Wi-Fi, and cellular networks are examples of wireless communication protocols that allow data to be transferred from sensors to backend systems seamlessly. Ensuring data integrity and confidentiality through secure communication protocols protects patient privacy and complies with legal requirements like Health Insurance Portability and Accountability Act (HIPAA).
3. Cloud-based data storage and processing:
   IoT devices generate enormous amounts of data, which calls for a reliable infrastructure for processing and storing data [13]. Platforms for cloud computing provide scalable and affordable options for handling, organizing, and interpreting medical data. To support clinical decision-making and individualized patient care, sophisticated analytics tools and machine learning algorithms process the data in real time, extracting relevant insights and patterns.
4. Integration with electronic health record systems:
   For IoT data to be seamlessly integrated with current healthcare systems, interoperability is essential. Bidirectional connection between IoT platforms

and electronic health record (EHR) systems is made possible by application programming interfaces (APIs). Healthcare professionals can obtain a thorough understanding of patient health state, history, and ongoing monitoring data by integrating IoT data with EHR. This allows for prompt interventions and continuity of care [14].

5. Predictive analytics and remote monitoring:
   IoT-enabled predictive analytics are critical for proactive healthcare management. Using both historical and present-day data, predictive models can identify potential health risks, predict how a disease will progress, and launch early therapies. By facilitating ongoing vital sign monitoring and adherence to treatment plans, remote monitoring technologies enable for patients to actively participate in their treatment, which lowers hospital readmission rates and improves patient outcomes.
6. Wearable devices and mobile applications:
   Wearables with IoT capabilities allow for customized health tracking and monitoring. These gadgets, which range from fitness trackers to smartwatches, constantly monitor sleep patterns, physical activity, and vital signs. Users can receive real-time feedback, actionable insights, and reminders for medication adherence and lifestyle changes through integrated mobile applications. Mobile apps and wearables encourage patient participation give them more control over their care, and make it easier for patients to consult with doctors remotely [15].
7. Measures for privacy and security:
   Ensuring the security and privacy of medical records is crucial for IoT implementations. Sophisticated authentication methods, access controls, and encryption strategies protect private patient data against hacking and illegal access. By reducing the legal and reputational risks associated with data privacy violations, compliance with industry best practices and regulatory requirements promotes confidence among patients and healthcare stakeholders [16].

   The IoT and its application into the healthcare sector represent a significant paradigm shift toward intelligent and efficient patient care. Healthcare providers can offer remote, proactive, and tailored monitoring solutions by utilizing sensor technologies, data analytics, and interoperable systems. The suggested approach places a strong emphasis on how IoT devices can be seamlessly integrated with the current healthcare infrastructure while maintaining patient privacy, data security, and interoperability. As IoT advances, it has the ability to completely transform healthcare by enhancing patient experiences, reducing expenses, and raising results (Figure 4.1).

In the following sections, this chapter illustrates how the proposed model can be implemented in various sections of the society, discussing in details about the methodology and the results that can be achieved using IoT integrated in the healthcare section to achieve better results.

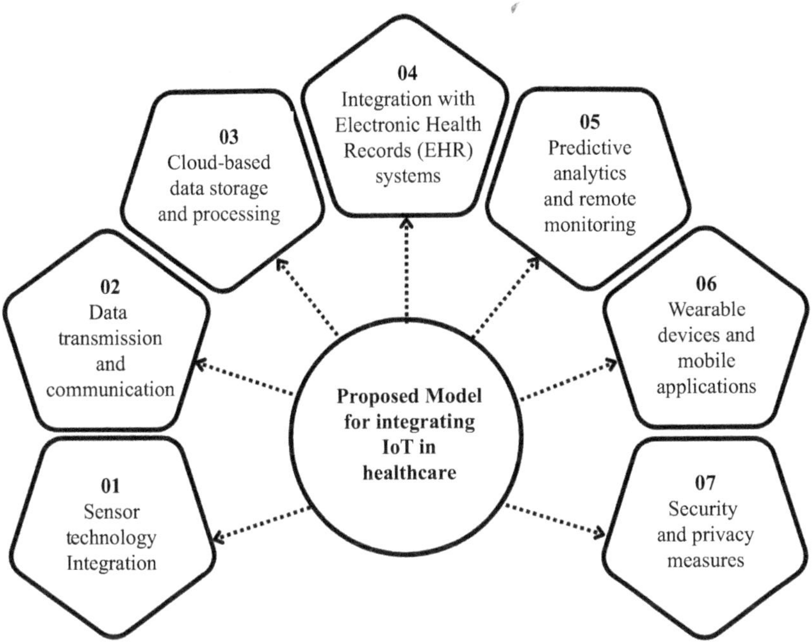

Figure 4.1 Proposed model for IoT in healthcare

4.4 Fall detection using IoT

An automated system that can monitor and notify us of accidents is necessary because falling accidents—such as construction workers falling from high floors or mountain climbers falling from high altitudes—occur frequently, and it can be difficult to provide emergency assistance in dire circumstances, specifically in view of the world's ageing population and associated issues such as inadequate medical services at the location of falls and ineffective fall detection systems. A device that recognizes incidents involving falls and instantly notifies rescue crews to come help [17]. In addition to sending notifications and precisely locating the victim using GPS technology, the device will be used to detect falling accidents. The technology immediately transmits the latitude and longitude coordinates of any falling mishap to the closest emergency service provider so that prompt action can be taken. The system uses both satellite navigation and mobile phone technology to find the car and alert the necessary parties—such as the police, family, and ambulance services—of an emergency. In particular, GSM is utilized to deliver alarm messages to the intended recipients that contain accurate position information, while GPS is used for location tracking. This approach may save the victim's life in isolated locations when incidents of this kind happen and no one is around to report them. This technology can facilitate a speedy reaction and possibly save lives in dire circumstances because of its capacity to identify accidents and notify emergency personnel promptly. The user's smartphone is connected to an accelerometer via GPS and GSM modules, and the system uses this information to make decisions and send messages to the device. Text messages containing the accident location are sent to the user's network connections and the nearby medical institutions by providing precise details about the victim's whereabouts.

4.4.1 Methodology

1. Arduino: The Arduino microcontroller is based on the ATmega328P. The Arduino UNO is a popular microcontroller board that uses an accelerometer sensor to detect any falling mishaps and a GSM module to notify emergency services or specified contacts. It is a well-liked option for embedded applications since it is effective, adaptable, and simple to use.
2. GSM module: Through the SIM900A GSM module, communication between GPS and a specified mobile number is enabled. With a frequency range of 900–1900 MHz, it uses a tri-band network that includes DSC 100 MHz, PCS 1900 MHz, and EGSM 900 MHz. It is made possible via the GPS module's sending pin and the GSM module's reception pin.
3. GPS module: The victim's location on Earth is determined using the SIM28ML GPS module, and coordinates are used to transfer the information to the Arduino board. The GPS module generates real-time position data in NMEA format and runs at 1575.42 MHz. The GSM module is then used to send the data to a specified contact.
4. Accelerometer: In accident detection systems, an accelerometer is a sensor that measures acceleration and recognizes changes in motion. The accelerometer can detect a large change in acceleration caused by the quick impact that occurs during an accident. After that, this information may sound an alarm or turn on safety features (Figure 4.2).

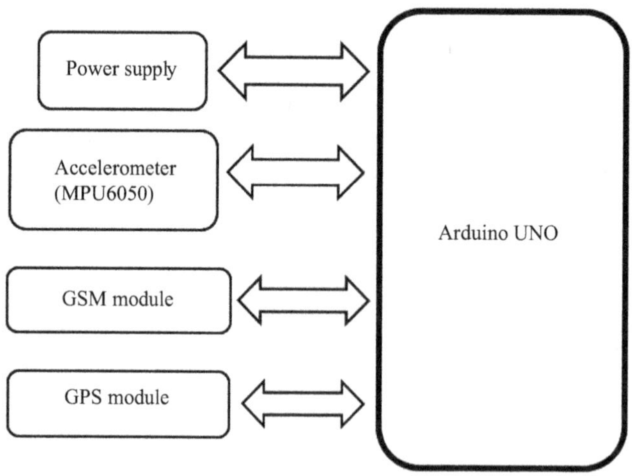

Figure 4.2 Block diagram of fall detection system

4.4.2 Working principle

1. Start: The procedure gets started. By doing this, the fall detection system is started and all required monitoring components are activated.
2. Calculate acceleration: One of the system's primary sensors, the accelerometer gauges the wearer's acceleration.
3. Check threshold: The recorded acceleration value is contrasted with a predefined threshold that is determined by user-specified criteria or empirical data. Should the recorded acceleration surpass the cutoff point, it implies an abrupt alteration in motion, which could potentially signify a fall incident.
4. Get GPS coordinates: When acceleration exceeds the threshold, the system initiates GPS coordinate retrieval. GPS coordinates give exact position information that is essential for emergency personnel to find the fall victim.
5. Send alert: Important details such as the GPS coordinates, the time of the fall, and any other pertinent information are included in an emergency alert message. Through suitable communication channels, including cellphone networks or internet-connected devices, the alarm is disseminated. Not only are the specified family members or caretakers receiving the message, but also adjacent hospitals and medical facilities. Several receivers guarantee the person who fell victim to timely support and aid.
6. End: After the alarm is sent, the procedure concludes. Upon successful transmission of the alert message, the fall detection system shuts off until the subsequent monitoring cycle. In standby mode, the system might continue to function, prepared to recognize and react to any further falls.

Fall detection systems are particularly beneficial for elderly people because they are at a higher risk of falls due to age-related factors, people with mobility issues like those with disabilities, or conditions that affect balance and movement. For building workers, fall detection devices can be quite helpful. Because falls from heights are so common, one area that stands to gain a great deal from this technology is the construction sector. The system can assist with real-time worker position and movement monitoring to identify falls early on. Prompt detection and alerting can result in quicker rescue and medical response, potentially saving lives. Automated alerts can expedite emergency protocols and guarantee that assistance is sent out right away. Incorporating fall detection systems into safety protocols on construction sites can greatly improve overall worker safety and reduce the incidence of serious injuries or fatalities due to falls. Also, this system can be used for mountain climbers. This type of system can monitor a climber's vitals, track their location via GPS, and detect falls, which is crucial in remote or high-altitude environments where quick rescue is essential. This technology enhances the safety of mountain climbers by ensuring that they can be located and assisted promptly if they experience a fall or other health issues while climbing (Figure 4.3).

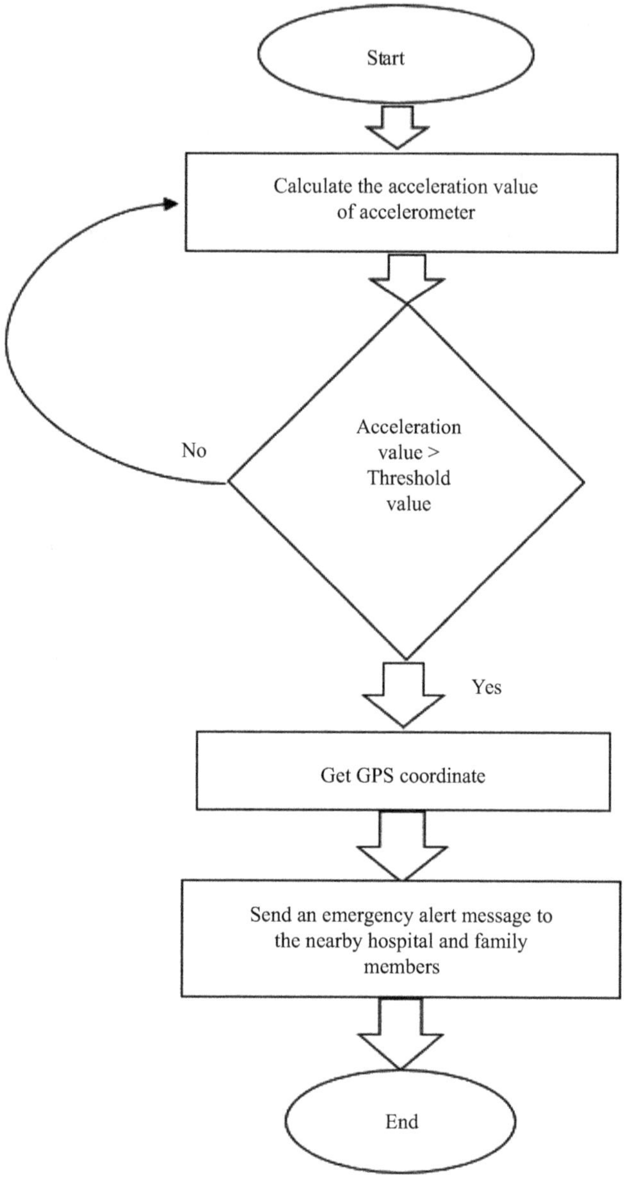

Figure 4.3 Flowchart of working of the fall detection model

4.5 IoT-powered smart bed

The main intention behind the development of smart beds is to improve patient outcomes, enhance comfort, and provide a safer environment for both patients and caregivers by leveraging IOT technology. Smart beds can collect real-time data automate various functions and provide timely inventions leading to better patient care and overall satisfaction. To improve patient outcomes, comfort, and safety through the integration of IoT technologies, smart beds represent a significant improvement in healthcare technology. Smart beds can improve patient care and satisfaction by utilizing IoT capabilities to gather data in real time, automate tasks, and deliver timely interventions [18]. By lightening the workload of medical personnel, smart beds significantly improve patient care in hospital settings. Routine duties like checking vital signs and modifying the bed's settings according to the patient's demands are automated by these beds. Caretakers can concentrate more on providing direct patient care by simplifying these procedures, which will boost productivity and improve resource allocation. Furthermore, by offering real-time alerts and notifications for any deviations from standard parameters, because they enable prompt interventions and prevent negative outcomes, smart beds contribute to the creation of a safer environment for both patients and carers. Smart beds have the benefit of continuous monitoring even when caregivers are not there for patients getting home care. Patients and their families may rest easy knowing that any anomalies or crises will be quickly identified and handled thanks to this continuous monitoring capacity. Caregivers can remotely access critical patient data, monitor trends, and receive alerts with smart beds that include remote monitoring functions. This allows caregivers to assist and support patients promptly even when they are far away. This raises the standard of care provided in home settings generally and promotes patient safety [19].

4.5.1 Methodology

The intelligent component of the smart bed is the IOT. It is a real-time data collection and processing network made up of seamlessly interacting sensors and networked devices. The centralized IOT platform receives continuous data collection from the sensors and processes and analyzes it. The IOT platform uses a complex algorithm to understand the data it has acquired in real time. This allows the system to identify abnormalities, and the smart bed can take action depending on the analysis. With the help of this clever technology, patients and healthcare professionals may communicate easily and remotely. All things considered, it increases effectiveness and makes fast interventions and modifications possible based on patient needs.

First, sensors are positioned within the bed in strategic ways to gather data in real time on patient movement, vital signs, and surroundings. Examples of these sensors are temperature, pressure, and accelerometer sensors. These sensors send data to an Arduino microcontroller, which is used to continuously monitor various parameters. After processing the incoming data, the Arduino microcontroller runs pre-programmed algorithms to decide what should be done. For instance, the microcontroller can activate linear actuators to help reposition the patient or to change the bed's posture for maximum comfort if a sensor notice extended inactivity or unusual vital signs that point to possible discomfort or distress. For remote access and control, the Arduino microcontroller can establish communication with other IoT devices or a central monitoring system. Caregivers can use this to remotely change the bed's settings as needed, keep an eye on the patient's condition, and get alerts for important events.

4.5.2 Working principle

1. The user lies on the bed: The user activates the monitoring system by lying down on the smart bed, which starts the procedure.
2. Is the patient's position normal: The system determines whether the user's initial bed position falls within predefined, regarded normal ranges.
3. Monitor position: The system begins to continuously monitor the user's position and motions if it determines that their posture is normal.
4. Position change detected: The user's posture is continuously monitored by the system to detect any changes that could indicate an odd movement or fall risk.
5. Activate airbag protection: The technology prevents the user from slipping off the bed by triggering the airbag protection mechanism when it senses a change in position that may indicate a risk.
6. Is user responsive: The system determines whether the user reacts to the situation after turning on the airbag protection.
7. Notify user: To guarantee that the user is aware of the safety precautions, the system alerts them if the airbag protection is turned on.
8. Alert caretaker: The device alerts the caregiver or medical personnel for assistance if the user does not reply to the activation of the airbag protection, suggesting possible distress or unconsciousness.
9. Adjust bed settings: The system will modify the bed settings by the user's saved preferences if the user's position is abnormal.
10. Check User's preferences: To guarantee individualized comfort, the system verifies the user's saved preferences, including preferred bed position and firmness.
11. Set bed position: The technology modifies the bed position to achieve the best possible comfort and support based on the user's preferences.
12. Activating smart alarm: To provide timely alerts, the system sets off a smart alarm by the prearranged wake-up time or in the event of an emergency.
13. Adjusting lights: The user's preferences are taken into account by the system, which brightens the lights when the user gets up and dims them while they sleep.
14. Monitor user: The system keeps track of the user's movements and sleeping habits to provide insights into the general health and quality of sleep.
15. User wakes up: Based on motions sensed or other cues, the system keeps track of whether the user has woken up from sleep.

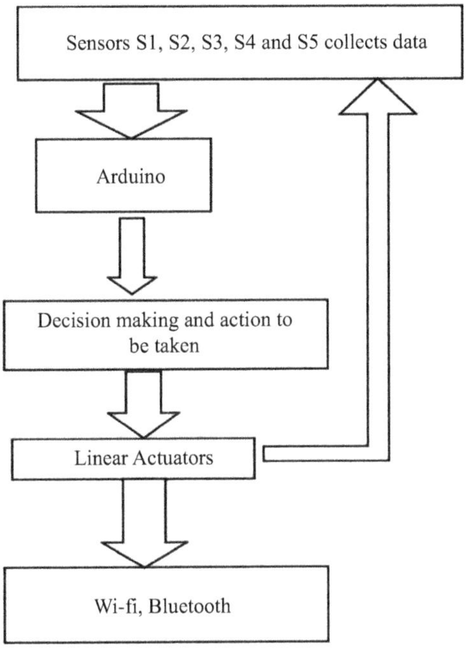

Figure 4.4 Block diagram of smart bed

16. Turn off the alarm: To avoid more disruption, the system silences the alert if the user is awake.
17. Adjusting the lights: Again, the system adjusts the room's illumination to make it comfortable for the user based on their level of awareness.
18. Continued monitoring: To ensure continued safety and comfort, the system keeps an eye on the user's actions and surroundings until a noticeable change is noticed (Figures 4.4 and 4.5).

By providing proactive fall prevention measures and individualized adjustments for the best possible comfort and safety, the use of IOT in smart beds allows for real-time monitoring of patient movement and vital signs. The IOT platform lowers risk and medical expenses, improving patient care. By automatically modifying bed settings, enabling a gentle wake-up routine with customized light and sound cues, and minimizing the need for continual bedside presence, IOT technology provides patients with impairments and illnesses with crucial help. Healthcare practitioners can obtain information from the IoT for individualized treatment programs and preventive care tactics. An IOT-enabled smart bed may easily interface with an EHR, allowing clinicians to share data and coordinate patient care.

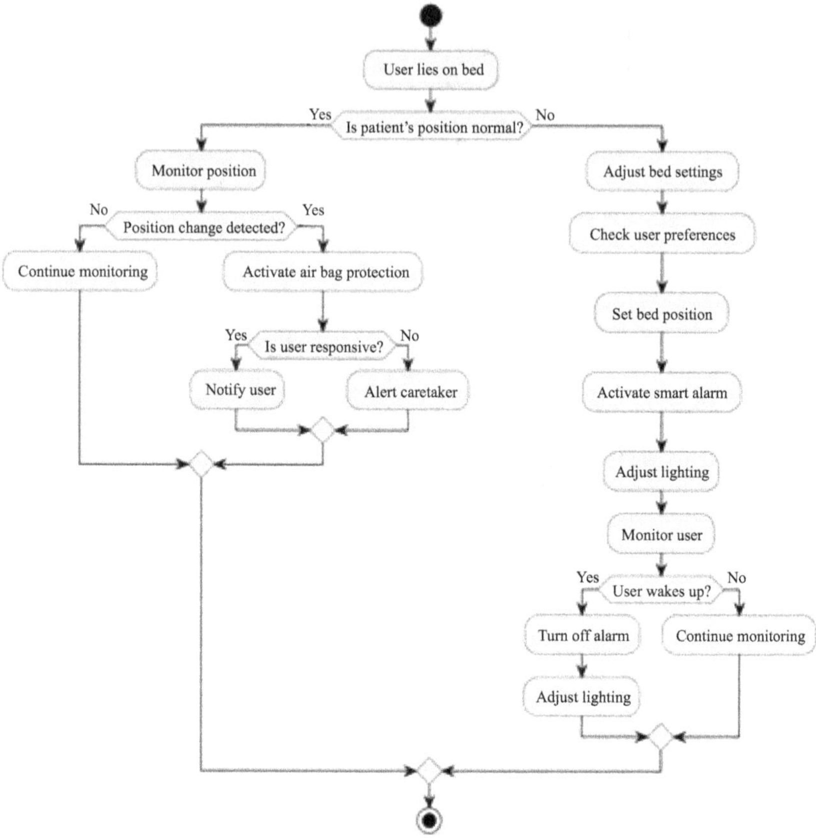

Figure 4.5 Flowchart of working of smart bed

4.6 IoT and AR-powered remote surgery assistance

The combination of the IoT and augmented reality (AR) technology has revolutionized the field of remote surgery assistance by providing surgeons with never-before-seen capabilities to support their work and improve patient outcomes. This novel method provides real-time instruction, data integration, and collaborative tools for remote surgical procedures by utilizing the immersive imagery of augmented reality and the connectivity of the IoT. AR and IoT-powered remote surgery assistance emerges as a promising solution to address challenges in delivering expert surgical expertise to underserved areas and optimizing surgical workflows in established healthcare facilities, as the demand for advanced medical care continues to rise beyond geographic barriers. AR technology improves situational awareness and precision during surgical procedures by superimposing virtual data and visuals onto the surgeon's field of view. Real-time data, including vital signs, imaging results, and instrument tracking, may be seamlessly integrated into the AR interface by incorporating IoT sensors into surgical tools and equipment. This allows surgeons to make well-informed judgments and modify their approaches in real time. Additionally, IoT connectivity makes it possible for surgical teams to collaborate remotely. This enables specialists to offer advice and support from a distance, increasing access to specialized treatment and encouraging knowledge exchange among medical professionals.

4.6.1 Methodology

Providing step-by-step insights into how AR and IOT are used in remote surgery, this aims to shed light on the benefits, limitations, and overall process of using AR to enhance the capabilities of qualified surgeons operating on distant sports patients.

1. Operating surgeon with HL: The surgeon operating wears a head-mounted display or AR software, which projects digital information onto their real-world view. HL likely refers to "Head-mounted Light" which might be part of the AR headset.
2. Video capture: It is the transcript or graphic displayed on the AR surgeon's view of the surgical site.
3. Voice (surgeon): It is through this equipment that the surgeon's voice messages will be passed first. Then, the surgeons resonate with their microphones with the remote consultant.
4. Capture of the environment: The camera captures the high-definition live video from the surgical field before it is broadcast back to the consultant.
5. Annotations and voice communication: The remote consultant receives the live video feed and can annotate it. Annotations may include instructions, guidance, or highlighting specific areas of interest. A speaker transmits the voice of a remote consultant, who could be a specialist surgeon or another medical professional observing the surgery remotely.
6. Access to medical images: Additional medical images, such as X-rays or CT scans, might be referenced during surgery and could be seen on the computer of the remote consultant or the surgeon's console.
7. Remote consultant with a computer: A specialist surgeon or other expert participates in the surgery remotely. They view a live video feed from the operating room on a computer monitor (Figure 4.6).

4.6.2 Working principle

1. Start: The process begins with activating the robot.
2. Initiate the robot: The robot is then initialized, presumably preparing its systems for operation.
3. Live stream the video: A live video feed is initiated, likely transmitting a view from a camera onboard the robot.
4. Move forward and follow the beacon system: The robot will start to move forward and will navigate by the beacon system it follows. A beacon sends the radio signals; the robot will then use them to locate itself and proceed on its path.
5. Obstacle detected: Sensors on the robot continuously scan for obstacles in its path.
6. Ultrasonic sensors: In case any obstruction is detected by ultrasonic sensors, further, the robot takes action. It works on high-frequency sound waves that are impossible for human ears and measures distance. In the given diagram is shown mounting front (F), left (L), and right (R) ultrasonic sensors.

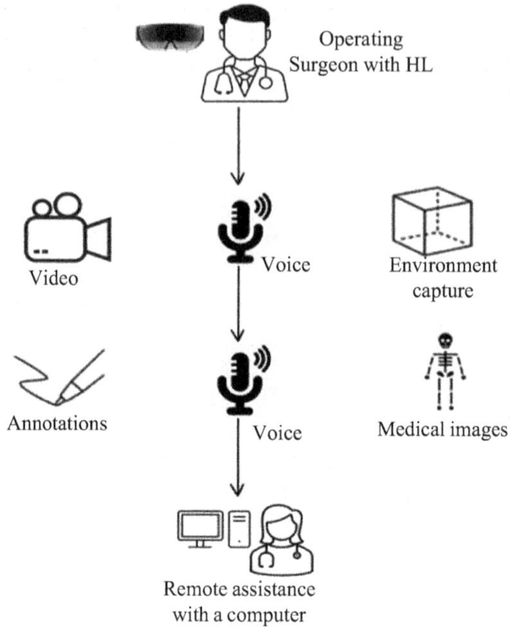

Figure 4.6 A presentation of HL using information sent by the remote consultant (speech, annotations, and medical pictures) and the operative surgeon (video, audio, and environment capture data)

7. Fuzzy-based optimal path: In case an obstacle is sensed, then there will be the operation of an optimal path of the robot with the assistance of the system of fuzzy logic. Fuzzy logic is an AI that provides an approach to approximate reasoning close to the capability of human beings' reasoning and judgment. It deals with imprecise and incomplete data and is fitted for applications where sensor readings could be inexact.
8. Turn toward the beacon direction: This implies that the robot will turn in a direction that would help it get closer to the beacon.
9. Play the message instructions through the Google-based voice entity: The message played by the robot on Google text-to-speech technology through a speaker when it arrives at a place.
10. Open/close door: A sensor detects whether the door is open or closed.
11. Reached the room status: This decision point indicates that the robot has reached its destination, likely a patient's room.
12. Enter the patient room and play the message: If the door is open, the robot enters the room and plays a message.
13. Deliver medicines to the patient: One possible task the robot performs in the room is delivering medication to a patient.
14. Take the patient's heart rate: The robot may also be used to measure the individual's heart rate.

15. Measure the patient's body temperature: Another vital sign the robot might measure is the patient's body temperature.
16. Update patient database: The collected patient data is likely stored in a database for medical staff to review.

This demonstrates the potential of AR in enhancing surgical precision and improving patient outcomes by providing real-time, three-dimensional information to surgeons. When access to surgical knowledge is critical in emergency crises and disaster response scenarios, AR and IoT-powered remote surgery aid have compelling advantages. First responders and remote medical teams can work together efficiently to perform life-saving interventions even in difficult circumstances with limited resources by utilizing wearable augmented reality gadgets and IoT-enabled surgical equipment. AR and IOT remote surgery assistant systems could greatly make positive impacts on the lives of humankind and could change parameters about the science of medicine and the delivery of healthcare (Figure 4.7).

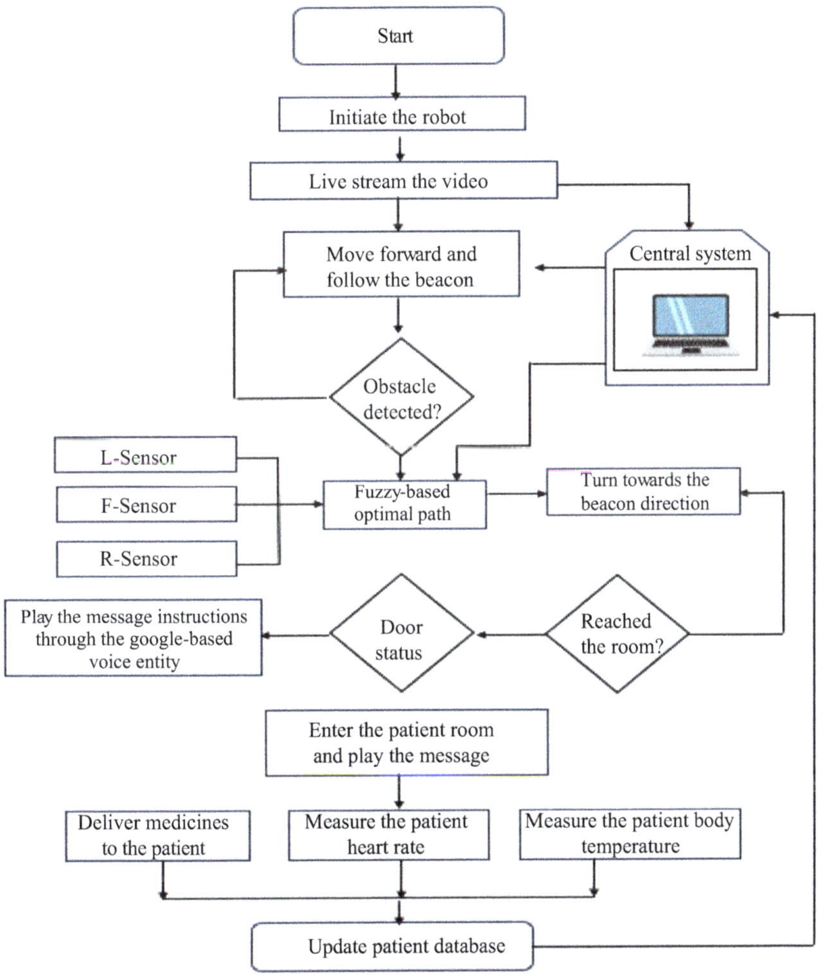

Figure 4.7 Flowchart to explain the working of remote surgery assistance

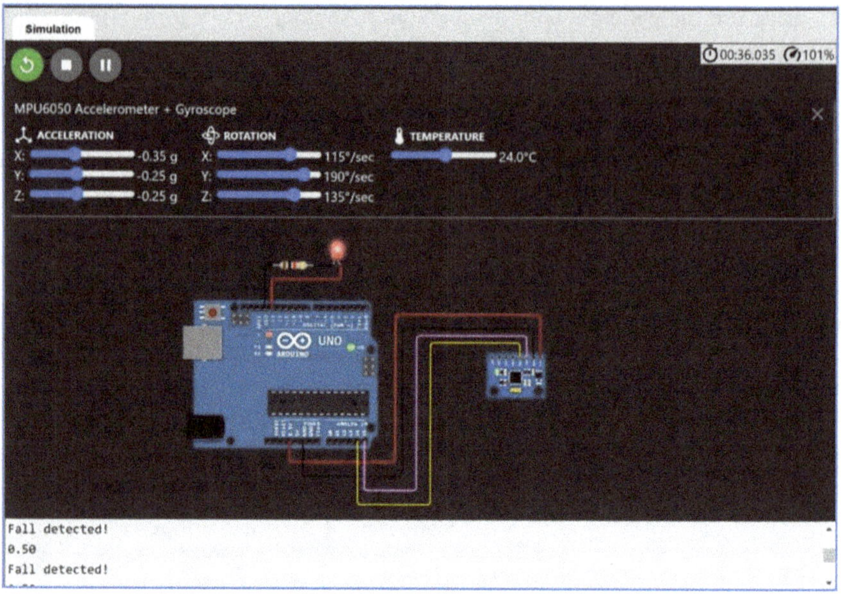

Figure 4.8 Screenshot of result obtained

4.7 Results and analysis

In Figure 4.8, a screenshot is shown which is the Virtual IOT Environment created for *fall detection*. The circuit includes Arduino Uno, MPU6050 Accelerometer + Gyroscope, resistor, and led. Once the acceleration and/or rotation is changed, fall is detected. If either of the *X* axis, *Y* axis, or *Z* axis value is changed, and if the value becomes <0.5, then fall is detected.

4.8 Future research direction

The upcoming research ought to concentrate on creating IoT solutions that are compatible with current healthcare devices and systems. Ensuring interoperability and data interchange across various platforms and suppliers would require standardization of communication protocols and data formats. In IoT healthcare research, addressing security and privacy issues will be crucial. It is important to create strong authentication procedures, data anonymization strategies, and encryption approaches to safeguard private patient data and reduce cybersecurity threats [20,21]. Research endeavors ought to concentrate on promoting IoT-based RPM solutions. To facilitate ongoing vital sign monitoring, medication adherence, and illness management outside of conventional healthcare settings, this comprises wearable sensors, home monitoring devices, and telemedicine platforms. By using IoT-generated data, healthcare analytics can gain an important understanding of disease outbreaks, population health patterns, and the efficacy of therapy. To extract useful insights from huge healthcare datasets, research in this field should investigate sophisticated analytics approaches including machine learning, predictive modeling, and data visualization.

For smart beds, improvement of the precision and dependability of the sensor to guarantee accurate tracking of the patient's movements and vital signs is needed. Cutting-edge AI algorithms need to be utilized to assess sleep trends and adjust bed settings for individual comfort. Using gesture- or voice-controlled interfaces to enhance accessibility and user experience for patients with limited mobility can be beneficial. In fall detection, increasing the specificity and sensitivity of the sensor to detect falls more precisely while reducing false alarms is needed. Machine learning algorithms use patterns in sensor data to distinguish between fall occurrences and regular activity. For quicker aid and intervention, improve integration with emergency response services and mobile alert systems. Investigating cutting-edge AR visualization methods to improve anatomical visualization and surgical guiding. IoT sensors are incorporated into surgical equipment to give real-time feedback on tissue properties, surgical results, and instrument location. Haptic feedback systems can enhance surgical dexterity during remote procedures by imitating tactile sensations. We can further advance the capabilities and effectiveness of IoT in healthcare by addressing these research directions and putting improvements into case studies such as smart beds, fall detection, and AR IoT remote surgery assistance. This will ultimately improve patient care outcomes and enhance healthcare delivery.

4.9 Conclusion

IoT integration in healthcare signifies a significant paradigm change toward intelligent and effective patient care. Healthcare providers may transform the delivery, monitoring, and management of patient care by utilizing IoT technology, including sensors, connectivity, and data analytics. IoT has the potential to totally change the healthcare industry by enabling preventative interventions, automating repetitive operations, and gathering data in real time. These benefits will ultimately result in better patient outcomes, more efficient use of resources, and improved healthcare results. Smart beds, fall detection devices, and remote surgical support are just a few examples of IoT-enabled products that have significant benefits for improving patient accessibility, comfort, and safety. With the help of these advances, healthcare professionals can now provide patients with prompt, individualized care in both hospital and home environments by providing them with actionable information, real-time monitoring capabilities, and collaborative tools.

IoT integration in healthcare also creates new opportunities for innovation and research in fields including security, interoperability, RPM, and healthcare analytics. Healthcare stakeholders may usher in a new era of linked healthcare, marked by enhanced efficiency, efficacy, and equity in healthcare delivery, by tackling these obstacles and fully utilizing IoT technologies. All things considered, the introduction of the IoT to the field of medicine signals a paradigm change toward a more intelligent, efficient, and patient-focused healthcare system. As we continue to study and advance in this rapidly developing field, the promise of IoT in healthcare offers a great opportunity to transform patient experiences and healthcare delivery, ultimately improving the health and well-being of people and their communities worldwide.

References

[1] Li C, Wang J, Wang S, and Zhang Y. A review of IOT applications in healthcare. *Neurocomputing*. 2024;565:127017. doi:10.1016/j.neucom.2023.127017.

[2] Dash, SP. The impact of IoT in healthcare: Global technological change & the roadmap to a networked architecture in India. *Journal of the Indian Institute of Science* 2020;100:773–85. doi:0.1007/s41745-020-00208-y.

[3] Rahman SM, Ibtisum S, Podder P, and Hossain SM. Progression and challenges of IoT in healthcare: A short review. 2023. arXiv preprint arXiv:2311.12869.

[4] Mohammed BG, and Hasan DS. Smart healthcare monitoring system using IoT. *International Journal of Interactive Mobile Technologies*. 2023;17(1):141–52.

[5] Asiminidis, C, Kokkonis, G, Kazanidis, I, and Kontogiannis, S. Design and implementation of wearable e-health monitoring system with medical and tactile sensors for remote patient surveillance. *International Journal of Science, Environment and Technology*. 2019: 8-1.

[6] Taiwo O, and Ezugwu AE. Smart healthcare support for remote patient monitoring during COVID-19 quarantine. *Informatics in Medicine Unlocked*. 2020;20:100428. doi:10.1016/j.imu.2020.100428.

[7] Balasundaram A, Routray S, Prabu AV, Krishnan P, Malla PP, and Maiti M. Internet of things (IoT)-based smart healthcare system for efficient diagnostics of health parameters of patients in emergency care. *IEEE Internet of Things Journal*. 2023;10(21):18563–70. doi:10.1109/jiot.2023.3246065.

[8] Juyal S, Sharma S, and Shankar Shukla A. Smart skin health monitoring using AI-enabled cloud-based IOT. *Materials Today: Proceedings*. 2021;46:10539–45. doi:10.1016/j.matpr.2021.01.074.

[9] Emelia Akashah PA, and Noor Shita A. An IOT platform for seizure alert wearable device. *IOP Conference Series: Materials Science and Engineering*. 2020;767(1):012012. doi:10.1088/1757-899x/767/1/012012.

[10] Pradhan B, Bhattacharyya S, and Pal K. IOT-based applications in healthcare devices. *Journal of Healthcare Engineering*. 2021;2021:1–18. doi:10.1155/2021/6632599.

[11] Babu BS, Srikanth K, Ramanjaneyulu T, and Narayana IL. IoT for healthcare. *International Journal of Science and Research*. 2016;5(2):322–6.

[12] Mohamad Jawad HH, Bin Hassan Z, Zaidan BB, Mohammed Jawad FH, Mohamed Jawad DH, and Alredany WH. A systematic literature review of

enabling IOT in healthcare: Motivations, challenges, and recommendations. *Electronics*. 2022;11(19):3223. doi:10.3390/electronics11193223.

[13] Selvaraj S, and Sundaravaradhan S. Challenges and opportunities in IOT healthcare systems: A systematic review. *SN Applied Sciences*. 2019;2:139. doi:10.1007/s42452-019-1925-y.

[14] Haghi Kashani M, Madanipour M, Nikravan M, Asghari P, and Mahdipour E. A systematic review of IOT in healthcare: Applications, techniques, and trends. *Journal of Network and Computer Applications*. 2021;192:103164. doi:10.1016/j.jnca.2021.103164.

[15] Shah R, and Chircu A. IoT and AI in healthcare: A systematic literature review. *Issues in Information Systems*. 2018;19(3).

[16] Azzawi, M, Hassan, R, and Abu Bakar KA. A review on Internet of Things (IoT) in healthcare. *International Journal of Applied Engineering Research*. 2016;11:10216–21.

[17] Yacchirema D, de Puga JS, Palau C, and Esteve M. Fall detection system for elderly people using IOT and big data. *Procedia Computer Science*. 2018; 130:603–10. doi:10.1016/j.procs.2018.04.110.

[18] Abdelmoghith A, Shaaban R, Alsheghri Z, and Ismail L. IOT-based healthcare monitoring system: Bedsores prevention. *2020 Fourth World Conference on Smart Trends in Systems, Security, and Sustainability (WorldS4)*, 2020. doi:10.1109/worlds450073.2020.9210319.

[19] Verdejo Espinosa Á, Lopez Ruiz JL, Mata Mata F, and Estevez ME. Application of IOT in healthcare: Keys to implementation of the sustainable development goals. *Sensors*. 2021;21(7):2330. doi:10.3390/s21072330.

[20] Yadav, AK, Singh, K, Amin, AH, Almutairi, L, Alsenani, TR, and Ahmadian, A. A comparative study on consensus mechanism with security threats and future scopes: Blockchain. *Computer Communications*.2023; 201:102–15.

[21] Praveen, G, Anand, M, Singh, PK, and Ranjan, P. An overview of blockchain consensus and vulnerability. *Proceedings of the Information and Communication Technology for Intelligent Systems – ICTIS 2020*, Volume 1, 459–68, 2021.

Chapter 5

Gradient-based optimization approach to solve fuzzy algebraic equations governed engineering problems

Paresh Kumar Panigrahi[1] *and Sukanta Nayak*[2]

Abstract

In this chapter, we explore a gradient-based optimization technique to tackle a fuzzy-valued unconstrained optimization problem in which the objective function is fuzzy. In this regard, the fuzzy system of linear equation (FSLE) can be converted into fuzzy-valued optimization problem. Using the concept of fuzzy center and fuzzy arithmetic, the proposed approach can deal with the uncertain system. Then, using the proposed method, both only fuzzy (where either the coefficient matrix or right-hand side vector is fuzzy) and fully fuzzy (where both coefficient matrix and right-hand side vector are fuzzy) systems are investigated. Convergence analysis is conducted to ensure the existence of a solution, followed by solving a variety of example problems using the proposed method across different scenarios. Then, the obtained solutions are compared with the other existing methods, and it is found to be a good agreement.

Keywords: Fuzzy set; triangular fuzzy number; fuzzy optimization problem; fuzzy-valued function; fuzzy gradient-based optimization technique

5.1 Introduction

In mathematical modeling, the system of linear equations plays an important role in identifying field variables and reveals the hidden relationships between the involved parameters. Additionally, the presence of uncertainties in the system needs modification in mathematical models that yield an uncertain system of linear equations. In many physical problems, epistemic uncertainties are found very often. These uncertainties occur due to physical inaccuracy, mechanical defects, experimental errors, vague data, and impreciseness. As such, fuzzy set theory [1] can be

[1]Department of Mathematics, School of Engineering, Dayananda Sagar University, India
[2]Department of Mathematics, School of Advanced Sciences, VIT-AP University, India

used as a tool to handle the same. The inclusion of fuzzy makes it challenging to investigate the system using mathematical models. However, to study the system with greater precision, uncertainties cannot be avoided. In 1965, Zadeh introduced the concept of fuzzy sets [1]. Then, various researchers used fuzzy sets for their modeled problems. As the system of equations is an integral part of quantifying the modeled problems, researchers presented different approaches to solving a fuzzy linear system (FLS) equation. A few of the relevant research works are discussed here. In [2–5], authors presented the analytical method to solve fuzzy linear and quadratic equations. To find the solution of the linear system of the equations, variables and coefficients of a parameter are presumed to be either real or complex fuzzy numbers. In this context, Friedman *et al.* [5] used an idea to transform the fuzzy $n \times n$ system into $2n \times 2n$ FSLE and then a simplified algorithm is presented to solve a FSLE. Recently, Abbasi and Allahviranloo [6] used the extension principle to solve FLS equations. These research works are carried out with an analytical approach.

However, due to the limitations of analytical approaches in the context of real-life problems, numerical methods can be adopted. In light of this, many relevant research works have been done with numerical methods to investigate the FSLE. Ma *et al.* [7] used an embedding method to solve the FLS equations with the concept of duality. Additionally, Dehghan and Hashemi [8] presented various iterative algorithms to solve the FSLE, whereas Dehghan *et al.* [9] employed an iterative algorithm to solve the fully FSLE. Further, the LU decomposition method is adopted by Abbasbandy *et al.* [10] to solve the FLS equations. Nayak and Chakraverty [11] considered the limit approach to defining fuzzy arithmetic and the same is adopted to solve the system. As intervals are the basis of fuzzy numbers, Karunakar and Chakraverty [12] introduced a technique to solve the interval system of equations. Besides, optimization techniques are helpful in the investigation of a system of linear equations. Hence, many contributions are reported to handle the system of equations under a fuzzy environment. Abbasbandy and Jafarian [13] have incorporated the steepest descent optimization method to solve the FLS equations. In addition, Nayak and Pooja [14] have investigated the interval system of the equation using an optimization method. Panigrahi and Nayak [15] proposed a fuzzy optimization technique to solve the system of equations. Panigrahi and Nayak [16] developed a derivative-free optimization technique to handle unconstrained optimization problems in a fuzzy environment. Furthermore, Panigrahi and Nayak [17] have applied the numerical method to solve the interval nonlinear system of equations.

The above literature review reveals that there are many analytical and numerical methods to solve the FLS equations. As such, the gradient-based optimization method can be used to solve the FLS equations. Furthermore, Abbasbandy and Jafarian [13] introduced the steepest descent optimization method to solve the FLS equations. Here, the authors investigate a $2n \times 2n$ system of linear equation. Therefore, the steepest descent gradient-based method is directly implemented for easy understanding to solve the same. One of the ideas is to use the parametric form for the FSLE to convert the crisp system. Then, the crisp system is computed by using the different cases. Besides this, the solution of this method can be discussed as the convergence theorem.

Therefore, we have extended the gradient descent optimization approach (GDOA) in a fuzzy environment. Here, we have used the proposed fuzzy gradient descent optimization (FGDO) to solve the linear system of equations for both only fuzzy and fully fuzzy by using different cases. The proposed gradient-based method works for the FSLE to give a better solution. To validate this method, convergence analysis is done here. Two example problems are studied, and the numerical solutions are discussed. Further, the solutions are shown graphically for easy interpretation. The solutions obtained are collected for three different cases and compared with two different approaches.

The chapter is organized as follows: Section 5.2 provides an overview of fuzzy numbers and their arithmetic operations. Section 5.3 categorizes FLS equations. Section 5.4 introduces the gradient descent optimization method within a fuzzy framework, along with its convergence theorems. Section 5.5 showcases the application of the FGDO in solving two example problems in a fuzzy environment, presenting the obtained solutions.

5.2 Preliminaries

This section includes the fundamentals of fuzzy numbers and its arithmetic [18].

A fuzzy number $\tilde{\mathcal{A}}$ is defined as a convex normalized fuzzy set $\tilde{\mathcal{A}}$ on the real line $\mathbb{R}$ with the membership function

$$\mu_{\tilde{\mathcal{A}}}(\tilde{x}) : \mathbb{R} \to [0,1],\ \forall \tilde{x} \in \mathbb{R}, \tag{5.1}$$

where $\mu_{\tilde{\mathcal{A}}}$ is piecewise continuous.

One of the fuzzy numbers, the triangular fuzzy number (TFN) is written as follows:

$$\tilde{\mathcal{A}} - [\tilde{a}_L, \tilde{a}_N, \tilde{a}_R] \tag{5.2}$$

where $\tilde{a}_L$, $\tilde{a}_N$, and $\tilde{a}_R$ are the left, center, and right values of the TFN $\tilde{\mathcal{A}}$ and $\tilde{a}_L \leq \tilde{a}_N \leq \tilde{a}_R$.

For example, if we take two TFNs, namely $\tilde{\mathcal{A}} = [\tilde{a}_L,\ \tilde{a}_N, \tilde{a}_R]$ and $\tilde{\mathcal{B}} = [\tilde{b}_L,\ \tilde{b}_N, \tilde{b}_R]$, then they are said to be equal if $\tilde{a}_L = \tilde{b}_L, \tilde{a}_N = \tilde{b}_N$ and $\tilde{a}_R = \tilde{b}_R$. The width of the TFN $\tilde{\mathcal{A}}$ is defined as $\tilde{w} = \tilde{a}_R - \tilde{a}_L$.

To compute TFNs, one may use the traditional arithmetic [19] operations. Let $\tilde{\mathcal{A}} = [\tilde{a}_L, \tilde{a}_N, \tilde{a}_R]$ may be converted into an interval by using $\alpha -$ cut as shown below

$$\tilde{\mathcal{A}} = [\tilde{a}_L, \tilde{a}_N, \tilde{a}_R] = [\tilde{a}_L + (\tilde{a}_N - \tilde{a}_L)\alpha,\ \tilde{a}_R - (\tilde{a}_R - \tilde{a}_N)\alpha], where\ \alpha \in [0,1].$$

Basically, the TFN is a collection of intervals, where each membership value $\alpha \in [0,\ 1]$ corresponds to an interval. Consequently, when computing with TFNs, operations are conducted pointwise. Thus, it is necessary to construct a function incorporating the variable α to compute TFNs [20] effectively. In the next section, we will investigate into solving fuzzy systems of linear equations utilizing these arithmetic operations.

In the following, we may rewrite the definition of TFN [21].

A fuzzy number $\tilde{x} = [\tilde{x}_L, \tilde{x}_N, \tilde{x}_R]$ is classified as a TFN when its membership values are defined as follows:

$$\mu_{\tilde{X}}(\tilde{x}) = \begin{cases} \tilde{f}_L, \tilde{x}_L \leq \tilde{x} \leq \tilde{x}_N \\ \tilde{f}_R, \tilde{x}_N \leq \tilde{x} \leq \tilde{x}_R \\ 0, \ otherwise \end{cases} \tag{5.3}$$

where, $\tilde{f}_L$ is the left monotonically increasing function and $\tilde{f}_R$ is the right monotonically decreasing function. These functions can be depicted as $\frac{\tilde{f}_L = \tilde{x} - \tilde{x}_L}{\tilde{x}_N - x_L}$ and $\frac{f_R = \tilde{x}_R - \tilde{x}}{\tilde{x}_R - \tilde{x}_N}$.

For computational convenience, an arbitrary TFN $\tilde{\mathcal{X}} = [\tilde{x}_L, \tilde{x}_N, \tilde{x}_R]$. It can be converted into a two-variable form by transforming TFN to interval form and then interval to crisp form.

Using α-cut, the TFN can be written as follows [22]:

$$\tilde{\mathcal{X}} = [\tilde{x}_L, \tilde{x}_N, \tilde{x}_R] \approx [\xi_L^\alpha, \xi_R^\alpha], \tag{5.4}$$

where

$$\xi_L^\alpha = \tilde{x}_L + \alpha(\tilde{x}_N - \tilde{x}_L) \text{ and } \xi_R^\alpha = \tilde{x}_R + \alpha(\tilde{x}_N - \tilde{x}_R); \ \alpha \in [0, 1].$$

The crisp representation of TFN $\tilde{\mathcal{X}}$ is obtained as follows:

$$\xi_L^\alpha - \beta(\xi_L^\alpha - \xi_R^\alpha), \beta \in [0, 1]. \tag{5.5}$$

To investigate a fuzzy system, utilize the aforementioned form in the following systems.

5.3 Fuzzy linear system equation

In this section, we have discussed FLS, which can be classified into three different areas cases [23].

Consider a matrix form of the system of linear equations

$$\tilde{M}\tilde{X} = \tilde{\mathcal{Y}} \tag{5.6}$$

Here, $\tilde{M}$ represents the coefficient matrix, $\tilde{b}$ denotes the right-hand side vector, and $\tilde{X}$ denotes the unknown vector to be determined. A brief presentation of this system is defined as

$$\tilde{M} = [\tilde{m}_{ij}], \tilde{\mathcal{Y}} = [\tilde{b}_1, \tilde{b}_2, \ldots, \tilde{b}_n]^T, \text{ and } \tilde{X} = [\tilde{x}_1, \tilde{x}_2, \ldots \tilde{x}_n]^T, \ i,j = 1, 2, \ldots n \tag{5.7}$$

Here, the entries $\tilde{a}_{ij}$ and $\tilde{b}_j$ are consider TFNs within the system of equations with n variables, that is,

$$\tilde{M}_{n\times n}\,\tilde{X}_{n\times 1}=\tilde{\mathcal{Y}}_{n\times 1}, \tag{5.8}$$

where

$$\tilde{M}_{n\times n} = \begin{bmatrix} \tilde{m}_{11} & \tilde{m}_{12} & \cdots & \tilde{m}_{1n} \\ \tilde{m}_{21} & \tilde{m}_{22} & \cdots & \tilde{m}_{2n} \\ \vdots & \vdots & \ddots & \vdots \\ \tilde{m}_{n1} & \tilde{m}_{n2} & \cdots & \tilde{m}_{nn} \end{bmatrix}, \tilde{X}_{n\times 1} = \begin{bmatrix} \tilde{x}_1 \\ \tilde{x}_2 \\ \vdots \\ \tilde{x}_n \end{bmatrix}, \text{ and } \tilde{\mathcal{Y}}_{n\times 1} = \begin{bmatrix} \tilde{b}_1 \\ \tilde{b}_2 \\ \vdots \\ \tilde{b}_n \end{bmatrix}. \tag{5.9}$$

The extended form of TFNs $\tilde{m}_{ij}$ and $\tilde{\mathcal{Y}}_j$ are

$$\tilde{m}_{ij} = \left[m_{L_{ij}}, m_{N_{ij}}, m_{R_{ij}}\right] \text{ and } \tilde{\mathcal{Y}}_j = \left[b_{L_{ij}}, b_{N_{ij}}, b_{R_{ij}}\right]; i,j = 1,2,\cdots,n.$$

The fully fuzzy system of equation (FSE) can be defined as

$$\tilde{M}_{n\times n} = \begin{bmatrix} \tilde{m}_{11} & \tilde{m}_{12} & \cdots & \tilde{m}_{1n} \\ \tilde{m}_{21} & \tilde{m}_{22} & \cdots & \tilde{m}_{2n} \\ \vdots & \vdots & \ddots & \vdots \\ \tilde{m}_{n1} & \tilde{m}_{n2} & \cdots & \tilde{m}_{nn} \end{bmatrix}, \quad \tilde{X}_{n\times 1} = \begin{bmatrix} \tilde{x}_1 \\ \tilde{x}_2 \\ \vdots \\ \tilde{x}_n \end{bmatrix}, \text{ and } \tilde{\mathcal{Y}}_{n\times 1} = \begin{bmatrix} \tilde{b}_1 \\ \tilde{b}_2 \\ \vdots \\ \tilde{b}_n \end{bmatrix}$$

where the entries of the coefficient matrix $\tilde{m}_{ij} = \left[m_{L_{ij}}, m_{N_{ij}}, m_{R_{ij}}\right]$ and the right-hand side vector $\tilde{\mathcal{Y}}_j = \left[b_{L_{ij}}, b_{N_{ij}}, b_{R_{ij}}\right]$ are TFN. Here, the above system of equations is a $n \times n$ FSE. Then, the subsequent system of equations is also $n \times n$ system, which is based on the parametric form of TFNs. The next section includes the GDOA to solve the FSE.

5.4 FGDO method for FSLE

Here, the FGDO technique is adopted to solve the FSE and then followed by presenting a modified algorithm.

Consider a FSE $\tilde{M}\tilde{X} = \tilde{\mathcal{Y}}$, where M is a symmetric positive definite matrix (SPDM). Then, we need to find the minimum of the quadratic function whose minimum is the solution of the system $\tilde{M}\tilde{X} - \tilde{\mathcal{Y}}$.

Take the quadratic function

$$F(X) = \frac{1}{2}\tilde{X}^T\tilde{M}\tilde{X} - \tilde{X}^T\tilde{\mathcal{Y}} \tag{5.10}$$

where

$$\tilde{M} = \begin{bmatrix} [\tilde{m}_{L_{11}}, \tilde{m}_{N_{11}}, \tilde{m}_{R_{11}}] & [\tilde{m}_{L_{12}}, \tilde{m}_{N_{12}}, \tilde{m}_{R_{12}}] & \cdots & [\tilde{m}_{L_{1n}}, \tilde{m}_{N_{1n}}, \tilde{m}_{R_{1n}}] \\ [\tilde{m}_{L_{21}}, \tilde{m}_{N_{21}}, \tilde{m}_{R_{21}}] & [\tilde{m}_{L_{22}}, \tilde{m}_{N_{22}}, \tilde{m}_{R_{22}}] & \cdots & [\tilde{m}_{L_{1n}}, \tilde{m}_{N_{1n}}, \tilde{m}_{R_{1n}}] \\ \vdots & \vdots & \ddots & \vdots \\ [\tilde{m}_{L_{n1}}, \tilde{m}_{N_{n1}}, \tilde{m}_{R_{11}}] & [\tilde{m}_{L_{n2}}, \tilde{m}_{N_{n2}}, \tilde{m}_{R_{n2}}] & \cdots & [\tilde{m}_{L_{nn}}, \tilde{m}_{N_{nn}}, \tilde{m}_{R_{nn}}] \end{bmatrix}, \tilde{X}_{n\times 1}$$

$$= \begin{bmatrix} [\tilde{x}_{L_1}, \tilde{x}_{N_1}, \tilde{x}_{R_1}] \\ [\tilde{x}_{L_2}, \tilde{x}_{N_2}, \tilde{x}_{R_2}] \\ \vdots \\ [\tilde{x}_{L_n}, \tilde{x}_{N_n}, \tilde{x}_{R_n}] \end{bmatrix}, \tilde{y} = \begin{bmatrix} [\tilde{b}_{L_1}, \tilde{b}_{N_1}, \tilde{b}_{R_1}] \\ [\tilde{b}_{L_2}, \tilde{b}_{N_2}, \tilde{b}_{R_2}] \\ \vdots \\ [\tilde{b}_{L_n}, \tilde{b}_{N_n}, \tilde{b}_{R_n}] \end{bmatrix}$$

The above-mentioned system is written as follows:

$$\sum_{i,j=1}^{n} \left[\tilde{m}^{\alpha}_{L_{ij}}, \tilde{m}^{\alpha}_{R_{ij}}\right]\left[\tilde{x}^{\alpha}_{L_j}, \tilde{x}^{\alpha}_{R_j}\right] = \left[\tilde{b}^{\alpha}_{L_j}, \tilde{b}^{\alpha}_{R_j}\right]. \tag{5.11}$$

Equation (5.11) can be expressed equivalently as follows:

$$M_L = \sum_{i,j=1}^{n} \tilde{m}^{\alpha}_{L_{ij}} \tilde{x}^{\alpha}_{L_j} = \tilde{b}^{\alpha}_{L_j} \tag{5.12}$$

$$M_R = \sum_{i,j=1}^{n} \tilde{m}^{\alpha}_{R_{ij}} \tilde{x}^{\alpha}_{R_j} = \tilde{b}^{\alpha}_{R_j} \tag{5.13}$$

Equations (5.12) and (5.13) can be written as in quadratic function

$$F(X) = [M_L, M_R] = \frac{1}{2}\left[\tilde{x}^{\alpha}_{L_j}, \tilde{x}^{\alpha}_{R_j}\right]^T \left[m^{\alpha}_{L_{ij}}, \tilde{m}^{\alpha}_{R_{ij}}\right]\left[\tilde{x}^{\alpha}_{L_j}, \tilde{x}^{\alpha}_{R_j}\right] - \left[\tilde{x}^{\alpha}_{L_j}, \tilde{x}^{\alpha}_{R_j}\right]^T \left[\tilde{b}^{\alpha}_{L_j}, \tilde{b}^{\alpha}_{R_j}\right]$$

By using the gradient descent method for (5.10). We get the gradient of the given function as $\left[\tilde{m}^{\alpha}_{L_{ij}}, \tilde{m}^{\alpha}_{R_{ij}}\right]$ is symmetric.

$$\nabla F(X) = \left[\tilde{m}^{\alpha}_{L_{ij}}, \tilde{m}^{\alpha}_{R_{ij}}\right]\left[\tilde{x}^{\alpha}_{L_j}, \tilde{x}^{\alpha}_{R_j}\right] - \left[\tilde{b}^{\alpha}_{L_j}, \tilde{b}^{\alpha}_{R_j}\right] \tag{5.14}$$

$$-\nabla F(X) = \left[\tilde{b}^{\alpha}_{L_j}, \tilde{b}^{\alpha}_{R_j}\right] - \left[\tilde{a}^{\alpha}_{L_{ij}}, \tilde{a}^{\alpha}_{R_{ij}}\right]\left[\tilde{x}^{\alpha}_{L_j}, \tilde{x}^{\alpha}_{R_j}\right] \tag{5.15}$$

Let

$$\left[\tilde{r}^{\alpha}_{L_j}, \tilde{r}^{\alpha}_{R_j}\right]_k = \left[\tilde{b}^{\alpha}_{L_j}, \tilde{b}^{\alpha}_{R_j}\right] - \left[\tilde{m}^{\alpha}_{L_{ij}}, \tilde{m}^{\alpha}_{R_{ij}}\right]\left[\tilde{x}^{\alpha}_{L_j}, \tilde{x}^{\alpha}_{R_j}\right]_k, \tag{5.16}$$

be the residual point of $\tilde{X}_k$. Then, $\tilde{X}_k$ is updated as follows:

$$\left[\tilde{x}^{\alpha}_{L_j}, \tilde{x}^{\alpha}_{R_j}\right]_{k+1} = \left[\tilde{x}^{\alpha}_{L_j}, \tilde{x}^{\alpha}_{R_j}\right]_k - \left[\tilde{\lambda}^{\alpha}_{L_j}, \tilde{\lambda}^{\alpha}_{R_j}\right]_k \left[\tilde{r}^{\alpha}_{L_j}, \tilde{r}^{\alpha}_{R_j}\right]_k \tag{5.17}$$

and $\tilde{X}_k$ are updated by using the gradient descent approach, where the matrix $\tilde{M}$ is a nonsingular matrix.

$$\left[\tilde{x}^{\alpha}_{L_j}, \tilde{x}^{\alpha}_{R_j}\right]_{k+1} = \left[\tilde{x}^{\alpha}_{L_j}, \tilde{x}^{\alpha}_{R_j}\right]_k - \left[\tilde{\lambda}^{\alpha}_{L_j}, \tilde{\lambda}^{\alpha}_{R_j}\right]_k \left[\tilde{r}^{\alpha}_{L_j}, \tilde{r}^{\alpha}_{R_j}\right]_k$$

where

$$\left[\tilde{\lambda}^{\alpha}_{L_j}, \tilde{\lambda}^{\alpha}_{R_j}\right]_k = \frac{\left[\tilde{r}^{\alpha}_{L_j}, \tilde{r}^{\alpha}_{R_j}\right]_k^T \left[\tilde{r}^{\alpha}_{L_j}, \tilde{r}^{\alpha}_{R_j}\right]_k}{\left[\tilde{r}^{\alpha}_{L_j}, \tilde{r}^{\alpha}_{R_j}\right]_k^T \left[\tilde{a}^{\alpha}_{L_{ij}}, \tilde{a}^{\alpha}_{R_{ij}}\right] \left[\tilde{r}^{\alpha}_{L_j}, \tilde{r}^{\alpha}_{R_j}\right]_k}. \tag{5.18}$$

The above formulation will be further illustrated and solved through example problems in next section. Additionally, the algorithm will address the application of gradient-based optimization techniques and their extension in a fuzzy environment.

5.4.1 Fuzzy gradient descent algorithm

The above GDOA can be used in the following cases to solve the FSE (Figure 5.1).

Case 1 In (5.9), the elements of the coefficient matrix are represented as TFNs. The entry $\tilde{m}_{ij}$ is defined as $\left[m_{L_{ij}}, m_{N_{ij}}, m_{R_{ij}}\right]$ while the right-hand side vector is $\mathcal{Y}_j$ and remains crisp.

The matrix representation of Case 1 of FSLE can be expressed as follows:

$$\begin{bmatrix} \left[\tilde{m}_{L_{11}}, \tilde{m}_{N_{11}}, \tilde{m}_{R_{11}}\right] & \left[\tilde{m}_{L_{12}}, \tilde{m}_{N_{12}}, \tilde{m}_{R_{12}}\right] & \cdots & \left[\tilde{m}_{L_{1n}}, \tilde{m}_{N_{1n}}, \tilde{m}_{R_{1n}}\right] \\ \left[\tilde{m}_{L_{21}}, \tilde{m}_{N_{21}}, \tilde{m}_{R_{21}}\right] & \left[\tilde{m}_{L_{22}}, \tilde{m}_{N_{22}}, \tilde{m}_{R_{22}}\right] & \cdots & \left[\tilde{m}_{L_{1n}}, \tilde{m}_{N_{1n}}, \tilde{m}_{R_{1n}}\right] \\ \vdots & \vdots & \ddots & \vdots \\ \left[\tilde{m}_{L_{n1}}, \tilde{m}_{N_{n1}}, \tilde{m}_{R_{11}}\right] & \left[\tilde{m}_{L_{n2}}, \tilde{m}_{N_{n2}}, \tilde{m}_{R_{n2}}\right] & \cdots & \left[\tilde{m}_{L_{nn}}, \tilde{m}_{N_{nn}}, \tilde{m}_{R_{nn}}\right] \end{bmatrix} \begin{bmatrix} \left[\tilde{x}_{L_1}, \tilde{x}_{N_1}, \tilde{x}_{R_1}\right] \\ \left[\tilde{x}_{L_2}, \tilde{x}_{N_2}, \tilde{x}_{R_2}\right] \\ \vdots \\ \left[\tilde{x}_{L_n}, \tilde{x}_{N_n}, \tilde{x}_{R_n}\right] \end{bmatrix} = \begin{bmatrix} b_1 \\ b_2 \\ \vdots \\ b_n \end{bmatrix}. \tag{5.19}$$

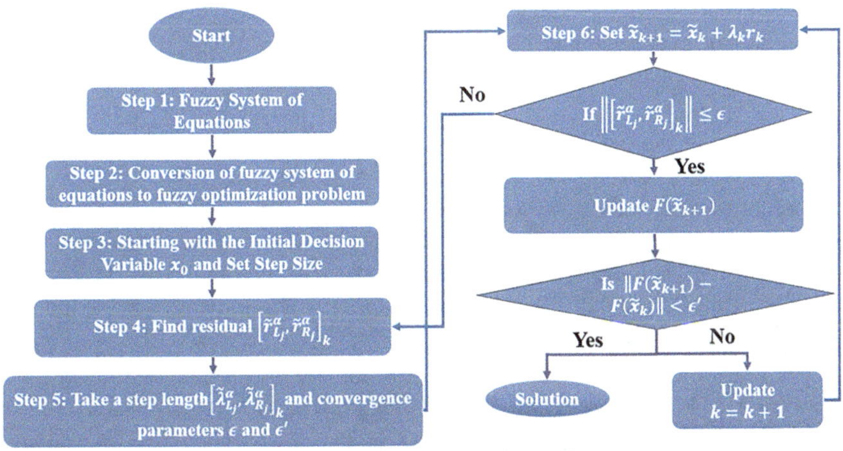

Figure 5.1 Flow chart of the proposed fuzzy gradient descent optimization algorithm

Case 2 In (5.9), the TFNs valued right-hand side vector is represent as $\tilde{\mathcal{Y}}_j$ and the same is defined as $\left[b_{L_{ij}}, b_{N_{ij}}, b_{R_{ij}}\right]$, whereas the coefficient of the left-hand side vector remains $\tilde{m}_{ij}$ and is crisp which is expressed as matrix representation of Case 2.

$$\begin{bmatrix} m_{11} & m_{12} & \cdots & m_{1n} \\ m_{21} & m_{22} & \cdots & m_{2n} \\ \vdots & \vdots & \ddots & \vdots \\ m_{n1} & m_{n2} & \cdots & m_{nn} \end{bmatrix} \begin{bmatrix} [\tilde{x}_{L_1}, \tilde{x}_{N_1}, \tilde{x}_{R_1}] \\ [\tilde{x}_{L_2}, \tilde{x}_{N_2}, \tilde{x}_{R_2}] \\ \vdots \\ [\tilde{x}_{L_n}, \tilde{x}_{N_n}, \tilde{x}_{R_n}] \end{bmatrix} = \begin{bmatrix} [\tilde{b}_{L_1}, \tilde{b}_{N_1}, \tilde{b}_{R_1}] \\ [\tilde{b}_{L_2}, \tilde{b}_{N_2}, \tilde{b}_{R_2}] \\ \vdots \\ [\tilde{b}_{L_n}, \tilde{b}_{N_n}, \tilde{b}_{R_n}] \end{bmatrix} \tag{5.20}$$

Case 3 In (5.9), the elements of both the coefficient matrix and the right-hand side vector are TFNs.

The matrix representation of Case 3 is shown as follows:

$$\begin{bmatrix} [\tilde{m}_{L_{11}}, \tilde{m}_{N_{11}}, \tilde{m}_{R_{11}}] & [\tilde{m}_{L_{12}}, \tilde{m}_{N_{12}}, \tilde{m}_{R_{12}}] & \cdots & [\tilde{m}_{L_{1n}}, \tilde{m}_{N_{1n}}, \tilde{m}_{R_{1n}}] \\ [\tilde{m}_{L_{21}}, \tilde{m}_{N_{21}}, \tilde{m}_{R_{21}}] & [\tilde{m}_{L_{22}}, \tilde{m}_{N_{22}}, \tilde{m}_{R_{22}}] & \cdots & [\tilde{m}_{L_{1n}}, \tilde{m}_{N_{1n}}, \tilde{m}_{R_{1n}}] \\ \vdots & \vdots & \ddots & \vdots \\ [\tilde{m}_{L_{n1}}, \tilde{m}_{N_{n1}}, \tilde{m}_{R_{11}}] & [\tilde{m}_{L_{n2}}, \tilde{m}_{N_{n2}}, \tilde{m}_{R_{n2}}] & \cdots & [\tilde{m}_{L_{nn}}, \tilde{m}_{N_{nn}}, \tilde{m}_{R_{nn}}] \end{bmatrix} \begin{bmatrix} [\tilde{x}_{L_1}, \tilde{x}_{N_1}, \tilde{x}_{R_1}] \\ [\tilde{x}_{L_2}, \tilde{x}_{N_2}, \tilde{x}_{R_2}] \\ \vdots \\ [\tilde{x}_{L_n}, \tilde{x}_{N_n}, \tilde{x}_{R_n}] \end{bmatrix}$$
$$= \begin{bmatrix} [\tilde{b}_{L_1}, \tilde{b}_{N_1}, \tilde{b}_{R_1}] \\ [\tilde{b}_{L_2}, \tilde{b}_{N_2}, \tilde{b}_{R_2}] \\ \vdots \\ [\tilde{b}_{L_n}, \tilde{b}_{N_n}, \tilde{b}_{R_n}] \end{bmatrix} \tag{5.21}$$

where the entries of the coefficient matrix $\tilde{m}_{ij} = \left[m_{L_{ij}}, m_{N_{ij}}, m_{R_{ij}}\right]$ and the right-hand side vector $\tilde{\mathcal{Y}}_j = \left[b_{L_j}, b_{N_j}, b_{R_j}\right]$ are TFNs.

The aforementioned cases will be explored and resolved through example problems. This algorithm is dedicated to addressing FSLE. To better understand the fuzzy gradient descent algorithm for FLS equations, we have discussed a few supporting theorems in the below paragraphs.

Definition 1 For the gradient descent method using TFN, the field variables must possess convergence to the center and width of the obtained TFN.

As our system deals with fuzzy, we get a FSE. So, to solve the FSE, we have extended the gradient descent method for the FSE. The proposed gradient-based technique for TFN has been shown to be convergent. According to Definition 1, we must first demonstrate that the center value of TFN shows convergence and then establish that the width of TFN converges.

Theorem 1 *If M is the real and SPDM, then solving* $\tilde{M}\tilde{X}_k = \tilde{\mathcal{Y}}$ *is equivalent to minimizing the quadratic function* $f(x_k) = \frac{1}{2}\tilde{X}_k^T\tilde{M}\tilde{X}_k - \tilde{X}_k^T\tilde{\mathcal{Y}}$.

Proof
The quadratic function $f(x_k) = \frac{1}{2}\tilde{X}_k^T \tilde{M} \tilde{X}_k - \tilde{X}_k^T \tilde{\mathcal{Y}}$
Furthermore, the minimum of $f(\tilde{X}_k)$ *is obtained by taking the above FSDA*
Considered an approximate solution $\tilde{X}_k$.
x_k *is updated by using the gradient descent method*

$$\tilde{X}_{k+1} = \tilde{X}_k + \lambda_k r_k \tag{5.22}$$

The gradient of $f(\tilde{X}_k)$ *is*

$$\nabla F(\tilde{X}) = \tilde{M}\tilde{X} - \tilde{b} \tag{5.23}$$

$$-\nabla F(\tilde{X}) = \tilde{\mathcal{Y}} - \tilde{M}\tilde{X} \tag{5.24}$$

The negative gradient is the residual of the point $\tilde{X}_k$, *that is,*

$$r_k = \tilde{\mathcal{Y}} - \tilde{M}\tilde{X}_k \tag{5.25}$$

where r_k *is the direction point of the* $f(X_k)$.
Let

$$f(\tilde{X}_{k+1}) = f(\tilde{X}_k + \lambda_k r_k) \tag{5.26}$$

$$f(x_{k+1}) = \frac{1}{2}(\tilde{X}_k + \lambda_k r_k)^T \tilde{M}(\tilde{X}_k + \lambda_k r_k) - (\tilde{X}_k + \lambda_k r_k)^T \tilde{\mathcal{Y}}$$

$$\frac{1}{2}\tilde{X}_k^T \tilde{M}\tilde{X}_k + \frac{\lambda^2}{2} r_k^T \tilde{M} r_k + \lambda_k \tilde{X}_k^T \tilde{M} r_k - \tilde{X}_k^T \tilde{\mathcal{Y}} - \lambda_k r_k \tilde{\mathcal{Y}}.$$

The function is minimum when $f'(\tilde{X}_k) = 0$ *in terms of* $\tilde{X}_k$ *and* r_k *by using (5.16) and (5.17),*

$$\lambda_k = \frac{r_k^T r_k}{r_k^T A r_k}, \text{ where } \tilde{X} = [\tilde{x}_1, \tilde{x}_2, \ldots, \tilde{x}_n]^T$$

$\tilde{X}_k$ *will be updated as a solution to the given system of linear equations.*

Hence, $\tilde{X}_k$ *will converge in n number of iterations. It can be shown by the next theorem that the error decreases iteration-wise.*

Theorem 2 *Suppose that M is a positive definite matrix (PDM), and* w_k *is the error vector generated by the proposed algorithm and* η *is the eigenvalue of matrix M, where* $\eta_{\max}$ *and* $\eta_{\min}$ *is the maximum and minimum eigenvalues. Then, by using the proposed algorithm of the* kth *iteration satisfies the following condition:*

$$\|w_{k+1}\|_{\tilde{M}} \leq \frac{\eta_{\max} - \eta_{\min}}{\eta_{\max} + \eta_{\min}} \|w_k\|_{\tilde{M}} \tag{5.27}$$

Proof
Let M be a positive definite matrix (PDM), then the error

$$w_k = x^* - x_k$$

where x^* *is the exact solution of* $\tilde{M}\tilde{X} = \tilde{\mathcal{Y}}$ *produced by using the gradient descent approach.*

Such that

$$\|w_{k+1}\|_M \leq \frac{\eta_{\max} - \eta_{\min}}{\eta_{\max} + \eta_{\min}} \|w_k\|_M$$

and the gradient descent approach convergence to any initial gauss x_0

$$\|w_k\|_A = w_k^T M w_k > 0$$

where M *is PDM.*

$$\|w_{k+1}\|_M \leq \frac{\eta_{\max} - \eta_{\min}}{\eta_{\max} + \eta_{\min}} \|w_k\|_M$$

$$= \frac{\frac{\eta_{\max}}{\eta_{\min}} - 1}{\frac{\eta_{\max}}{\eta_{\min}} + 1} \|w_k\|_M$$

here the spectral condition number always $\frac{|\lambda_{\max}|}{|\lambda_{\min}|} > 1$*, then after a few steps,* w_{k+1} *converges.*

Hence, the convergence rate depends on the spectral condition number. If the spectral condition number is low, then the convergence rate is fast.

Theorem 3 *The gradient descent optimization method satisfies the given condition*

$$\|x^* - x_{k+1}\|_2 < \|x^* - x_k\|_2 \tag{5.28}$$

unless $x_k = x^*$

Proof

The above statement can be proved by using the theorem given in [24].

Using Theorem 3, a unique solution of the linear system of equations can be obtained.

As the center solution of TFN means solution at $\alpha-$cut unity, Theorem 3 gives the center is convergence. Next, we need to prove the width of TFN. Theorem 3 shows the proof of the gradient descent method when the membership function is one. However, Theorem 4 is presented, when the membership function is zero.

Theorem 4 [15] *Suppose* $\{\tilde{x}_n\}$ *and* $\{\tilde{y}_n\}$ *are two convergent sequences for the left and right values of the TFN, respectively, such that* $\lim_{n\to\infty} \tilde{x}_n \to \tilde{a}$ *and* $\lim_{n\to\infty} \tilde{y}_n \to \tilde{b}$*. Suppose,* $\tilde{\eta}_n$ *and* $\tilde{\xi}$ *be the widths of the TFNs which are defined as* $\tilde{\eta}_n = |\tilde{x}_n - \tilde{y}_n|$ *and* $\tilde{\xi} = |\tilde{a} - \tilde{b}|$*. Then,* $\lim_{n\to\infty} \tilde{\eta}_n \to \tilde{\xi}$*.*

Proof: Given, $\lim_{n\to\infty} \tilde{x}_n \to \tilde{a}$ and $\lim_{n\to\infty} \tilde{y}_n \to \tilde{b}$.

Let there exist $\epsilon > 0$ and integers N_1 and N_2 such that

$$d(\tilde{x}_n, \tilde{a}) < \frac{\epsilon}{2} \text{ for } n_1 > N_1 \text{ and } d(\tilde{y}_n, \tilde{b}) < \frac{\epsilon}{2} \text{ for } n_2 > N_2.$$

Then, we need to prove that $\left| d(\tilde{x}_n - \tilde{y}_n, \tilde{a} - \tilde{b}) \right| < \epsilon$.

Using triangle inequality,

$$d(\tilde{x}_n, \tilde{y}_n) \leq d(\tilde{x}_n, \tilde{a}) + d(\tilde{y}_n, \tilde{b}) + d(\tilde{a}, \tilde{b}). \tag{5.29}$$

Equation (5.29) gives

$$d(\tilde{x}_n, \tilde{y}_n) - d(\tilde{a}, \tilde{b}) \leq d(\tilde{x}_n, \tilde{a}) + d(\tilde{y}_n, \tilde{b}) = \frac{\epsilon}{2} + \frac{\epsilon}{2} = \epsilon. \tag{5.30}$$

Similarly,

$$d(\tilde{a}, \tilde{b}) \leq d(\tilde{x}_n, \tilde{a}) + d(\tilde{y}_n, \tilde{b}) + d(\tilde{x}_n, \tilde{y}_n). \tag{5.31}$$

Equation (5.31) can be represented as

$$d(\tilde{a}, \tilde{b}) - d(\tilde{x}_n, \tilde{y}_n) \leq d(\tilde{x}_n, \tilde{a}) + d(\tilde{y}_n, \tilde{b}) = \frac{\epsilon}{2} + \frac{\epsilon}{2} = \epsilon. \tag{5.32}$$

From (5.30) and (5.32), we get

$$\left| d(\tilde{x}_n, \tilde{y}_n) - d(\tilde{a}, \tilde{b}) \right| \leq d(\tilde{x}_n - \tilde{a}) + d(\tilde{y}_n - \tilde{b}) < \epsilon \forall n > \max\{N_1, N_2\}. \tag{5.33}$$

This proves

$$\left| d(\tilde{x}_n, \tilde{y}_n) - d(\tilde{a}, \tilde{b}) \right| \leq \epsilon. \tag{5.34}$$

Therefore, we conclude that both the center and width of TFNs converge. The next section discussed two example problems to demonstrate the proposed algorithm and its effectiveness.

Example problem Here we discuss the proposed optimization approach to solve the FSLE using the following example problems.

Example 1 Take the TFN linear system of equation [25]

$$\begin{aligned} \tilde{a}_{11}\tilde{x}_1 + \tilde{a}_{12}\tilde{x}_2 &= \tilde{\chi}_1 \\ \tilde{a}_{21}\tilde{x}_1 + \tilde{a}_{22}\tilde{x}_2 &= \tilde{\chi}_2 \end{aligned} \tag{5.35}$$

where $\tilde{a}_{11} = [2.5,\ 3,\ 3.5]$, $\tilde{a}_{12} = [-1.5,\ -1,\ -0.5]$, $\tilde{a}_{21} = [-1.5,\ -1,\ -0.5]$, $\tilde{a}_{22} = [1.5,\ 2,\ 2.5]$, $\tilde{\chi}_1 = [0,\ 3,\ 5]$, $\tilde{\chi}_2 = [1,\ 2,\ 7]$. Assume the initial approximation $x^0 = (0, 0)^T$, and tolerance value $\epsilon = 10^{-3}$.

Case 1 Here, we consider the left-hand side elements that consist TFNs, i.e., the coefficient matrix is taken as fuzzy. Accordingly, the following coefficients are used for the investigation: $\tilde{a}_{11} = [2.5, 3, 3.5]$, $\tilde{a}_{12} = [-1.5, -1, -0.5]$, $\tilde{a}_{21} = [-1.5, -1, -0.5]$, $\tilde{a}_{22} = [1.5, 2, 2.5]$ $\tilde{\chi}_1 = [3]$, $\tilde{\chi}_2 = [2]$.

Applying the FGDO algorithm for an FSLE, the obtained solution vectors $\tilde{x}_1$, and $\tilde{x}_2$ are listed in Tables 5.1 and 5.2. The fuzzy solution is shown graphically in Figure 5.2.

In Table 5.1, the width of $\tilde{x}_1$ is defined as the difference between the right and the left values of the TFN.

In Table 5.2, the width of $\tilde{x}_2$ is defined as the difference between the right the left values of the TFN.

The obtained fuzzy solution for Case 1 of Example 1 is present in Figure. 5.2 and width-wise convergence is shown in Figure 5.3.

Case 2 Here, the right-hand side is treated as TFNs, Thus, the right-hand side vector is observed as fuzzy. Consequently, the following vector is used for the investigation: a_{ij} are crisp $(i,j = 1, 2)$, $\tilde{\chi}_1 = [0, 3, 5]$, $\tilde{\chi}_2 = [1, 2, 7]$ becomes fuzzy.

Table 5.1 TFNs solution component of $\tilde{x}_1$ for Case 1 of Example 1

Iteration	$\tilde{x}_{L_1}$	$\tilde{x}_{N_1}$	$\tilde{x}_{R_1}$	**Width**
1	0	0	0	0
2	1.0986	1.6957	3.7143	2.6157
3	0.9521	1.3996	2.9088	1.9567
4	1.0047	1.612	4.4623	3.4576
5	0.9977	1.5749	4.1254	3.1277
6	1.0002	1.6015	4.7751	3.7749
7	0.9999	1.5969	4.6342	3.6343
8	0.9999	1.6002	4.9059	3.906
9	0.9999	1.5996	4.847	3.8471
End Point	0.9999	1.5996	4.9992	3.9993

Table 5.2 TFNs solution component $\tilde{x}_2$ for Case 1 of Example 1

Iteration	x_{L_2}	x_{N_2}	x_{R_2}	**Width**
1	0	0	0	0
2	0.7324	1.1304	2.4762	1.7438
3	0.9521	1.5745	3.6845	2.7324
4	0.9872	1.7161	4.7201	3.7329
5	0.9977	1.7718	5.2255	4.2278
6	0.9994	1.7895	5.6586	4.6592
7	0.9999	1.7965	5.87	4.8701
8	0.9999	1.7987	6.0511	5.0512
9	0.9999	1.7996	6.1395	5.1396
End point	0.9999	1.7996	6.3323	5.3324

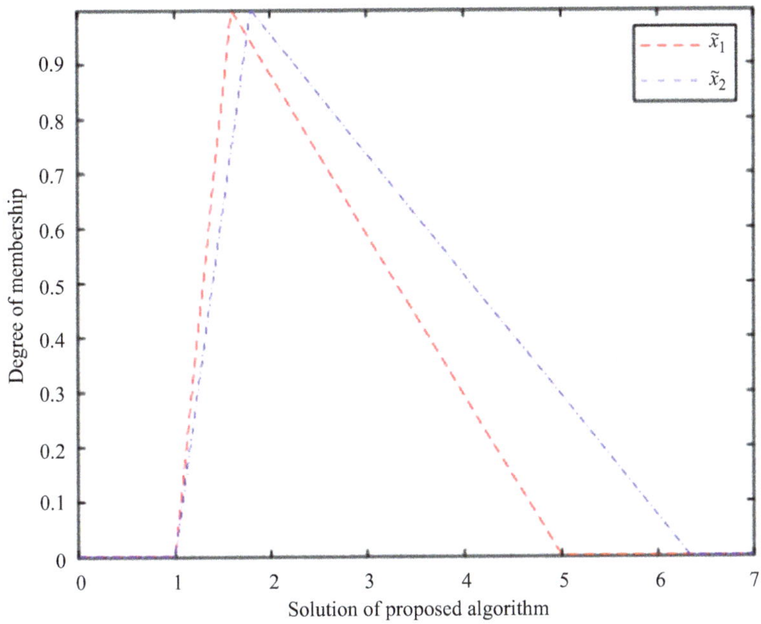

Figure 5.2 Solution components $\tilde{x}_1$ and $\tilde{x}_2$ for Case 1 of Example 1

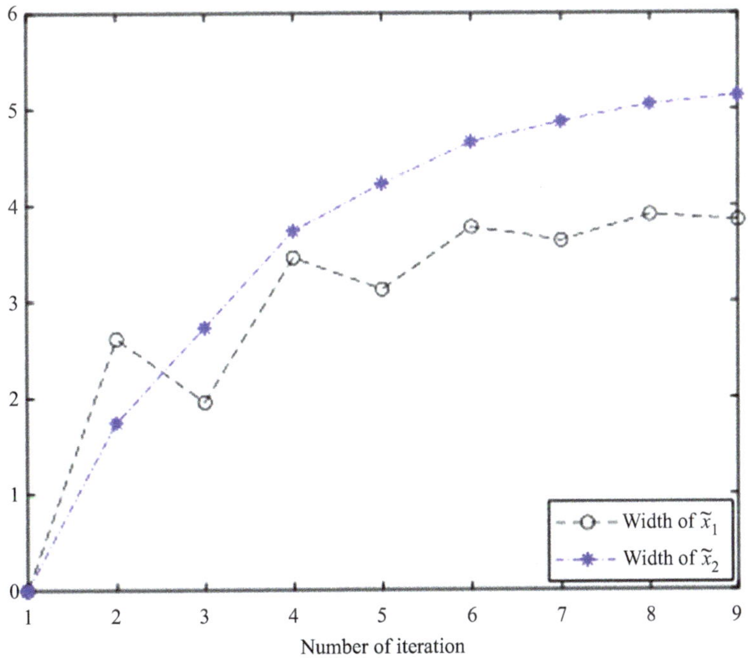

Figure 5.3 Uncertain width of $\tilde{x}_1$ and $\tilde{x}_2$ for Case 1 of Example 1

By applying the modified gradient descent algorithm for the FSLE in (5.35), the obtained solution sets of $\tilde{x}_1$, and $\tilde{x}_2$ are depicted in Tables 5.3 and 5.4. The fuzzy solution is shown graphically in Figures 5.4 and 5.5.

In Table 5.3, the width of $\tilde{x}_1$ is defined as the difference between the right and the left values of the TFN.

In Table 5.2, the width of $\tilde{x}_2$ is defined as the difference between the right and the left values of the TFN.

The obtained fuzzy solution for Case 2 of Example 1 is presented in Figures 5.3 and 5.4, and width-wise convergence is shown in Figure 5.6.

To understand the effectiveness of the proposed algorithm, obtained results are compared with [25]. The graphical representation of the solution $\tilde{x}_1$and $\tilde{x}_2$ components in comparison with [25] is shown in Table 5.5 as well as Figures 5.3 and 5.4.

Case 3 In this case, both the left-right and right-hand sides are treated as TFNs, i.e., both the coefficient matrix and the right-hand side vectors are considered fuzzy. The following elements of coefficients matrix and right-hand side vectors are used for investigation:

Table 5.3 TFN solutions component $\tilde{x}_1$ for Case 2 of Example 1

Iteration	x_{L_1}	x_{N_1}	x_{R_1}	Width
1	0	0	0	0
2	0	1.6957	3.5922	3.5922
3	0.1667	1.3996	3.385	3.2183
4	0.1667	1.612	3.4008	3.2341
5	0.1944	1.5749	3.3999	3.2055
6	0.1944	1.6015	3.3999	3.2055
7	0.1991	1.5969	3.3999	3.2008
8	0.1991	1.6002	3.3999	3.2008
9	0.1998	1.5996	3.3999	3.2001
End point	0.1998	1.5996	3.3999	3.2001

Table 5.4 TFN solutions component $\tilde{x}_2$ for Case 2 of Example 1

Iteration	x_{L_2}	x_{N_2}	x_{R_2}	Width
1	0	0	0	0
2	0.5	1.1304	5.0291	4.5291
3	0.5	1.5745	5.1771	4.6771
4	0.5833	1.7161	5.1992	4.6159
5	0.5833	1.7718	5.1999	4.6166
6	0.5972	1.7895	5.1999	4.6027
7	0.5972	1.7965	5.1999	4.6027
8	0.5995	1.7987	5.1999	4.6004
9	0.5995	1.7996	5.1999	4.6004
End point	0.5995	1.7996	5.1999	4.6004

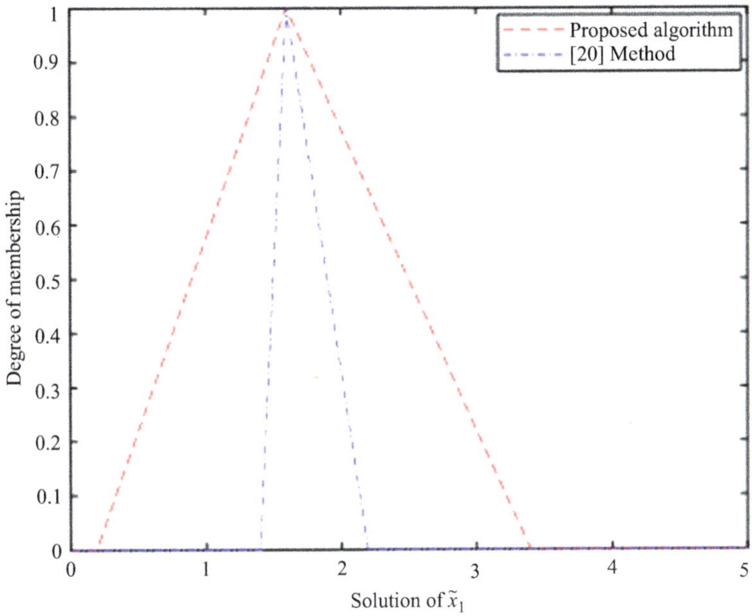

Figure 5.4 *Comparison solution component $\tilde{x}_1$ for the proposed algorithm and [25] method for Case 2 of Example 1*

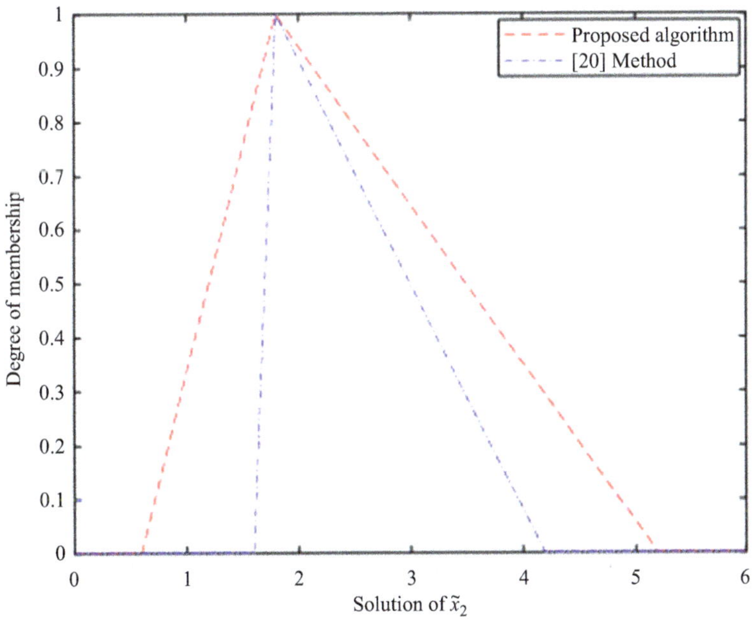

Figure 5.5 *Comparison solution component $\tilde{x}_2$ for the proposed algorithm and [25] method for Case 2 of Example 1*

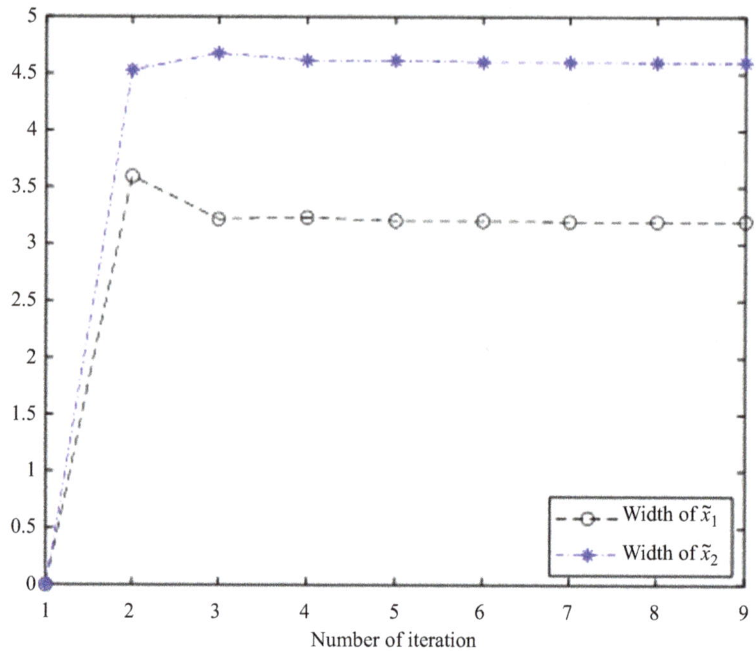

Figure 5.6 Uncertain width of $\tilde{x}_1$ and $\tilde{x}_2$ for Case 2 of Example 1

Table 5.5 Solution component $\tilde{x}_1$ and $\tilde{x}_2$ and its comparison with [25]

Method	$\tilde{x}_1$	$\tilde{x}_2$
Ref. [25] method	[1.4, 1.6, 2.2]	[1.6, 1.8, 4.2]
Proposed algorithm	[0.1998, 1.5996, 3.3999]	[0.5995, 1.7996, 5.1999]

$\tilde{a}_{11} = [2.5,\ 3,\ 3.5]$, $\tilde{a}_{12} = [-1.5,\ -1,\ -0.5]$, $\tilde{a}_{21} = [-1.5,\ -1,\ -0.5]$, $\tilde{a}_{22} = [1.5,\ 2,\ 2.5]$, $\tilde{\chi}_1 = [0,\ 3,\ 5]$, $\tilde{\chi}_2 = [1,\ 2,\ 7]$.

Applying the modified gradient descent approach to the system of linear equations in (5.35), the obtained solution $\tilde{x}_1$ and $\tilde{x}_2$ are depicted in Tables 5.6 and 5.7, respectively. The fuzzy solution is shown graphically in Figures 5.7 and width-wise convergence is shown in Figure 5.8.

In Table 5.6, the width of $\tilde{x}_1$ is defined as the difference between the right and the left values of the TFN.

In Table 5.7, the width of $\tilde{x}_2$ is defined as the difference between the right and the left values of the TFN.

The obtained fuzzy solution for Case 3 of Example 1 is presented in Figure 5.7 and width-wise convergence is shown in Figure 5.8.

Table 5.6 TFN solutions component $\tilde{x}_1$ for Case 3 of Example 1

Iteration	x_{L_1}	x_{N_1}	x_{R_1}	Width
1	0	0	0	0
2	0	1.6957	2.1143	2.1143
3	0.4	1.3996	1.8612	1.4612
4	0.4	1.612	1.885	1.485
5	0.64	1.5749	1.8821	1.2421
6	0.64	1.6015	1.8824	1.2424
7	0.784	1.5969	1.8824	1.0984
8	0.784	1.6002	1.8824	1.0984
9	0.8704	1.5996	1.8824	1.012
End point	0.9992	1.5996	1.8824	0.8832

Table 5.7 TFN solutions component $\tilde{x}_2$ for Case 3 of Example 1

Iteration	x_{L_2}	x_{N_2}	x_{R_2}	Width
1	0	0	0	0
2	0.6667	1.1304	2.96	2.2933
3	0.6667	1.5745	3.1408	2.4741
4	1.0667	1.7161	3.174	2.1073
5	1.0667	1.7718	3.1761	2.1094
6	1.3067	1.7895	3.1764	1.8697
7	1.3067	1.7965	3.1764	1.8697
8	1.4507	1.7987	3.1764	1.7257
9	1.4507	1.7996	3.1764	1.7257
End point	1.6654	1.7996	3.1764	1.511

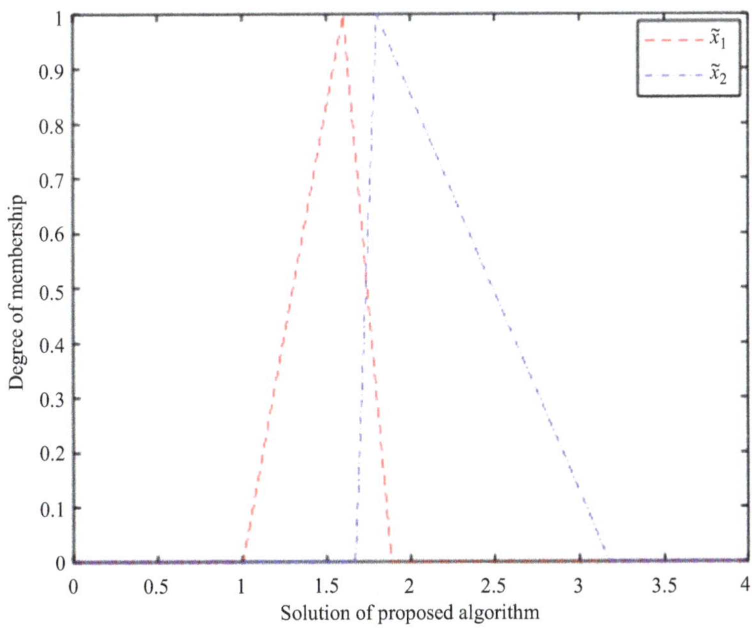

Figure 5.7 Solution components $\tilde{x}_1$ and $\tilde{x}_2$ for Case 3 of Example 1

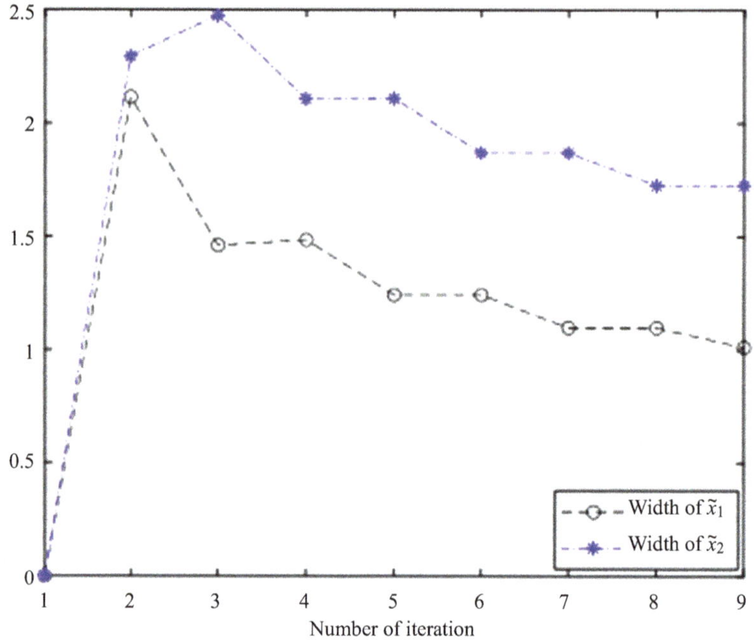

Figure 5.8 Uncertain width of $\tilde{x}_1$ *and* $\tilde{x}_2$ *for Case 3 of Example 1*

Example 2 Take a FSLE

$$\begin{aligned}\tilde{a}_{11}\tilde{x}_1 + \tilde{a}_{12}\tilde{x}_2 + \tilde{a}_{13}\tilde{x}_3 &= \tilde{\chi}_1\\ \tilde{a}_{21}\tilde{x}_1 + \tilde{a}_{22}\tilde{x}_2 + \tilde{a}_{23}\tilde{x}_3 &= \tilde{\chi}_2\\ \tilde{a}_{31}x_1 + \tilde{a}_{32}\tilde{x}_2 + \tilde{a}_{33}\tilde{x}_3 &= \tilde{\chi}_3\end{aligned} \tag{5.36}$$

where $\tilde{a}_{11} = [3.5,\ 4,\ 4.5]$, $\tilde{a}_{12} = [1.5,\ 2,\ 2.5]$, $\tilde{a}_{13} = [-1.5,\ -1,\ -0.5]$, $\tilde{a}_{21} = [1.5, 2, 2.5]$, $\tilde{a}_{22} = [6.5,\ 7,\ 7.5]$, $\tilde{a}_{23} = [5.5,\ 6,\ 6.5]$, $\tilde{a}_{31} = [-1.5,\ -1,\ -0.5]$, $\tilde{a}_{32} = [5.5,\ 6,\ 6.5]$, $\tilde{a}_{33} = [9.5,\ 10,\ 10.5]$, $\tilde{\chi}_1 = [-27, -20,\ -7]$, $\tilde{\chi}_2 = [1,\ 16,\ 40]$, and $\tilde{\chi}_3 = [26,\ 44,\ 47]$. Assume the initial approximation $x^0 = (0, 0, 0)^T$, and tolerance value $\epsilon = 10^{-3}$.

Case 1 Here, we consider the left-hand side elements that consist of TFNs, i.e., the coefficient matrix is taken as fuzzy. Accordingly, the following coefficients are used for the investigation: $\tilde{a}_{11} = [3.5,\ 4,\ 4.5]$, $\tilde{a}_{12} = [1.5,\ 2,\ 2.5]$, $\tilde{a}_{13} = [-1.5,\ -1,\ -0.5]$, $\tilde{a}_{21} = [1.5,\ 2,\ 2.5]$, $\tilde{a}_{22} = [6.5,\ 7,\ 7.5]$, $\tilde{a}_{23} = [5.5,\ 6,\ 6.5]$, $\tilde{a}_{31} = [-1.5, -1,\ -0.5]$, $\tilde{a}_{32} = [5.5,\ 6,\ 6.5]$, $\tilde{a}_{33} = [9.5,\ 10,\ 10.5]$, $\tilde{\chi}_1 = -20$, $\tilde{\chi}_2 = 16$, $\tilde{\chi}_3 = 44$.

Applying the modified gradient descent algorithm for FLS equations the obtained solution vectors $\tilde{x}_1$, $\tilde{x}_2$, and $\tilde{x}_3$ are listed in Tables 5.8, 5.9, and 5.10. The obtained fuzzy solution is presented in Figure 5.9.

Table 5.8 TFN solutions component $\tilde{x}_1$ for Case 1 of Example 2

Iteration	x_{L_1}	x_{N_1}	x_{R_1}	Width
1	0	0	0	0
2	−1.6788	−1.6364	−1.5961	0.0827
3	−3.2207	−3.201	−3.1634	0.0573
4	−3.4889	−3.4829	−3.4745	0.0144
5	−3.7816	−3.7684	−3.7329	0.0487
6	−3.8454	−3.8241	−3.7816	0.0638
7	−3.9103	−3.8836	−3.8418	0.0685
8	−3.9261	−3.8989	−3.8571	0.069
9	−3.9456	−3.9209	−3.8782	0.0674
10	−3.9509	−3.9268	−3.8882	0.0627
11	−3.9591	−3.9379	−3.9031	0.056
12	−3.962	−3.9425	−3.911	0.051
End point	−3.9695	−3.9495	−3.9194	0.0501

Table 5.9 TFN solutions component $\tilde{x}_2$ for Case 1 of Example 2

Iteration	x_{L_2}	x_{N_2}	x_{R_2}	Width
1	0	0	0	0
2	1.2768	1.3091	1.343	0.0662
3	−0.1881	−0.1324	−0.0877	0.1004
4	0.0247	0.1009	0.1636	0.1389
5	−0.1882	−0.1344	−0.0907	0.0975
6	−0.1444	−0.0861	−0.037	0.1074
7	−0.166	−0.1184	−0.0801	0.0859
8	−0.1394	−0.0992	−0.064	0.0754
9	−0.1347	−0.0986	−0.0685	0.0662
10	−0.1122	−0.0815	−0.0568	0.0554
11	−0.1073	−0.0778	−0.0555	0.0518
12	−0.0893	−0.0642	−0.0456	0.0437
End point	−0.0906	−0.0606	−4.85e−02	0.0421

Table 5.10 TFN solutions component $\tilde{x}_3$ for Case 1 of Example 2

Iteration	x_{L_3}	x_{N_3}	x_{R_3}	Width
1	0	0	0	0
2	3.5113	3.6	3.6933	0.182
3	3.2951	3.413	3.5491	0.254
4	3.9647	4.0238	4.094	0.1293
5	3.9174	3.982	4.0569	0.1395
6	4.0427	4.0824	4.1338	0.0911
7	4.0282	4.065	4.108	0.0798
8	4.0506	4.0815	4.1214	0.0708
9	4.0402	4.0605	4.088	0.0478
10	4.0453	4.0672	4.0975	0.0522
11	4.0334	4.0477	4.0701	0.0367
12	4.0365	4.0529	4.0776	0.0411
End point	4.0304	4.0445	4.0505	0.0201

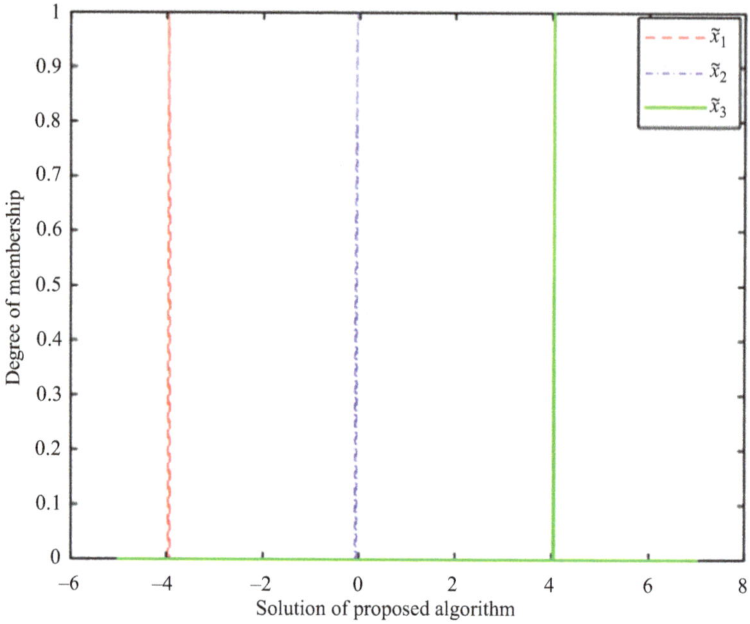

Figure 5.9 Solution components $\tilde{x}_1$, $\tilde{x}_2$ and $\tilde{x}_3$ for Case 1 of Example 2

In Table 5.8, the width of $\tilde{x}_1$ is defined as the difference between the right and the left values of the TFN.

In Table 5.9, the width of $\tilde{x}_2$ is defined as the difference between the right and the left values of the TFN.

In Table 5.10, the width of $\tilde{x}_3$ is defined as the difference between the right and the left values of the TFN.

In Tables 5.8, 5.9, and 5.10, computationally after 12 iterations the width of TFNs gets converged to the end point. The obtained fuzzy solution for Case 1 of Example 2 is presented in Figure 5.9, and width-wise convergence is shown in Figure 5.10.

Case 2 Here, we consider the right-hand side elements consisting of TFNs, i.e., the right-hand side vector is taken as fuzzy. Accordingly, the following vectors are used for the investigation: $\tilde{a}_{ij}$ are crisp, $(i, j = 1, 2)$, $\tilde{\chi}_1 = [-27, -20, -7]$, $\tilde{\chi}_2 = [1, 16, 40]$ and $\tilde{\chi}_3 = [26, 44, 47]$.

Applying the modified gradient descent method for the FSLE in (5.36), the obtained solution sets of $\tilde{x}_1$, $\tilde{x}_2$, and $\tilde{x}_3$ are presented in Tables 5.11, 5.12, and 5.13, respectively.

In Table 5.11, the width of $\tilde{x}_1$ is defined as the difference between the right and the left values of the TFN.

In Table 5.12, the width of $\tilde{x}_2$ is defined as the difference between the right and the left values of the TFN.

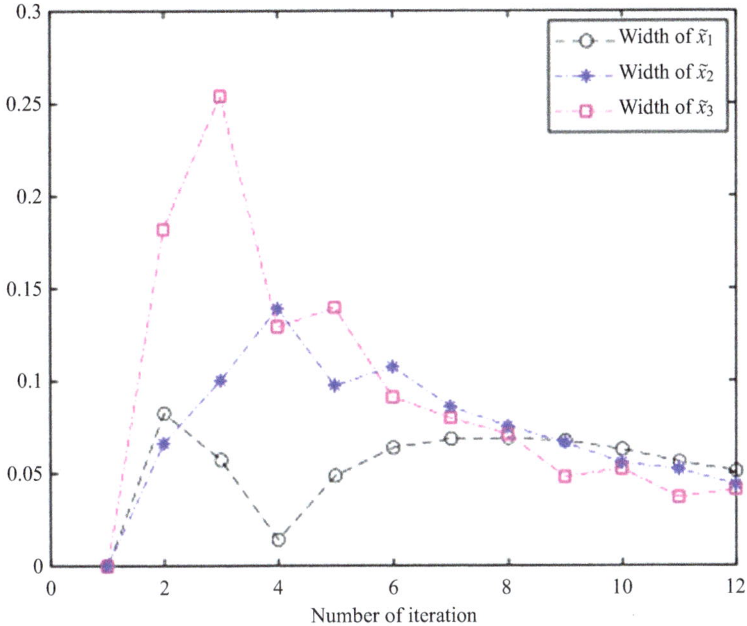

Figure 5.10 *Uncertain width of* $\tilde{x}_1$, $\tilde{x}_2$ *and* $\tilde{x}_3$ *for Case 1 of Example 2*

Table 5.11 *TFN solutions component* $\tilde{x}_1$ *for Case 2 of Example 2*

Iteration	x_{L_1}	x_{N_1}	x_{R_1}	Width
1	0	0	0	0
2	−3.3621	−1.6364	−0.4859	2.8762
3	−4.2445	3.5727	−3.201	1.0435
4	−5.0288	−3.4889	−3.3212	1.7076
5	−5.29	−3.7684	−3.4283	1.8617
6	−5.5826	−3.8241	−3.6123	1.9703
7	−5.7068	−3.8836	−3.8488	1.858
8	−5.8685	−4.0476	−3.8989	1.9696
9	−5.9643	−.2512	−3.9209	2.0434
10	−6.0903	−4.4153	−3.9268	2.1635
11	−6.1659	−4.58	−3.9379	2.228
12	−6.2654	−4.712	−3.9425	2.3229
End point	−6.9225	−5.9226	−3.9995	2.923

In Table 5.13, the width of $\tilde{x}_3$ is defined as the difference between the right and the left values of the TFN.

In Tables 5.11–5.13, computationally after 12 iterations, the width of the TFN is converged to the end point. The obtained fuzzy solution for Case 2 of Example 2 is presented in Figures 5.11–5.13, and width-wise convergence is shown in

Table 5.12 TFN solutions component $\tilde{x}_2$ for Case 2 of Example 2

Iteration	x_{L_2}	x_{N_2}	x_{R_2}	Width
1	0	0	0	0
2	0.1245	1.3091	2.7763	2.6518
3	−0.9258	−0.1324	3.6016	4.5274
4	−0.6828	0.1009	4.6746	5.3574
5	−0.813	−0.1344	4.7677	5.5807
6	−0.4359	−0.0861	5.4195	5.8554
7	−0.4703	−0.1184	5.44	5.9103
8	−0.1361	−0.0992	5.9376	6.0737
9	−0.1621	−0.0986	5.9472	6.1093
10	−0.0815	0.1027	6.3446	6.4261
11	−0.0778	0.0822	6.3517	6.4295
12	−0.0642	0.2913	6.6706	6.7348
End point	−5.00e−04	0.9994	7.9994	7.9999

Table 5.13 TFN solutions component $\tilde{x}_3$ for Case 2 of Example 2

Iteration	x_{L_3}	x_{N_3}	x_{R_3}	Width
1	0	0	0	0
2	3.2376	3.2622	3.6	0.3624
3	2.1001	2.3617	3.413	1.3129
4	2.194	2.8604	4.0238	1.8298
5	1.4175	2.513	3.982	2.5645
6	1.521	2.5917	4.0824	2.5614
7	0.9719	2.2947	4.065	3.0931
8	1.076	2.3237	4.0815	3.0055
9	0.6414	2.088	4.0605	3.4191
10	0.727	2.1101	4.0672	3.3402
11	0.3785	1.924	4.0477	3.6692
12	0.4475	1.9414	4.0529	3.6054
End point	0.6919	1.3081	4.0004	3.3085

Figure 5.14. The obtained solutions are $[x_L, x_N, x_R]$ and the value of uncertain solutions lies between them. Further, to understand the effectiveness of the proposed algorithm, the obtained results are compared with [13] in Table 5.14. The graphical representation of the solution $\tilde{x}_1$, $\tilde{x}_2$, and $\tilde{x}_3$ components in comparison with [13] is shown in Figures 5.11, 5.12, and 5.13, respectively.

Case 3 In this case, both the left-hand and right-hand sides are treated as TFNs, i.e., both the coefficient matrix and the right-hand side vectors are considered fuzzy. The elements of coefficients matrix and right-hand side vectors are used for investigation: $\tilde{a}_{11} = [3.5, 4, 4.5]$, $\tilde{a}_{12} = [1.5, 2, 2.5]$, $\tilde{a}_{13} = [-1.5, -1, -0.5]$, $\tilde{a}_{21} = [1.5, 2, 2.5]$, $\tilde{a}_{22} = [6.5, 7, 7.5]$, $\tilde{a}_{23} = [5.5, 6, 6.5]$, $\tilde{a}_{31} = [-1.5, -1, -0.5]$, $\tilde{a}_{32} = [5.5, 6, 6.5]$, $\tilde{a}_{33} = [9.5, 10, 10.5]$, $\tilde{\chi}_1 = [-27, -20, -7]$, $\tilde{\chi}_2 = [1, 16, 40]$, and $\tilde{\chi}_3 = [26, 44, 47]$.

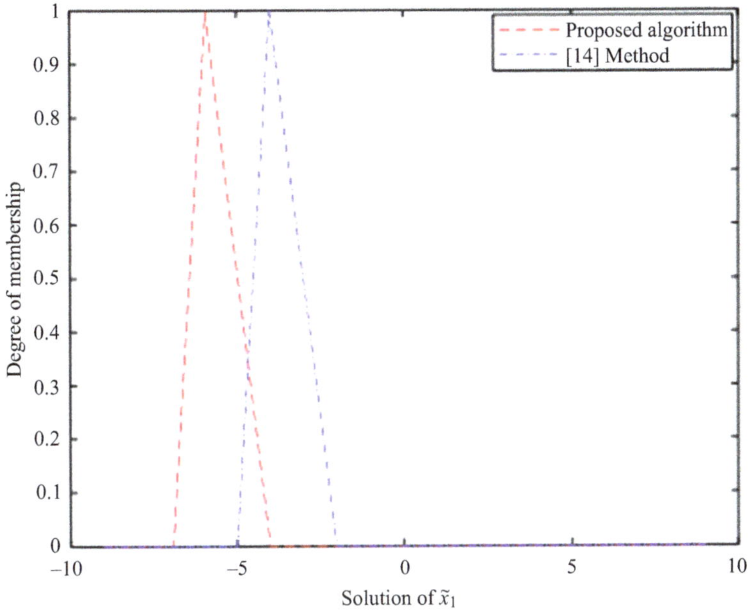

Figure 5.11 Comparison of solution component $\tilde{x}_1$ for the proposed algorithm and [13] method for Case 2 of Example 2

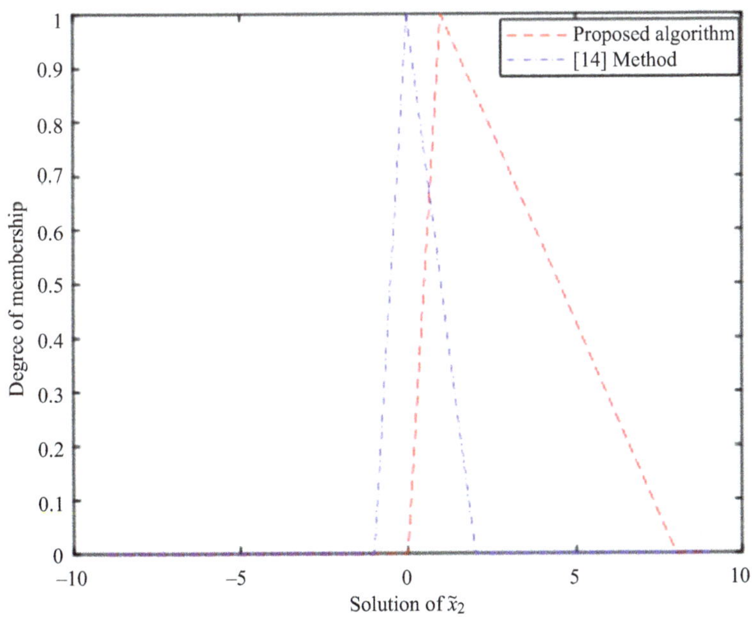

Figure 5.12 Comparison of solution components of $\tilde{x}_2$ for the proposed algorithm and [13] method for Case 2 of Example 2

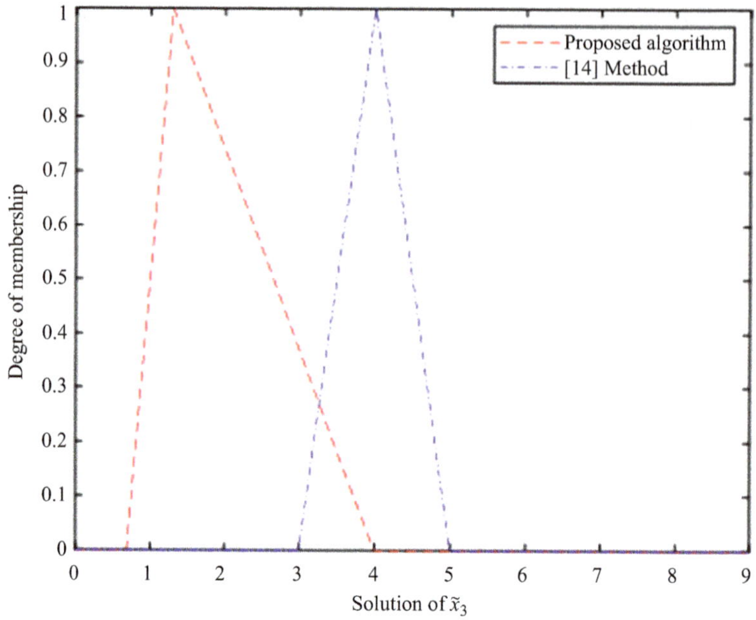

Figure 5.13 Comparison of solution component $\tilde{x}_3$ for the proposed algorithm and [13] method for Case 2 of Example 2

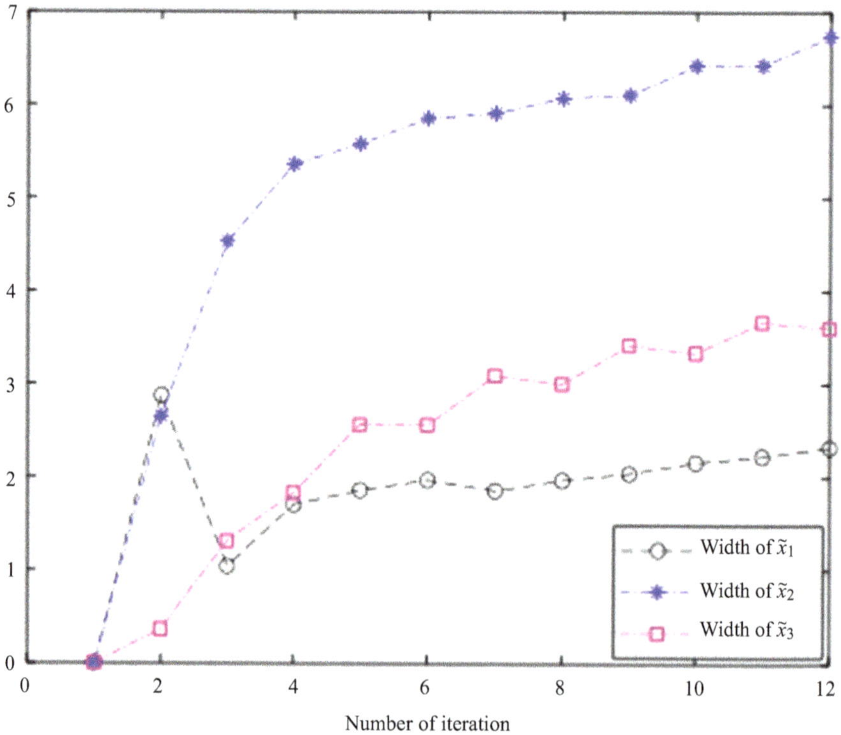

Figure 5.14 Uncertain width of $\tilde{x}_1$, $\tilde{x}_2$, and $\tilde{x}_3$ for Case 2 of Example 2

Table 5.14 Comparison obtained solution components $\tilde{x}_1$, $\tilde{x}_2$, and $\tilde{x}_3$ with [13]

Method	$\tilde{x}_1$	$\tilde{x}_2$	$\tilde{x}_3$
Ref. [13] method	[−5, −4, −2]	[−1, 0, 2]	[3, 4, 5]
Proposed algorithm	[−6.9225, −5.9226, −3.9995]	[−5.00e−04, 0.9994, 7.9994]	[0.6919,1.3081,4.0004]

Table 5.15 TFN solutions component $\tilde{x}_1$ for Case 3 of Example 2

Iteration	x_{L_1}	x_{N_1}	x_{R_1}	Width
1	0	0	0	0
2	−3.3621	−1.6364	−0.4594	2.9027
3	−4.3699	−3.2203	−3.201	1.1689
4	−5.2941	−3.4889	−3.1026	2.1915
5	−5.6489	−3.7684	−3.3633	2.2856
6	−6.0422	−3.8241	−3.5464	2.4958
7	−6.2216	−3.8836	−3.8801	2.3415
8	−6.4684	−4.0535	−3.8989	2.5695
9	−6.6053	−4.3315	−3.9209	2.6844
10	−6.8	−4.4725	−3.9268	2.8732
11	−6.91	−4.6958	−3.9379	2.9721
12	−7.0669	−4.8088	−3.9425	3.1244
End point	−8.1813	−6.1548	−3.9995	4.1818

Applying the modified gradient descent approach to the system of linear equations in (5.36), the obtained solution vector components $\tilde{x}_1$, $\tilde{x}_2$, and $\tilde{x}_3$ are depicted in Tables 5.15, 5.16, and 5.17, respectively.

In Table 5.15, the width of $\tilde{x}_1$ is defined as the difference between the right and the left values of the TFN.

In Table 5.16, the width of $\tilde{x}_2$ is defined as the difference between the right and the left values of the TFN.

In Table 5.17, the width of $\tilde{x}_3$ is defined as the difference between the right the left values of the TFN.

In Tables 5.15–5.17, computationally after 82 iterations, the width of TFN is converged to the end point. The obtained fuzzy solution for Case 3 of Example 2 is presented in Figure 5.15 and width-wise convergence is shown in Figure 5.16.

From this study, it may be observed that in Case 1, the width of TFN convergence occurs at fewer iterations.

In the results and discussion, we highlighted the efficacy and significance of our proposed gradient-based optimization approach in solving fuzzy-valued

Table 5.16 TFN solutions component $\tilde{x}_2$ for Case 3 of Example 2

Iteration	x_{L_2}	x_{N_2}	x_{R_2}	Width
1	0	0	0	0
2	0.1245	1.3091	2.6252	2.5007
3	−1.075	−0.1324	3.0162	4.0912
4	−0.7681	0.1009	4.0885	4.8566
5	−0.9351	−0.1344	4.2242	5.1593
6	−0.4602	−0.0861	4.9559	5.4161
7	−0.4978	−0.1184	5.0192	5.517
8	−0.0992	−0.0695	5.593	5.6922
9	−0.0986	−0.0949	5.639	5.7376
10	−0.0815	0.2535	6.0971	6.1786
11	−0.0778	0.2334	6.1335	6.2113
12	−0.0642	0.5146	6.4999	6.5641
End point	−5.00e−04	1.6055	8.1113	8.1118

Table 5.17 TFN solutions component $\tilde{x}_3$ for Case 3 of Example 2

Iteration	x_{L_3}	x_{N_3}	x_{R_3}	Width
1	0	0	0	0
2	3.0846	3.2376	3.6	0.5154
3	2.2373	2.3407	3.413	1.1757
4	2.4675	2.8002	4.0238	1.5563
5	1.5622	2.3087	3.982	2.4198
6	1.7245	2.4313	4.0824	2.3579
7	1.0626	2.0015	4.065	3.0024
8	1.205	2.067	4.0815	2.8765
9	0.6808	1.7168	4.0605	3.3797
10	0.7958	1.7677	4.0672	3.2714
11	0.3769	1.4849	4.0477	3.6708
12	0.469	1.5259	4.0529	3.5839
End point	−0.8381	0.5156	4.0004	4.8385

optimization problems, particularly in addressing FLS equations. The proposed method demonstrates notable improvements in solution quality compared with existing approaches shown in Table 5.5. Through rigorous convergence analysis, we validate the reliability and robustness of this method, supporting its suitability for tackling fuzzy systems. For instance, in the case of solely fuzzy systems, where either the coefficient matrix or the right-hand side vector is fuzzy, our method showcases superior convergence properties and solution gives a good agreement, which is shown in Figure. 5.2–5.16. This is particularly evident when compared with traditional methods that may struggle to effectively handle uncertainty inherent in fuzzy systems. To provide concrete evidence of the

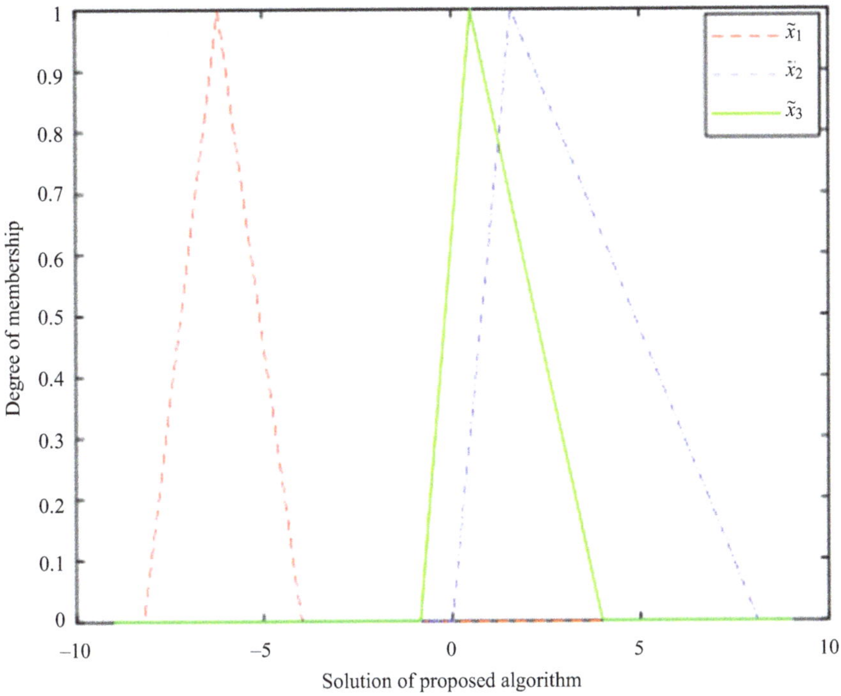

Figure 5.15 Solution components $\tilde{x}_1$, $\tilde{x}_2$, and $\tilde{x}_3$ for Case 3 of Example 2

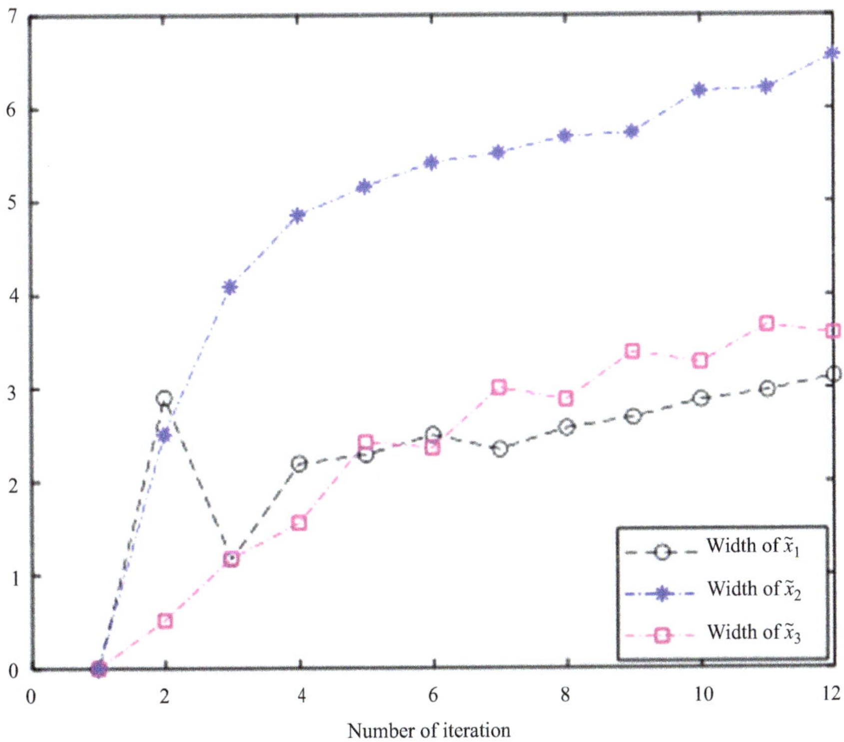

Figure 5.16 Uncertain width of $\tilde{x}_1$, $\tilde{x}_2$, and $\tilde{x}_3$ for Case 3 of Example 2

present method's effectiveness, we present numerical solutions for two example problems, accompanied by graphical representations for intuitive interpretation. These results not only validate the applicability of our approach but also underscore its practical utility in real-world scenarios. In summary, the proposed gradient-based optimization approach represents a significant advancement in the field of fuzzy-valued optimization. Its ability to effectively handle uncertainty is a valuable tool for addressing complex fuzzy systems across various domains.

5.5 Conclusion

In this chapter, the GDOA is extended to address the FLS equations. To study the efficiency of the algorithm, a convergence theorem is presented. The GDOA is applied with two example problems of a system of linear equations in a fuzzy environment. The fuzzy system was divided into three cases, namely, only fuzzy (either the coefficient matrix or the right-hand side vector being fuzzy) and fully fuzzy system. Then, the same is solved through the proposed algorithm and the fuzzy solutions are reported. The obtained solutions are compared with other methods which found a good agreement. Finally, both the numerical and graphical solutions are discussed case-wise. The present study in different cases shows the uncertainty propagation and its nature for the three different cases. The study also suggest the sensitiveness of the system with respect to the place where uncertainty occurs.

The future scope of the present research in fuzzy optimization technique is promising and diverse. It includes integration with machine learning, multi-objective optimization, dynamic case, different types of fuzzy environment, and exploration of quantum computing. These directions offer opportunities for handling the uncertain field variable and addressing complex challenges across various domains.

References

[1] L. A. Zadeh, "Fuzzy sets," *Information and Control*, vol. 8, no. 3, pp. 338–353, 1965.

[2] J. J. Buckley and Y. Qu, "Solving linear and quadratic fuzzy equations," *Fuzzy Sets and Systems*, vol. 38, no. 1, pp. 43–59, 1990.

[3] J. J. Buckley and Y. Qu, "Solving systems of linear fuzzy equations," *Fuzzy Sets and Systems*, vol. 43, no. 1, pp. 33–43, 1991.

[4] K. Peeva, "Fuzzy linear systems," *Fuzzy Sets and Systems*, vol. 49, no. 3, pp. 339–355, 1992.

[5] M. Friedman, M. Ming and A. Kandel, "Fuzzy linear systems," *Fuzzy Sets and Systems*, vol. 96, no. 2, pp. 201–209, 1998.

[6] F. Abbasi and T. Allahviranloo, "Solving fully fuzzy linear system: A new solution concept," *Information Sciences*, vol. 589, pp. 608–635, 2022.

[7] M. Ma, M. Friedman and A. Kandel, "Duality in fuzzy linear systems," *Fuzzy Sets and Systems*, vol. 109, no. 1, pp. 55–58, 2000.

[8] M. Dehghan and B. Hashemi, "Iterative solution of fuzzy linear systems," *Applied Mathematics and Computation*, vol. 175, no. 1, pp. 645–674, 2006.

[9] M. Dehghan, B. Hashemi and M. Ghatee, "Solution of the fully fuzzy linear systems using iterative techniques," *Chaos, Solitons & Fractals*, vol. 34, no. 2, pp. 316–336, 2007.

[10] S. Abbasbandy, R. Ezzati and A. Jafarian, "LU decomposition method for solving fuzzy system of linear equations," *Applied Mathematics and Computation*, vol. 172, no. 1, pp. 633–643, 2006.

[11] S. Nayak and S. Chakraverty, "A new approach to solve fuzzy system of linear equations," *Journal of Mathematics and Computer Science*, vol. 7, no. 3, pp. 205–212, 2013.

[12] P. Karunakar and S. Chakraverty, "Solving fully interval linear systems of equations using tolerable solution criteria," *Soft Computing*, vol. 22, no. 14, pp. 4811–4818, 2018.

[13] S. Abbasbandy and A. Jafarian, "Steepest descent method for system of fuzzy linear equations," *Applied Mathematics and Computation*, vol. 175, no. 1, pp. 823–833, 2006.

[14] S. Nayak and J. Pooja, "Numerical optimization technique to solve imprecisely defined nonlinear system of equations with bounded parameters," *International Journal of Mathematics in Operational Research*, 2021.

[15] P. K. Panigrahi and S. Nayak, "Numerical investigation of non-probabilistic systems using inner outer direct search optimization technique," *AIMS Mathematics*, vol. 8, no. 9, pp. 21329–21358, 2023.

[16] P. K. Panigrahi and S. Nayak, "Numerical approach to solve imprecisely defined systems using inner outer direct search optimization technique," *Mathematics and Computers in Simulation*, vol. 215, pp. 578–606, 2024.

[17] P. K. Panigrahi and S. Nayak, "Fuzzy Levenberg Marquart optimization algorithm with inexact line search technique to solve imprecisely defined nonlinear unconstrained optimization problems," *International Journal of Machine Learning and Cybernetics*, vol. 16, pp. 5527–5551, 2025.

[18] H. J. Zimmermann, Fuzzy set theory—and its applications, New York: Springer Dordrecht, 2001.

[19] S. Nayak, "Uncertain quantification of field variables involved in transient convection diffusion problems for imprecisely defined parameters," *International Communications in Heat and Mass Transfer*, vol. 119, p. 104894, 2020.

[20] D. Chakraborty and D. Guha, "Addition of two generalized fuzzy numbers," *International Journal of Industrial Mathematics*, vol. 2, no. 1, pp. 9–20, 2010.

[21] P. K. Panigrahi and S. Nayak, "Conjugate gradient with Armijo line search approach to investigate imprecisely defined unconstrained optimisation problem," *International Journal of Computational Science and Engineering*, vol. 27, pp. 458–471, 2024.

[22] S. Priyadarshini and S. Nayak, "A new hybrid approach to study heat and mass transfer in porous medium influenced by imprecisely defined parameters," *Case Studies in Thermal Engineering*, vol. 51, p. 103619, 2023.

[23] S. Priyadarshini and S. Nayak, "A numerical approach to study heat and mass transfer in porous medium influenced by uncertain parameters," *International Communications in Heat and Mass Transfer*, vol. 139, p. 106411, 2022.

[24] J. M. Ortega, *Introduction to Parallel and Vector Solution of Linear Systems*, New York: Springer, 1988.

[25] H. S. Najafi and S. A. Edalatpanah, "H-matrices in fuzzy linear systems," *International Journal of Computational Mathematics*, vol. 2014, p. 6, 2014.

Chapter 6

A convolutional neural network-based biomarkers for Alzheimer's diagnosis and prognosis

H. Meenal[1], C. Kishor Kumar Reddy[2], Reddaboina Rajini[3], Pendyala Loka Priya[3] and Marlia Mohd Hanafiah[4]

Abstract

Convolutional neural networks (CNNs) are used by biomarkers to diagnose Alzheimer's disease (AD). To enable prompt intervention and improve patient outcomes, the project aims to improve early identification and precise prediction of disease development. With the help of Open Access Series of Imaging Studies (OASIS), the data is prepared for the model testing for prediction. The model train consists of a few steps like preprocessing including standardizing and augmenting imaging data. Functional and structural biomarkers suggestive of AD were trained into the CNN. Different metrics like accuracy, Roc curve is used for predicting model performance. To make sure the findings could be applied to a wider audience, cross-validation methods were used. When it came to differentiating between AD, MCI, and healthy control participants, the CNN model showed excellent accuracy. The model found distinct patterns of hippocampal shrinkage and alterations in cortical thickness as key biomarkers. By monitoring the course of the disease in MCI patients, the model's predictive abilities were confirmed and those who converted to AD were correctly predicted. The accuracy of CNN model is 99.95%. This work opens the door for the prediction of AD using the CNN model. The model's strong predictive abilities and high accuracy demonstrate its usefulness for early AD detection and monitoring in clinical settings. To further improve the model's therapeutic applicability, future studies should concentrate on integrating multimodal data and testing it across bigger and more diverse populations.

[1]Department of Computer Science and Engineering, Methodist College of Engineering and Technology, Osmania University, India

[2]Department of Computer Science and Engineering, Stanley College of Engineering and Technology for Women, India

[3]Department of Computer Science and Engineering, Keshav Memorial Institute of Technology, Jawaharlal Nehru Technological University, Hyderabad (JNTUH), India

[4]Faculty of Science and Technology, University Kebangasaan Malaysia, Malaysia

Keywords: Alzheimer's disease; biomarkers; prognosis; neuroimaging; early detection; precision medicine; neurodegenerative disorders; diagnostics cognitive decline

6.1 Introduction

Alzheimer's disease (AD) is basically a neurodegenerative disease that affects memory loss, behavior change, and lowering brain health overall [1]. It is the most prevalent cause of dementia in the early stage and is distinguished by a steadily declining ability to think [2]. The illness progresses gradually, starting with mild memory loss and possibly leading to major impairment in day-to-day activities and fundamental physiological functions [3]. AD can cause a wide range of symptoms, but the most typical ones include memory loss, confusion, difficulty making decisions and solving problems, disorientation in time and space, and linguistic difficulties. People may find it difficult to remember recent discussions or occurrences in the early stages [4]. They may have behavioral and emotional changes as the illness worsens such as increased agitation, anxiety, or depression. Severe memory loss develops in later stages, accompanied by a reduction in cognitive function, an inability to recognize loved ones, and a decline in motor functions such as walking and swallowing.

AD is more common in older age groups; hence, elderly people are more vulnerable to it. Another important factor is genetics. People who have a family history of AD are more vulnerable [5]. Some genetic factors, such as the APOE $\varepsilon4$ allele, are linked to an increased risk of acquiring the illness.

Understanding these genetic predispositions is critical to identifying at-risk individuals and initiating early therapy [6]. AD risk also increases with other cardiovascular disorders such as diabetes, high blood pressure, and high cholesterol. This risk is further increased by lifestyle choices such as low cognitive engagement, poor eating habits, and physical inactivity. These elements emphasize the significance of preventing AD holistically, addressing lifestyle and genetic variables to slow the illness's start and progression. Through deep learning (DL)-based biomarkers, the prediction of Alzheimer's may be efficient and accurate, allowing for earlier intervention and better patient outcomes [7]. In order to create these markers, which will eventually help with AD early detection and therapy, it is critical to understand how genetic, cardiovascular, and lifestyle variables interact.

Table 6.1 provides a breakdown of all fatalities by year from 2010 to 2023, including numbers specific to men and women. The overall death toll has risen over this time, rising from 500,000 in 2010 to 760,000 in 2023. The data shows a steady trend wherein each year, there are more deaths among women than among men. 2010 saw 300,000 fatalities among women and 200,000 deaths among men. These numbers increased to 304,000 for males and 456,000 for women by 2023, continuing the long-term trend of women dying at a higher rate than men, graphically shown in Figure 6.1.

Table 6.1 Death rate for Alzheimer's disease in 2010–2023 across the globe

Sl. no	Year	Total deaths	Men	Women
1	2010	500,000	200,000	300,000
2	2011	520,000	208,000	312,000
3	2012	540,000	216,000	324,000
4	2013	560,000	224,000	336,000
5	2014	580,000	232,000	348,000
6	2015	600,000	240,000	360,000
7	2016	620,000	248,000	372,000
8	2017	640,000	256,000	384,000
9	2018	660,000	264,000	396,000
10	2019	680,000	272,000	408,000
11	2020	700,000	280,000	420,000
12	2021	720,000	288,000	432,000
13	2022	740,000	296,000	444,000
14	2023	760,000	304,000	456,000

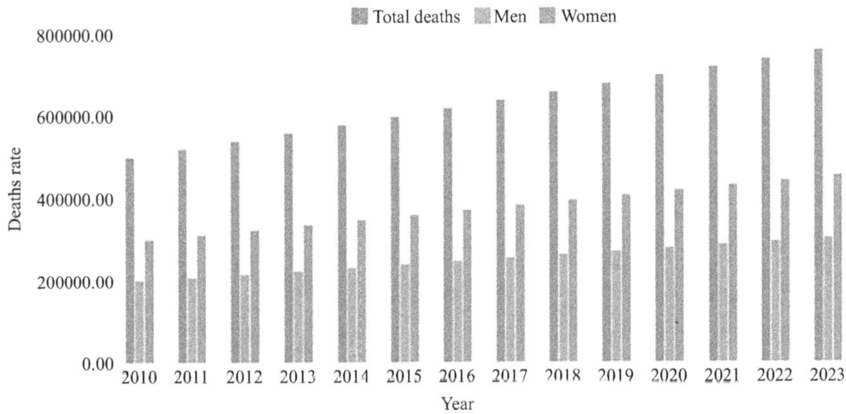

Figure 6.1 Mortality rates for Alzheimer's disease across the world from 2010 to 2023

Table 6.2 shows the number of deaths from 2010 to 2023, along with a breakdown by gender. The overall death toll has steadily increased during this time, going from 10,000 in 2010 to 17,500 in 2023. The statistics regularly indicates that there are more deaths among women than among men each year. 2010 saw 6,000 fatalities among women and 4,000 deaths among men. These figures rose to 10,500 deaths for women and 7,000 deaths for men by 2023. This trend shows a persistent increase in mortality rates for both sexes throughout the course of the measured years, with a larger frequency among women, graphically shown in Figure 6.2.

Numerous techniques, including DL and machine learning (ML) techniques, are put forth. These are the Random Forest classifiers [1], Decision Tree classifiers

Table 6.2 Death rate for Alzheimer's disease in 2010–2023 across India

Sl. no	Year	Total deaths	Men	Women
1	2010	10,000	4,000	6,000
2	2011	10,500	4,200	6,300
3	2012	11,000	4,400	6,600
4	2013	11,500	4,600	6,900
5	2014	12,000	4,800	7,200
6	2015	13,000	5,200	7,800
7	2016	13,500	5,400	8,100
8	2017	14,000	5,600	8,400
9	2018	14,500	5,800	8,700
10	2019	15,000	6,000	9,000
11	2020	16,000	6,400	9,600
12	2021	16,500	6,600	9,900
13	2022	17,000	6,800	10,200
14	2023	17,500	7,000	10,500

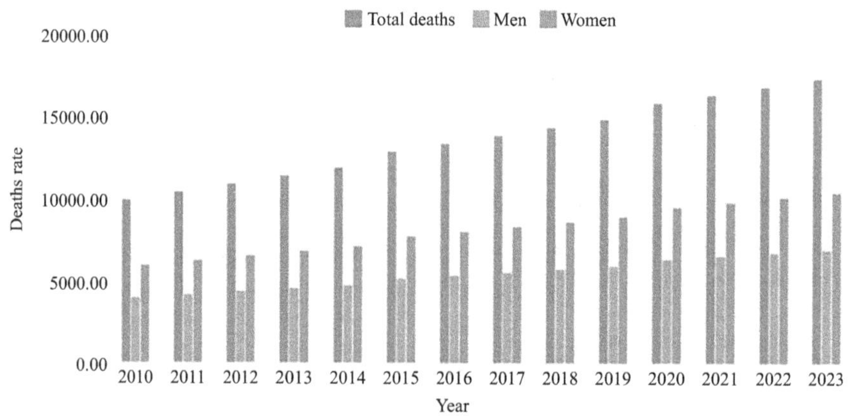

Figure 6.2 Mortality rates for Alzheimer's disease across the world from 2010 to 2023

[1], Support Vector Machine (SVM) [1], Artificial Neural Network (ANN) [8], Generative Adversarial Network (GAN) [8], Stacked Auto-encoder (SAE) [3], Deep Neural Network (DNN), and Recurrent Neural Network (RNN) [4], Bidirectional encoder [4], Single Value Decomposition (SVD) [9], Principal Component Analysis(PCA) [5], Logistic regression [10], Stochastic Gradient Descent(SGD) [11], Gradient Boosting Classifier (GBC) [11], K-Nearest Neighbors (KNN) [11], Multilayer Perception (MLP) [11], PRISMA [11], NeAE-Eye [12], Transfer Learning, and the Grad-CAM technique [7].

Existing approaches to AD diagnosis and prognosis face several significant limitations. The sensitivity and precision required by current approaches, including

Table 6.3 Comparison of diagnostic methods for Alzheimer's disease

Diagnostic method	Invasiveness	Cost	Accuracy	Accessibility	Novel biomarkers
Clinical assessments	Low	Low	Moderate	High	No
Neuroimaging	Moderate	High	High	Moderate	Limited
CSF biomarkers	High	High	High	Low	Yes
Genetic testing	Low	Moderate	Moderate	High	No
Deep learning-based	Low	Low	High	High	Yes

neuroimaging and biomarker assays, are sometimes lacking, especially in the early phases when treatments could have the greatest potential impact. The cost and discomfort of invasive treatments such as PET scans and cerebrospinal fluid (CSF) analysis prevent widespread usage. Subjective clinical evaluations can cause inconsistent diagnosis, which makes it more difficult to provide patients with consistent and trustworthy care. Furthermore, these methods might not be able to accurately anticipate the course of a disease or customize a course of therapy for a given patient, which emphasizes the need for more sophisticated and accurate diagnostic instruments. Integrating DL-based biomarkers could address these problems and enhance patient care and outcomes in AD by offering increased accuracy, non-invasiveness, and the potential for early diagnosis. By providing improved accuracy, non-invasiveness, and the possibility of early identification, by using DL-based biomarkers it is easy for the prediction of AD for the patient treatment at an early stage.

Table 6.3 shows the analysis of several diagnostic procedures based on important factors such as cost, accuracy, simplicity of use, intrusiveness, and creative use of biomarkers.

The Alzheimer's convolutional neural network (CNN) is the main topic of this chapter. The field of DL-based biomarkers, which have revolutionized determining AD diagnosis and prognosis, is led by CNNs. CNNs' unparalleled ability to analyze and comprehend complex imaging data has allowed them to analyze and detect indications of AD through PET and MRI images from the dataset. Through automation of the feature extraction process, CNNs are able to detect minute but significant alterations to the brain's composition and capabilities that may occur before the clinical symptoms. This DL method increases diagnosis accuracy and aids in prognostic assessments by predicting the disease's trajectory. This chapter's primary focus is the Alzheimer's CNN. The field of DL-based biomarkers, which have revolutionized the diagnosis and prognosis of AD, is led by CNNs. Because of CNN decipher and evaluation of image data, it became easier to predict Alzheimer's using PET and MRI data. CNN is able to detect minute but significant alterations in the composition and functioning of the brain that take place prior to the clinical symptoms by the model extraction. By forecasting the course of the illness, this DL technique improves prognostic assessments and raises diagnosis accuracy. The chapter consists of Section 6.2 about the Literature Survey followed by Section 6.3 which is about the model used for Alzheimer's prediction while Section 6.4 talks about the result and conclusion followed by Reference.

6.2 Literature survey

Koga *et al.* [1] used phosphorylated-tau (CP-13) pictures, CBD, and PSD databases. They suggested a model for AD that obtained performance values higher than 92% using decision tree and random forest classifiers. These results show the adaptability of the model for predicting AD diagnosis across various datasets. The Alzheimer's Disease Neuroimaging Initiative (ADNI) and the Open Access Series of Imaging Studies (OASIS) were two of the numerous datasets used by Arya *et al.* [2]. They proposed CNN, RNN, and ANN with the following accuracies: SVM 85.71%, CNN 98.6%, and RNN 91.2%. These results demonstrate the adaptability and potential of DL models for accurate AD diagnosis across a range of datasets. The ADNI and the OASIS were two of the few datasets used by Zhao *et al.* [8]. They suggested utilizing CNN, ANN, and radio frequency (RF), which produced a wide range of accuracy of 90.3%.

Many datasets like ADNI are used by Aqeel *et al.* [3]. They suggested employing DL techniques, which produced a wide range of results, including 88.2% accuracy, 88.64% precision rate, 88.39% F1-score, and 11.84% false negative rate. Error rate: 0.12% and area under the curve (AUC): 0.92%. These findings present the DL models' versatility and promise for precise AD diagnosis on a variety of datasets. Abdelwahab *et al.* [13] utilized multiple datasets including microarray gene expression dataset. They proposed using Preprocessing Gene Selection (SVD, CNN, PCA) which resulted in an accuracy of 96.09% and an error rate of less than 1%. These findings highlight the DL models' versatility and promise for precise AD diagnosis on a variety of datasets. Numerous datasets like ADNI are used by Sellappan *et al.* [4]. They suggested a DL method known as Variational Bayesian Network, which produced results with an error rate of less than 1% and an accuracy of 98.57%. These findings highlight the DL models' versatility and promise for precise AD identification across many datasets. Magnetic Resonance Imaging and the OASIS datasets were among the most used datasets used by Mohammed *et al.* [9]. They proposed a classification approach that yielded results with an error rate of less than 6% and 94% accuracy rate, 93% precision, 98% recall, and 96% accuracy, along with AUC values of 94.8%, 93%, 97.75%, and 99.7%.

Hu *et al.* used datasets of ADNI [5]. They recommended DL techniques that produced a performance metric of 91.8% and an error rate of less than 9%. These findings highlight the DL models' versatility and promise for precise Alzheimer's on a variety of datasets. Odusami *et al.* [10] used datasets such as pet photographs from MRI and the ADNI. They recommended DL techniques, which produced results with a 73.90% accuracy rate and a 26.1% error rate. These findings highlight the DL models' versatility and promise for precise AD diagnosis on a variety of datasets. The MRI report is one of the datasets used by Dara *et al.* [6]. They recommended DL techniques, which produced precision: Genetic algorithms with SVM: 96.80 SVM with 93.30 image processing; Fisher: 96.32 SVM: 92.48, 95 Procedures for image processing and weight extraction; 91.40 Multimodal Neuroimaging arbitrary. Stochastic forest: 93 PCA using SVM: 95 CNN with various architectures: 98.08, less than 7.52% error rate on average. These findings show how versatile and promising DL models are for providing precise diagnoses of Alzheimer's across a variety of datasets. Syed *et al.* [14] employed datasets, including the Training cum-validation dataset and the independent test dataset. They suggested ML methods that yielded results with less than 1% error rate and 0.9799% accuracy.

Korean-based aging study for the diagnosis initiated the prediction of AD (KBASE; kbase.kr) dataset was one of the datasets used by Kim *et al.* [15] from the Severance Hospital. They recommended applying DL techniques; the results produced are accuracy with 78.05, sensitivity with 76.0, and F1-score with 74.0, respectively, with an error rate of 10–20%. These findings show how adaptable and efficient DL models are for predicting AD using datasets. The dataset OASIS was utilized by Bangyal *et al.* [16]. It was suggested to SVM, SGD, GBC, KNN, DT, RF, MLP. CNN provided accuracy of 94.61% for CNN and 92.12% for MLP with a 5.39 error rate range. The data collecting and preparation methods suggested by Sun *et al.* [11] yielded an accuracy of 85% and an error rate of 15%. These outcomes highlight the DL models' versatility and promise for precise AD diagnosis on various datasets. Battineni *et al.* [17] used the OASIS dataset. CNN was suggested by them, and it produced results with an accuracy rate of 80% and an error rate of less than 20%. This shows the DL models' versatility and promise for precise AD diagnosis on various datasets. Agarwal *et al.* [18] used datasets from OASIS, ADNI, IXI, AIBL, MIRIAD, and MILAN. They advised using the PRISMA approach, which produced results with an error rate of less than 8.30% and an accuracy of 91.30%. Sun *et al.* [11] used eye tracking data. It was advised to use the NeAE-Eye approach, which produced results with an accuracy of 85% and an error rate of 15%.

Bringasa *et al.* [12] received information from the AFAC daycare center. They suggested using CNN, which produced results with an error rate of less than 9% and an accuracy of 90.91%. Raju *et al.* [7] employed two-dimensional coronal view MRI image slices that they obtained from Kaggle. They suggested using transfer learning in conjunction with the Grad-CAM approach, which produced results with a 1% error rate and 99% accuracy. These findings highlight the DL models' versatility and promise for precise AD diagnosis on various datasets. Table 6.4 shows the approaches for predicting AD using different datasets and models. These works leverage the MRI, ADNI, and OASIS datasets in addition to ML and DL methods, including CNN, RNN, SVM, and GAN. Performance evaluations demonstrate how well these techniques work to achieve high accuracy. Results are greatly enhanced by methods such as data augmentation and transfer learning. Table 6.4 shows the approaches to AD prediction.

Table 6.4 Analysis of the existing approaches to Alzheimer's prediction

Reference no	Dataset used	Algorithm proposed	Demerits
[1]	Images of phosphorylated-tau (CP13), (PSP), (CBD)	DT RF	A lone researcher trained the diagnostic model using data from a single neuropathology laboratory, which may affect generalization
[2]	ADNI OASIS	CNN SVM RNN	Significant overlap with the high cost of acquiring fMRI data and restricted access to other datasets
[8]	ADNI	CNN SVM RF	The datasets in the AD sector are still small compared with datasets in computer vision tasks since medical data privacy is an issue. Due to the complexity of AD-related tasks, researchers need a large-scale dataset in order to develop more powerful and efficient models
[3]	ADNI	RNN NM MRI Biomarkers MLP	Complexity in feature learning Data dependency
[13]	Microarray gene expression data	Preprocessing Gene Selection (SVD, CNN, PCA)	High-dimensional data
[4]	ADNI	VBN CNN RNN LR	Over fitting ML is less effective than deep learning
[9]	OASIS MRI Data	SVM DT RF KNN	Susceptibility to overfitting, High computational cost for prediction
[5]	ADNI NIFD	SGD	In the loser datasets, the multiclassification performance is not up to the par
[10]	MRI and PET images ADNI	DL methods	Lower sensitivity, complex architecture
[6]	MRI reports	SVM CNN PCA	Lower sensitivity, higher specificity, complex processes

(Continues)

Table 6.4 (Continued)

Reference no	Dataset used	Algorithm proposed	Demerits
[14]	Training cum-validation on dataset Independent test dataset	ML methods	Obtaining all of these biomarkers for multiple marker examinations from a single patient is both costly and impractical. Our novel CSF protein combination offers the unique advantage of being economical, as it can precisely classify patients with AD in its early stages by profiling
[15]	The Alzheimer's Disease Neuroimaging Initiative includes the Korean Brain Aging Study for the Early diagnosis and prediction of Alzheimer's disease (KBASE; kbase.kr) dataset from Severance Hospital.	Deep learning framework technique	Large datasets 10% of false negative instances can be attributed to amyloid PET sensitivity
[16]	The Imaging Studies Open Access Series (OASIS)	MLP CNN	MLP gives less accuracy than CNN
[11]	3D T1-weighted structural MRI data collected from two open-accessible databases: NIFD and ADNI	CNN SGD	Sensitivity to feature scaling, computational cost
[17]	(OASIS-3) Dataset	CNN	Early stopping of the model at the 80th iteration out of 100 epochs
[18]	ADNIOASIS IXI AIBL MIRIAD MILAN	PRISMA methodology	Data heterogeneity, variability in pre-processing, limited generalizability techniques
[11]	Eye-tracking datasets	NeAE-Eye	Lack of large-scale eye-tracking datasets, current models and algorithms are mostly applied to 2D displays
[12]	Data from (AFAC) daycare center	CNN	Few sequences are obtained for each patient, seeking to increase the size of the original dataset
[7]	Two-dimensional MRI image slices in the coronal view, sourced from Kaggle	Transfer learning and the Grad-CAM technique	Limited dataset, overfitting risk

6.3 Materials and methods dataset

The Washington University Knight Alzheimer Disease Research Center collected clinical data on MRI and PET imaging results from 1,098 people over a 15-year period. This data is known as the OASIS dataset. The ages of the participants range from 42 to 95 years old. The data comprises 493 persons in varying stages of cognitive decline and 605 adults with full cognitive functioning. The OASIS collection contains around 2,000 MR sessions with different functional and structural sequences. Together with raw imaging scans from the PET metabolic and amyloid imaging, about 1,500 images from the PET unified pipelines are viewable. Another benefit provided by OASIS is access to imaging data which was post-processed, including 3D image analysis and positron emission tomography investigations. The imaging data includes APOE, mental state, and long-term thinking, and behavioral effects. The OASIS dataset is made available to the scientific community for use in studying dementia and healthy aging-related topics. In addition to its longitudinal dataset, OASIS contains a cross-sectional dataset.

Researchers and other interested parties can access brain imaging data from the organization OASIS dataset. The OASIS Longitudinal [4] and OASIS Cross-sectional [13] datasets were obtained using MRI image neuroanatomical atlases. OASIS is a neuroimaging biomarker that can be used in both healthy individuals and AD patients, both cross-sectionally and longitudinally. Clinical and demographic data are supplied in the file in XML format. Data is obtained from magnetic resonance imaging scans. For additional details on the picture properties and terminology, visit http://www.oasisbrains.org/longitudinal_facts.html (accessed on May 25, 2021). Fifteen features from 374 patients of whom 37 are converted, 190 are demented, and 147 are not included in the OASIS longitudinal dataset. The age range of patients is 60–90 years old.

6.3.1 Data preprocessing

There are several data pre-processing techniques utilized to create clean data collection for the investigations. To increase the precision and caliber of the analytical findings, clinical datasets related to AD can be pre-processed using a variety of techniques. Pre-processing techniques that researchers most frequently employ include data cleaning which is the process of eliminating and rectifying any erroneous, faulty, or absent information from a dataset. This method normally starts with looking for anomalies, discrepancies, and missing values in the data. then thereafter altering or getting rid of them. Feature selection is the process of choosing the most important features that are expected to have a considerable influence on the analysis' findings. from the dataset. This can be accomplished using a variety of statistical methods, including principal component analysis, mutual information, and correlation analysis. The process of transforming the data into an analysis-ready format is called as data transformation. Applying techniques such as normalization, logarithmic transformation, and standardization will help achieve this.

The amputation process entails substituting approximate values for the blanks. Many imputation techniques, including mean imputation, regression imputation, and multiple imputations, can be used to achieve this. Imaging datasets for AD include PET and MRI scans. The following pre-processing techniques were applied to the research articles that this review article reviewed. The practice of aligning each intensity value in an image with every other image is known as image normalization. To do this, techniques including histogram equalization, contrast stretching, and normalization can be leveraged to enhance the images' visual presentation. Cropping and resizing images include resizing the photos to a uniform size and removing any extraneous components from the picture. This could increase the approaches' efficiency and lower the analysis' computing complexity. Enhancement of images: This technique uses a number of modifications, including rotation, flipping, and scaling, to generate additional images from the original dataset. By doing so, the dataset may be increased, and the analysis resilience can be strengthened. Extraction of features: During this stage, pertinent features from the photos that can be utilized for analysis are extracted. Techniques such as edge detection, texture analysis, and form analysis can be used for this.

An important step in data mining is data cleaning, which entails replacing missing values and eliminating anomalies. We have now analyzed the distribution of categorical and numerical columns to determine which features align with our research goals; these attributes should not have a substantial relationship with the target trait. In order to eliminate them, we also looked at characteristics with distinct values. Missing values and incorrect data types were the next topics we covered. No correlation was observed between the subject ID, MRI ID, and the target feature. The hand feature was also non-informative, having a constant value of R. Therefore, these columns were removed from the dataset to reduce dimensionality.

6.3.2 Processing the unbalance of the OASIS dataset

Three imbalanced classes make up the 373 rows in the OASIS dataset. Three classifications, representing changing frequencies within the dataset, are non-demented (190 instances, or 51% of the dataset), demented (147 instances, or 39%), and converted (37 occurrences, or 10%). In this study, we applied the SMOTE algorithm to reduce the impact of class imbalance. SMOTE is a useful method for achieving dataset equilibrium. The SMOTE technique finds the minority class nearest neighbors, creates new samples for the minority class at random intervals, and chooses the minority class ranks at random.

The performance of classification is significantly impacted by the size of the training set. In each of the previously described datasets, there is a cap on the quantity of imaging scans that were recovered from individuals suffering from AD and MCI. Pre-processing is usually required in investigations before data tampering. Pre-processing is the term used to describe a set of image processing processes carried out on the resulting image scans. Many MRI software packages, such as Statistical Parametric Mapping (SPM), Computational Anatomy Toolbox (CAT12), FMRIB Software Library (FSL) [19], Free Surfer [20], and ANTS [21], have well-encapsulated pre-processing approaches accessible for them. Temporal filtering, covariates removal, registration, normalization, smoothing, segmentation, skull-stripping, and noise reduction are pre-processing techniques that are commonly used. This review will address intensity normalization, registration, tissue segmentation, skull-stripping, and class balancing.

When dealing with datasets that exhibit significant range variability, normalization is essential to the data preparation phase. This is especially true after turning nominal values (1 and 2) into real numbers. Due to a lack of standardization, some traits might be more prevalent than others with smaller numerical ranges, which could induce biases in the analysis. Normalization not only minimizes these biases but also facilitates algorithm performance by restricting the use of broad datasets. This guarantees a more exhaustive and unbiased inquiry, supporting the overall stability of the ensuing data-driven procedures. Here, values are kept between 0 and 1, in accordance with (6.1), by data scaling. Min a denotes the attribute's lowest value, whereas max a denotes its highest value, the scaled value by x normalized, and the attribute's initial value by x in this equation. By transforming the values of each characteristic into a predetermined range, this scaling method facilitates consistency and comparison across the various features in the dataset. The normalization method ensures that the initial scale of the attributes will not affect later research, allowing for a more thorough and precise analysis of the data.

$$X_{\text{normalized}} = (X - \min)/(\max - \min) \tag{6.1}$$

6.3.3 CNN for Alzheimer's prediction

To address complex learning tasks, a specialized subset of ML called DL uses a hierarchical architecture made up of several layers. DL techniques have acquired a lot of popularity in recent years and are now frequently used in different brain research projects. CNNs are particularly popular among these techniques. CNNs can effectively recognize and analyze patterns in input data by extracting pertinent features through the use of convolutional procedures. The field has rapidly grown and extended as a result of CNN's outstanding performance in a variety of applications, including object identification, classification, and segmentation. CNNs are closely watched by researchers because they have yielded state-of-the-art results in various domains such as medical imaging, natural language processing, and computer vision. The remarkable efficiency of CNNs in classifying and dividing reality-based photos is utilized in application development.

The input layer provides data to the network. The input for image data in this layer is a multidimensional array representing the image. Convolutional layers perform convolution operations to input data through a sequence of learnable filters to extract features such as edges, shapes, and textures. It is made up of filters and kernels. Little matrices are slid over the input image to build feature maps. For example, a 3×3 filter only captures a small percentage of the input image. The Mechanism of Activation ReLU (Rectified Linear Unit) is usually added after the convolution procedure to add non-linearity. By keeping only the most important data, the pooling layer—more especially, the max pooling technique—down samples the feature maps. Through this procedure, the spatial dimensions are lowered, which lowers the number of factors and processing demands. Max pooling is a useful technique for capturing the most prominent features in a feature map by choosing the maximum value from each patch. Dense layer: The dense layer performs higher-level reasoning and classification using the features that have been gathered by the convolutional and pooling layers. Neurons in a dense layer in a neural network are connected to all other neurons in the layer above it. This allows the network to learn complex representations. Activation function: Common activation functions (in classification issues) are SoftMax for the output layer and ReLU complex for hidden layers. The output layer, which normally consists of as many neurons as there are classes in a classification issue, produces the final predictions.

Algorithm for Alzheimer's prediction using CNN:

1. Input dataset, dataset true labels, word2vec matrix.
2. Finding the output score for the model using test dataset.
3. Set 3d matrix using f as feature.
4. For j in i.
5. Set feature set matrix for sample i.
6. For j in i.
7. Vectorize v_j.
8. Append f from f_i.
9. Set ***ftrain***, ***ftest***, ***ltrain*** into train and test subsets.
10. Evaluate score using ***ftrain*** and ***ltrain***.
11. At last, the return score for the model.

After multiplying two Maricely represented images, a new matrix is created from which a CNN retrieves attributes. Two stages are carried out by CNNs: first, they automatically extract pertinent characteristics from the input data, and then they classify these features into predetermined categories. CNNs' capacity to learn and generalize characteristics from big datasets is one of their main advantages. Using the set of input images, the convolution operation creates a series of feature maps that highlight significant structures and patterns. A pooling layer is used to minimize the dimensions of these feature maps and highlight the most significant features, allowing the model to concentrate on the most pertinent data. The offsets of the subregion in the x and y directions are indicated by the variables "a" and "b," respectively. The bottom-right corner of the subregion is indicated by these offsets, which are binary numbers (0 or 1) that represent its location in relation to the origin ($2x$, $2y$).

$$fx, (s) = \max a,\ b = 0s2x+, 2y + b \tag{6.2}$$

Convolutional architectures are used by CNNs, a class of feed-forward neural networks, to extract features from input automatically and without the need for human involvement. However, manual feature engineering is the foundation of conventional feature extraction techniques. CNN's activation functions mimic the biological neurons' threshold-based signal propagation mechanism, while the CNN's kernels represent a variety of receptors with varying patterns and features recognition abilities. CNN outperforming other generic ANN Local Domains Lowering parameters is facilitated by the fact that only a tiny fraction of neuron layer is connected to the layers above them. And then those accelerate the convergence, Weight sharing.

Parameters can be lowered by distributing in various factors. Down sampling is a key technique in DL that reduces dimensionality while maintaining essential information in images. Pooling layers can efficiently down sample images by using local correlation to remove unnecessary information and keep just the most important characteristics. This procedure lowers the model's parameter count while simultaneously increasing its effectiveness. Convolution is a crucial step in feature extraction that is highly relied upon by CNNs, the foundation of DL, to produce feature maps that effectively capture the essence of the input image. The size of the convolution kernel, however, needs to be carefully considered since it can result in information loss at the image borders, highlighting the necessity of wise kernel selection. Alzheimer's disease using DL was proposed by different researchers [22–26].

6.4 Results and discussion

Accuracy is the degree to which a diagnostic test correctly finds or excludes a condition. A high true positive rate indicates that a test is accurate; a high true negative rate indicates that it is not accurate in identifying those who do not have the ailment. It is necessary for precise diagnosis and therapy planning. A few elements that impact accuracy are the caliber of the test, the technology used, and the experience of healthcare professionals. Achieving high diagnosis accuracy in medicine is essential to patient safety and effective treatment outcomes. The accuracy of several current models is displayed in Table 6.4. Table 6.5 illustrates the accuracy. The graphical representation of the different current model accuracy rates is shown in Figure 6.3.

$$\text{Accuracy} = (\text{TP} + \text{TN}) * 100/(\text{TP} + \text{TN} + \text{FP} + \text{FN}) \tag{6.3}$$

Table 6.5 shows the accuracy of various existing models compared with proposed model. From Table 6.5, CNN has high accuracy compared with other existing models, that is, 99.95%. The graphical representation of accuracy for different models is shown in Figure 6.4.

One essential statistic in ML that is used to assess a classification model's performance is its error rate. It shows the percentage of inaccurate predictions the model produced out of all the predictions. The frequency of the model predictions does not match with the actual labels in the dataset. It is directly measured by the error rate. Understanding the error rate is essential to determining how effective an ML model is. Better performance, or fewer errors made by the model, is indicated by a lower error rate. However, if the error rate is more, the model prediction is not accurate.

$$\text{Error rate} = \frac{\text{Number of incorrect prediction}}{\text{Total number of prediction}} \tag{6.4}$$

Another popular indicator for assessing model performance is accuracy, which has a tight relationship with the error rate. The percentage of accurate forecasts among all predictions is known as accuracy. There is a clear correlation between accuracy and mistake rate. Table 6.6 shows the error rate of various models. From

Table 6.5 Comparison of accuracy of proposed models on various existing models

Sl. no	Model name	Accuracy (%)
1	CNN	99.95
2	DTRF	95
3	SVM	85.71
4	RF	90.3
5	RNN	88.24
6	Preprocessing Gene Selection (SVD))	96.09
7	VBN	98.57
8	KNN	99.7
9	SGD	91.83
10	DL methods	73.9
11	PCA	95
12	ML methods	97
13	Deep learning model technique	75
14	MLP	92.12
15	SGD	85
16	GBC	80
17	PRISMA methodology	91.30
18	NeAE-Eye	85
19	PCA	90.91
20	Transfer learning and the Grad-CAM technique	99

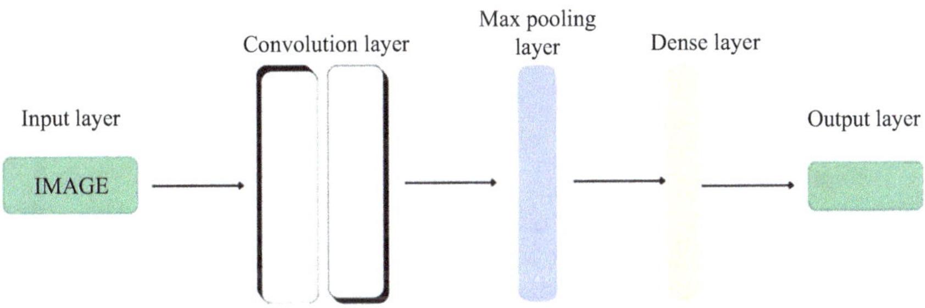

Figure 6.3 Convolutional neural network for Alzheimer's prediction

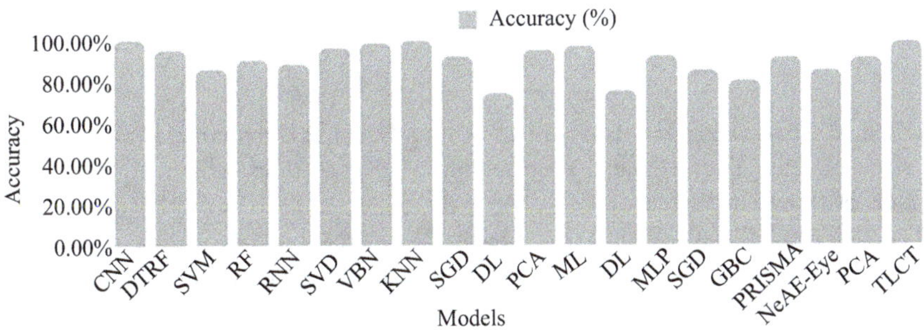

Figure 6.4 Comparison of accuracy of proposed model with various existing models

Table 6.6 Comparison of error rate of the proposed model with various existing models

Sl. no	Model name	Error rate (%)
1	CNN	0.05
2	DTRF	5
3	SVM	14.26
4	RF	9.7
5	RNN	11.76
6	Preprocessing gene selection (SVD)	3.91
7	VBN	1.43
8	KNN	0.3
9	SGD	8.17
10	DL methods	26.1
11	PCA	5
12	ML methods	3
13	Deep learning model technique	25
14	MLP	7.88
15	SGD	15
16	GBC	20
17	PRISMA methodology	8.7
18	NeAE-Eye	15
19	PCA	9.09
20	Transfer learning and the Grad-CAM technique	1

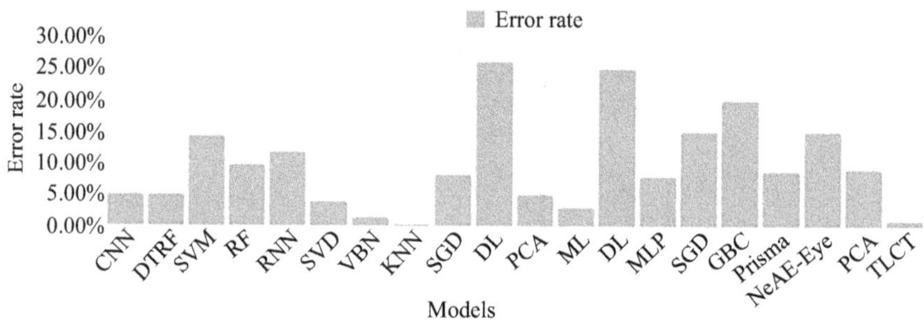

Figure 6.5 Comparison of error rate of the proposed model on various existing models

Table 6.6, we conclude that the least error rate model is CNN compared with other existing models. The graphical representation of error rates with various existing models is shown in Figure 6.5.

As shown in (6.6), when the model has low false positive value, we say that positive predictions are accurate. Recall is usually sacrificed for precision. Precision is a critical factor in the creation of reliable and trustworthy classification models for a variety of real-world applications. Table 6.7 shows the precision value of the proposed model on various existing models. Figure 6.6 shows the graphical representation of various existing models for recall value.

Table 6.7 Comparison of precision of the proposed model on various existing models

Sl. no	Model name	Precision (%)
1	CNN	93.80
2	LR	82.49
3	RNN	88.24
4	GAN	72.12
5	PCA	93.01
6	SVM	91
7	KNN	72

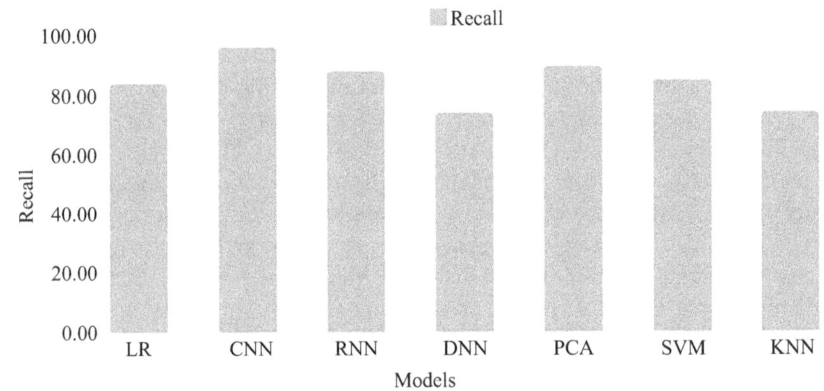

Figure 6.6 The graphical representation of various existing models for recall value

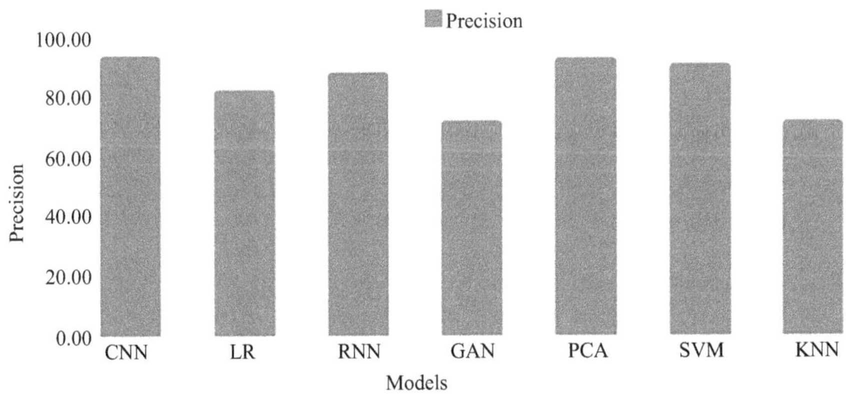

Figure 6.7 Comparing the proposed model's precision rates with several other models

$$\text{Precision} = \text{TP} * 100/(\text{TP} + \text{FP}) \tag{6.5}$$

Table 6.7 shows the precision values for the various models. From Table 6.7, it is concluded that CNN model has a high precision value, that is, 93.80% among other existing models. The graphical representation of precision valve for the models is shown in Figure 6.7.

6.5 Conclusion

AD is a dangerous neurological condition that progressively goes worse over time, affecting cognitive function, leading to severe memory loss and the inability to do even basic tasks. It is a common cause of dementia that affects millions of people globally, especially the elderly. AD places a tremendous strain on all those involved, including family members and caregivers. In this study, we concentrate on applying cutting-edge ML methods to enhance AD diagnosis and prognosis. To be more precise, we use a CNN in conjunction with the extensive brain imaging dataset from OASIS to find and examine disease-related biomarkers. Using this dataset, we train CNN and compare its results with those of other available methods.

The research utilized a range of assessment criteria, including accuracy, recall, error rate, precision, and several other elements that are critical to assessing the performance of diagnostic models. Comprehensive testing and analysis show that CNN outperforms traditional methods in each and every one of these characteristics. The CNN model shows great potential as a powerful tool for early diagnosis and prognosis as it has been demonstrated to be a highly accurate and reliable way of identifying biomarkers associated with dementia. A proficient way for timely diagnosis of AD can lead to improved management and treatment strategies, thereby shortening the illness's course and improving patients' quality of life. Furthermore, the use of CNNs in medical imaging and diagnostics creates new opportunities in clinical and research uses, which makes this a noteworthy advancement in the medical field. The study concludes by demonstrating how well CNN scan diagnose and prognosticate AD. Our work highlights the possible use of these ML techniques in the fight against AD by showing the superiority of CNNs over traditional methods using the OASIS dataset.

References

[1] Shunsuke Koga, Akihiro Ikeda, and Dennis W. Dickson, "Deep learning-based model for diagnosing Alzheimer's disease and tauopathies", *Neuropathol Appl Neurobiol.* 2022;48(1):e12759. https://doi.org/10.1111/nan.12759.

[2] Akhilesh Deep Arya, Sourabh Singh Verma, Prasun Chakarabarti, *et al.*, "A systematic review on machine learning and deep learning techniques in the effective diagnosis of Alzheimer's disease", *Brain Inf.* 2023;10:17. https://doi.org/10.1186/s40708-023-00195-7.

[3] Anza Aqeel, Ali Hassan, Muhammad Attique Khan, *et al.*, "A long short-term memory biomarker-based prediction framework for Alzheimer's disease", *Sensors.* 2022;22(4):1475. https://doi.org/10.3390/s22041475.

[4] Suganyadevi Sellappan, Shiny Pershiya Anand, Finney Daniel Shadrach, Balasamy Krishnasamy, Renu Karra, Umamaheshwari Annamalai, "A survey of Alzheimer's disease diagnosis using deep learning approaches", *J Autonom Intell.* 2024;7(3). https://doi.org/10.32629/jai.v7i3.660.

[5] Jingjing Hu, Zhao Qing, Renyuan Liu, *et al.*, "Deep learning based classification", *Front Neurosci.* 2021;14:626154. https://doi.org/10.3389/fnins.2020.626154.

[6] Omer Asghar Dara, Jose Manuel Lopez-Guede, Hasan Issa Raheem, Javad Rahebi, Ekaitz Zulueta, and Unai Fernandez-Gamiz, "Alzheimer's disease diagnosis using machine learning", *Appl Sci.* 2023;13(14):8298. https://doi.org/10.3390/app13148298.

[7] Manu Raju, M Thirupalani, S. Vidhyabharathi and S. Thilagavathi, "Deep learning based multilevel classification of Alzheimer's disease using MRI scans", *IOP Conf Ser: Mater Sci Eng.* 2021;1084:012017. https://doi.org/10.1088/1757-899X/1084/1/012017.

[8] Zhen Zhao, Joon Huang Chuah, Khin Wee Lai, *et al.*, "Conventional machine learning and deep learning in Alzheimer's disease diagnosis using neuroimaging: A review", *Front Comput Neurosci.* 2023;17:1038636. https://doi.org/10.3389/fncom.2023.1038636.

[9] Badiea Abdulkarem Mohammed, Ebrahim Mohammed Senan, Taha H. Rassem, *et al.*, "Multimethod analysis of medical records and MRI images for early diagnosis of dementia and Alzheimer's disease based on deep learning and hybrid methods", *Electronics.* 2021;10(22):2860. https://doi.org/10.3390/electronics10222860.

[10] Modupe Odusami, Rytis Maskeliūnas, Robertas Damasevicius, and Sanjay Misra, "Deep-learning-based diagnosis of Alzheimer's disease using multimodal input fusion of PET and MRI images", *J Med Biol Eng.* 2023;43:291–302. https://doi.org/10.1007/s40846-023-00801-3.

[11] Jinglin Sun, Yu Liu, Hao Wu, Peiguang Jing and Yong Ji, "A novel deep learning approach for diagnosing Alzheimer's disease based on eye-tracking data", *Front Hum Neurosci.* 2022;16:972773. https://doi.org/10.3389/fnhum.2022.972773.

[12] Santos Bringasa, Sergio Salomónb, Rafael Duquec, Carmen Laged, and José Luis Montañac, "Alzheimer's disease stage identification using deep learning models", *J Biomed Inform.* 2020;109:103514.

[13] Mahmoud M. Abdelwahab, Khamis A. Al-Karawi and Hatem E. Semary, "Deep learning-based prediction of Alzheimer's disease using microarray gene expression data", *Biomedicines.* 2023;11(12):3304. https://doi.org/10.3390/biomedicines11123304.

[14] Asif Hassan Syed, Tabrej Khan, Atif Hassan, Nashwan Alromema, Muhammad Binsawad, and Alhuseen Omar Alsayed, "An ensemble-learning based application to predict the earlier stages of Alzheimer's disease", *IEEE Access.* 2020;8:222126–222143. https://doi.org/10.1109/ACCESS.2020.3043715.

[15] Suhong Kim, Peter Lee, Kyeong Taek Oh, *et al.*, "Deep learning-based amyloid PET positivity classification model in the Alzheimer's disease continuum by using 2-FDG PET", *EJNMMI Res.* 2021;11(1):56. https://doi.org/10.1186/s13550-021-00798-3.

[16] Waqas Haider Bangyal, Najeeb Ur Rehman, Asma Nawaz, *et al.*, "Constructing domain ontology for Alzheimer disease using deep learning based approach", *Electronics*. 2022;11(12):1890. https://doi.org/10.3390/electronics11121890.

[17] Gopi Battineni, Nalini Chintalapudi, Francesco Amenta and Enea Traini, "Deep learning type convolution neural network architecture for multiclass classification of Alzheimer's disease", In *Proceedings of the 14th International Joint Conference on Biomedical Engineering Systems and Technologies (BIOSTEC)*, pp. 209–215, 2021.

[18] Deevyankar Agarwal, Gonçalo Marques, Isabel de la Torre-Díez, Manuel A. Franco, Begoña García Zapiraín and Francisco Martín Rodríguez, "Transfer learning for Alzheimer's disease through neuroimaging biomarkers: A systematic review", *Sensors*. 2021;21(21):7259. https://doi.org/10.3390/s21217259.

[19] Vijeeta Patil, Manohar Madgi and Ajmeera Kiran, "Early prediction of Alzheimer's disease using convolutional neural network", *Egypt J Neurol Psychiatry Neurosurg*. 2022;58:130. https://doi.org/10.1186/s41983-022-00571-w.

[20] Tausifa Jan Saleem, Syed Rameem Zahra, Fan Wu, *et al.*, "Deep learning-based diagnosis of Alzheimer's disease", *J Pers Med*. 2022;12(5):815. https://doi.org/10.3390/jpm12050815.

[21] Changxing Qu, Yinxi Zou, Yingqiao Ma, *et al.*, "Diagnostic performance of generative adversarial network-based deep learning methods for Alzheimer's disease", *Front Aging Neurosci*. 2022;14:841696. https://doi.org/10.3389/fnagi.2022.841696.

[22] Deevyankar Agarwal, Manuel Álvaro Berbís, Antonio Luna, Vivian Lipari, Julien Brito Ballester and Isabel de la Torre-Díez1, "Automated medical diagnosis of Alzheimer's disease using an efficient net convolutional neural network", *J Med Syst*. 2023;47(1):57. https://doi.org/10.1007/s10916-023-01941-4.

[23] Gowhar Mohi ud din dar, Avinash Bhagat, Syed Immamul Ansarullah, *et al.*, "A novel framework for classification of different Alzheimer's disease stages using CNN model", *Electronics*. 2023;12(2):469. https://doi.org/10.3390/electronics12020469.

[24] Mark Jenkinson, Christian F Beckmann, Timothy E J Behrens, Mark W Woolrich, and Stephen M Smith, "FSL", *Neuroimage*. 2012;62(2):782–790. https://doi.org/10.1016/j.neuroimage.2011.09.015.

[25] Bruce Fischl, "FreeSurfer", *Neuroimage*. 2012;62(2):774–781. https://doi.org/10.1016/j.neuroimage.2012.01.021.

[26] Brian B Avants, Nicholas J Tustison, Gang Song, Philip A Cook, Arno Klein, and James C Gee, "A reproducible evaluation of ANTs similarity metric performance in brain image registration", *Neuroimage*. 2011;54(3):2033–2044. https://doi.org/10.1016/j.neuroimage.2010.09.025.

Chapter 7

An adaptive approach to EEG-based seizure onset detection

G. Yogarajan[1], V. Angayarkanni[2] and Ben Othman Soufiane[3]

Abstract

Epilepsy is a significant brain disorder diagnosed after two seizures not attributable to known medical conditions such as alcohol or sugar levels. Individuals with epilepsy experience uncontrollable movements and loss of consciousness, leading to serious injury risks and potential fatalities. To mitigate these dangers, it is important to develop a computerized seizure detection technique to safeguard epileptic patients during seizures and promptly alert caregivers upon detection. In this chapter, a seizure onset detection method based on electroencephalogram (EEG) data has been discussed. It utilizes a pre-processing unit responsible for band-limiting with a bandpass filter, amplification, voltage level detection, and an adaptive signal rejection algorithm. The adaptive algorithm identifies abnormalities in EEG signals, often manifested as hypersynchronous pulses, at the onset of seizures and removes unwanted pulses using an adaptive technique. Upon detecting seizure onset, the system notifies medical staff for immediate intervention. This methodology demonstrates superior accuracy, sensitivity, and specificity compared with existing state-of-the-art methods. This approach is suitable for developing wearable devices leveraging fabrication technologies within the Internet of Healthcare platform, enabling rapid seizure detection and potentially saving the lives of critical patients.

Keywords: Internet of Healthcare; seizure onset detection; EEG signal processing; hypersynchronous pulses

[1]Department of Information Technology, Mepco Schlenk Engineering College, India
[2]Department of Computing Technologies, Faculty of Engineering and Technology, SRM Institute of Science and Technology, India
[3]Department of Computer Science, Applied College, King Faisal University, Saudi Arabia

7.1 Introduction

The foundation of smart healthcare systems lies in the Internet of Medical Things (IoMT) and the Internet of Healthcare Things. Providing IoMT-based smart healthcare to detect abnormalities (seizure) in the electroencephalogram (EEG) signals can help detect seizures before their onset. Severe epilepsy not only irreparably harms the human body but also poses a life-threatening risk. Hence, investigations based on the diagnosis and management of epilepsy possess critical clinical importance [1]. The EEG works on the principle of detecting even minute abnormalities in human brain signals from the wires and sensors that are attached to the head's scalp even when the subjects are asleep. Nowadays, by allowing patients to take preventive measures or medicine when a seizure is detected, advances in automated seizure detection systems are able to improve the management of epilepsy [2].

With the help of this smart healthcare system, the treatment gap between the people with less awareness and knowledge of the anti-epileptic drugs and anticonvulsant medications can be filled. In today's era of smart healthcare, there is a demand for intelligent detection systems that have the capability of accurately identifying seizures and delivering rapid results to users for remote healthcare monitoring. Epilepsy is one of the common neurological disorders [3] that are caused by the trigger of recurrent seizures, and these seizures may be caused due to any acquired head injury or due to genetic traits. Having more than two seizures within a daytime (24 h) without any identifiable reason or cause would be considered epilepsy. The seizures may range from 30 s to 2 min or more depending on the person's health condition and if it exceeds 5 min, it becomes a life threat and requires immediate medical response. Seizure disorders can be treated with medications but some people remain refractory to these anti-epileptic drugs and medicines. Thus, the introduction of smart healthcare systems might help people with easy tracking of their medical records in their day-to-day lives. Seizures can be broadly categorized into two main types: generalized seizures and focal (or partial) seizures. Generalized seizures impact the entire brain, while focal seizures affect only a specific part of the brain. Essentially, seizures are sudden and uncontrolled electrical disturbances in the brain that can alter a person's behavior, movements, and consciousness [4].

EEG signals are initially processed and filtered and then made into hypersynchronous pulses for the detection of seizures using the voltage level detector (VLD) and the signal rejection algorithm (SRA) with adaptive approach. The preprocessing or the filtering unit along with the amplification process is done to the signal feed into the VLD. The major contributions from this chapter are discussed below:

- This chapter presents a methodology to detect the onset of seizures in epileptic patients using EEG data, providing a significant advancement in automated health monitoring.
- The algorithm developed for detecting seizure onset is capable of identifying abnormalities in EEG signals, particularly hypersynchronous pulses, and

removing unwanted pulses using an adaptive technique. This innovation contributes to the robustness and reliability of the detection system.
- The approach is suitable for integration into wearable devices, leveraging fabrication technologies within the Internet of Healthcare platform.

The remainder of this chapter is organized as follows: Section 7.2 discusses the related works on abnormality detection using EEG signal processing. Section 7.3 illustrates the system architecture on seizure onset detection using adaptive threshold approach. The results and discussion are shown in Section 7.4 and Section 7.5 provides the conclusion and directions for future work.

7.2 Literature review

An epileptic seizure detection real-time system [5] using a less power implantable integrated device with CMOS to detect epileptic seizures, utilizing a unique seizure onset feature and algorithm is developed. The approach described in [6] generates a feature vector that encapsulates the morphology and spatial distribution of an EEG epoch using a wavelet decomposition algorithm. After that, a support-vector machine classification method is used to examine this vector in order to ascertain whether it depicts a seizure or non-seizure EEG signal for the patient. To identify seizures, the system uses an SRA in conjunction with a hypersynchronous signal detecting circuit. It monitors EEG signals, eliminates unwanted pulses, and determines seizure onset time. Tested on the Bern-Barcelona EEG Database, this energy-efficient algorithm is accurate across various datasets [7].

An implantable CMOS-integrated device in a system is described in [8] to identify partial-onset seizures, which are the beginning of epileptic episodes. By obtaining information about the beginning of seizures from brain impulses and tracking them over a predetermined duration, it improves the management of epilepsy.

A non-invasive method of measuring brain electrical activity is employed to detect seizures and classify significant features using EEG signals. It selects affected channels from CHB-MIT EEG datasets, extracts relevant features, and evaluates seven classifiers using machine learning techniques for efficient output as discussed in [9].

The chapter proposes a co-design approach for hardware and software for EEG-based seizure detection in an IoMT edge-based system, addressing issues such as high latency and privacy. The hardware-friendly method uses EMA-GHE for ictal and interictal EEG features extraction RF classifier and SMOTE for binary classification [10].

A study proposes seizure detection model in real time [11] that applies the standard kriging technique to classify fractal features from patients' EEG signals. An article introduces wearable technology for smart healthcare, seizures using EEG in the Internet of Things (IoT), discrete wavelet transform (DWT), Hjorth parameters (HPs), statistical features, and machine learning classifier [12].

Deep neural networks and the binary dragonfly algorithm are used in an EEG-based seizure detection system to improve seizure identification's precision and effectiveness [13]. Using a variety of statistical and Hjorth parameters obtained from wavelet-decomposed EEG signals, the deep neural network is able to understand the basic properties of EEG signals. A detailed study has been conducted to bridge the gap between edge computing and the Internet of Healthcare solutions [14] with a discussion on the various associated challenges and future trends.

EEG signals along with ECG signals have been used to study the effect of drowsiness on vehicle drivers [15]. A deep learning-based data fusion approach has been used to process the EEG and ECG signals to detect drowsiness, which helps to improve the safety of vehicle drivers. Multichannel EEG signals are used to study brain disorders like schizophrenia which often lead to messy speech and hallucinations in patients [16]. Convolutional neural networks (CNN) and temporal convolution networks have been used to develop a smart healthcare framework. An IoT-based EEG monitoring system has been developed that monitors the multichannel EEG signals and detects the seizures. It used CNN with several derived features to perform EEG seizure detection with an average accuracy of 98.48% [17].

A real-time seizure onset detection method using high spatial frequencies has been developed in [18]. But this method does not explicitly reveal the underlying dynamics and require larger datasets for training. An intelligent system has been developed for identifying the presence and onset time of seizures in intracranial EEG recordings from the responsive neurostimulation. It does not explore the impact of potential artifacts or noise in the intracranial EEG recordings on the accuracy of seizure detection that affects its reliability in real-world clinical settings [19].

A single-channel seizure detection system using brain-rhythmic recurrence biomarkers and an optimized model (ONASNet) has been developed to analyze nonlinear features from phase-space representations using a deep neural network, which provides new insights for EEG decoding. It focuses on single-channel seizure detection, limiting the utilization of multichannel EEG data for a more comprehensive analysis [20].

7.3 Proposed system

A seizure is an abrupt, fleeting disruption of brain activity that causes loss of consciousness due to the burst of electrical impulses in the brain. Patients are even prone to death because of ignorance of these seizures. Thus, early detection helps in recognizing patients with seizures and might even notify the medical team in terms of any emergency.

This section represents the logic and principles of the proposed system and the procedural working of the system with the help of the dataset acquired from the CHB-MIT Scalp EEG database [21].

7.3.1 System architecture

Figure 7.1 illustrates the various processes involved in the proposed system architecture. The proposed system helps in recognizing the seizure onset of the patient

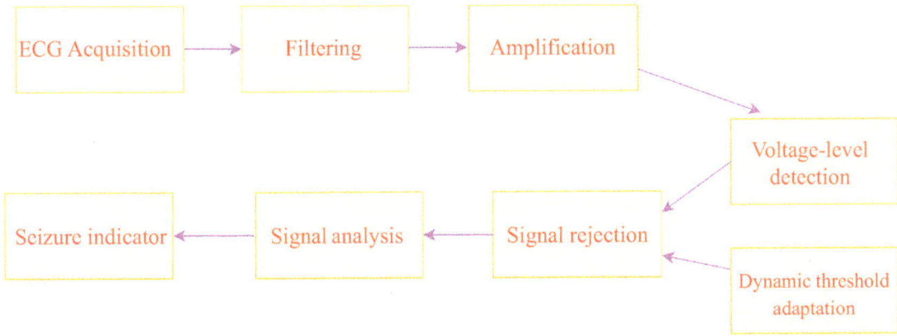

Figure 7.1 Proposed system architecture

by continuously monitoring the brain signals emitted by the neurons with the help of the EEG device. Initially, the EEG signals are extracted and then passed through the bandpass filter (BPF) later amplified and passed on to the VLD which is then revolutionized by the SRA that states if a seizure is present or not in a patient from their EEG signals. The patient's EEG signals are recorded and subsequently extracted from the databases. This process is called data acquisition. The filtration process is mainly focused on removing unwanted signals and noise from the EEG signals and then filtering it to the desired range or frequency.

The amplification process is done by an adjustable gain amplifier to amplify the signal to a particular range which is then passed on to the VLD. It is responsible for converting the amplified signals into hypersynchronous pulses and thus subjecting those pulses to SRA. The main aim of the proposed architecture is to detect the seizure from those hypersynchronous pulses after fine filtering of the pulses obtained from the VLD which is done by removing the spurious and unwanted noises generated during the VLD and then monitoring the EEG signals continuously for respective periods.

7.3.2 System design

The proposed system consists of three units: a sensor unit, a transmission and storage unit, and an access unit.

7.3.2.1 Sensor unit

The pre-processing unit examines the EEG signal that is given as input. Seizures are constantly being monitored by the seizure detector, and the information is automatically transferred wirelessly to a remote storage location.

7.3.2.2 Transmission and storage unit

Cloud storage is preferred since the storage unit's job is to keep and manage each patient's data. Here, the transmission unit serves as an interface between the access unit and the sensor unit, facilitating the transfer of data to the intended location.

7.3.2.3 Access unit

The access unit is responsible for allowing the respective persons to access data and information that is been stored in the cloud. With this provision, the patient's medication history can be viewed anytime by the medical team and can help in checking the required dosage and medication required for the patients for the treatment of epilepsy.

7.3.2.4 Pre-processing unit

This unit is responsible for the acquisition of data from the patient/dataset and then processing it with the BPF so as to remove all the unwanted noise from the EEG signals that may be caused by defect on arrival (DOA) or other effects. The CHB-MIT scalp EEG database that consists of EEG recordings from pediatric subjects is fed as the input to the BPF. The BPF used in this work is said to have a desired frequency range and the range is:

1. EEG minimum frequency: 3 Hz
2. EEG maximum frequency: 29 Hz

The BPF also helps in the smoothening of the sharp transition signals that are later passed on to amplifier unit.

7.3.2.5 Amplification unit

After the signals are outsourced from the filtration process, that is, after the BPF method is applied on the EEG signals, they are subjected to pass through the amplification unit. The signals retrieved from the pre-processing unit are of low amplitude neural signals that need to be amplified prior to analysis. The amplification method here is subjected to boost the signal that is obtained from the filtration process. The modulated signal is amplified with the help of adjustable or variable gain amplifier. The variable gain amplifier can be used to amplify up to a desired value. In this system, the low power neural signal is amplified up to five units using the adjustable gain amplifier.

7.3.2.6 Voltage level detection

The amplified signals are obtained and passed on to the VLD to generate the hypersynchronous pulses. The maximum and minimum voltages of the VLD detect theses hypersynchronous signals. Hypersynchronous pulses are determined from the heuristic analysis of the amplified signal and calculated with the help of (7.1)–(7.4).

The minimum and maximum voltage are already predefined, and the values of average maximum voltage is 450 mV and average minimum voltage is 150 mV. The optimal values are obtained using a trial and error method and is then applied to the unknown seizure and non-seizure instances. Thus, the amplified signals are now modified into hypersynchronous pulses and is now proceeded to signal rejection block.

It is detected using the following equation:

$$V_{VLD}(n) = \begin{cases} 1\,for\ V\max > V\text{mod}(n) > V\min \\ 0, otherwise \end{cases} \tag{7.1}$$

7.3.2.7 Signal rejection algorithm

Hypersynchronous signals that are obtained from the VLD unit are now analyzed using the SRA. The seizure detector continuously monitors the neural signals and detects the hypersynchronous pulses. Since a large number of good signals also pass through the VLD phase, there is a requirement to remove those unwanted and unnecessary noise and signals caused by external factors and thus this algorithm is applied. The unwanted pulses are removed if they fall below the defined threshold. The choice of threshold should be carefully tuned based on the characteristics of the EEG signals and the desired sensitivity of the seizure detection algorithm.

The statistical measure of the EEG signals is used to calculate an adaptive value that helps to remove unwanted hypersynchronous pulses. The adaptive thresholding formula using the Z-score for seizure onset detection is given by:

$$Z_{Score} = \frac{DP - MV}{SD} \tag{7.2}$$

where DP refers to the most recent EEG data point, MV refers to the current mean of the EEG signals, and SD refers to the current standard deviation.

The seizure onset can be detected by comparing the absolute value of the Z-score to a threshold level:

$$\text{Seizure\_Onset} = |\ \text{Z}_{\text{score}}\ | > \text{Threshold} \tag{7.3}$$

7.4 Results and discussion

7.4.1 Performance metrics

The performance analysis is done based on the parameters that have a significant role in the proposed system when compared with other existing systems considered:

1. Sensitivity
2. Specificity
3. Accuracy

The metrics are analyzed and calculated using the following equations:

$$Sensitivity = \frac{True\ positive}{True\ positive + True\ negative} \tag{7.4}$$

Sensitivity is defined as the minimal and smallest amount of change that takes place in any reading or measurement.

$$Specificity = \frac{True\ negative}{True\ negative + False\ positive} \tag{7.5}$$

Specificity is defined as the extent to which the system is specific or particular in nature with respect to the constraints and conditions that are being applied to it.

$$Accuracy = \frac{True\ positive + True\ negative}{True\ positive + True\ negative + False\ positive + False\ negative} \tag{7.6}$$

Accuracy is defined as the correctness or the exactness of the result that is been produced from calculation of this formula. The terms used in the above equations are defined below:

- True positive: correctly predicts positive class
- True negative: correctly predicts negative class
- False positive: wrongly predicts positive class
- False negative: wrongly predicts negative class

7.4.2 Simulation results

In this work, the datasets consist of the EEG recordings from pediatric subjects and it has about 23 datasets that are collected from about 22 subjects. The EEG recordings are obtained at 256 samples per second with 16-bit resolution. Figures 7.2(a) and (b) show the transient analysis on the EEG data for two different cases. Each EEG data consists of 23 channels and each channel consists of 921,600 samples. The x-axis refers to the channel number and the y-axis indicates the amplitude of the EEG samples.

Sample segments of EEG signals from different subjects, highlighting variations in brain activity over specific time intervals, have been illustrated in Figures 7.3(a) and (b). Figure 7.3(a) depicts the input EEG signal recorded between 2,800 and 3,200 s from the file chb01_03.edf. This segment captures a particular period of brain activity, which is critical for understanding the temporal dynamics of the subject's neural patterns. Figure 7.3(b) shows the input EEG signal recorded between 290 and 310 s from the file chb11_82.edf. This segment offers a different temporal snapshot, allowing for the examination of variations in neural activity between different subjects.

The different EEG signals after being filtered by the BPF along with the filter characteristics are depicted in Figures 7.4(a) and (b).

The filtered signal after the amplification using a variable gain amplifier is illustrated in Figures 7.5(a) and (b) for two different EEG segments. The x-axis indicates the time instance, and y-axis indicates the amplitude of the EEG sample.

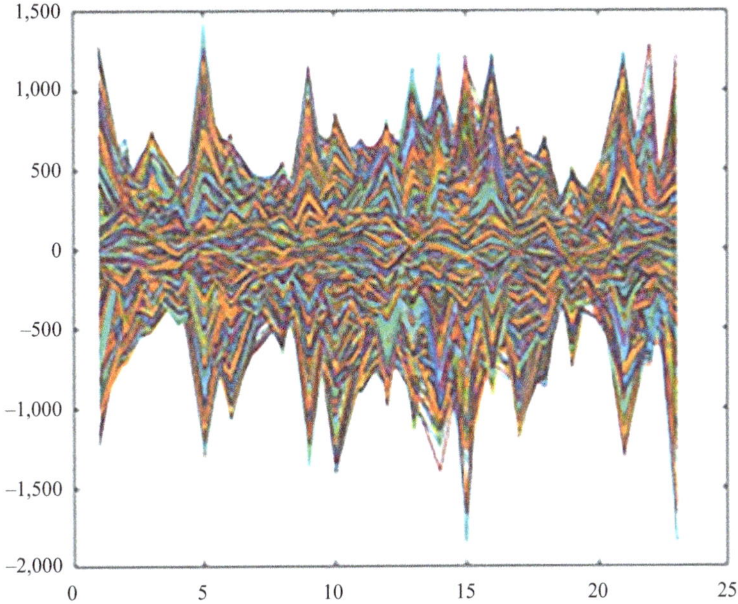

(a)

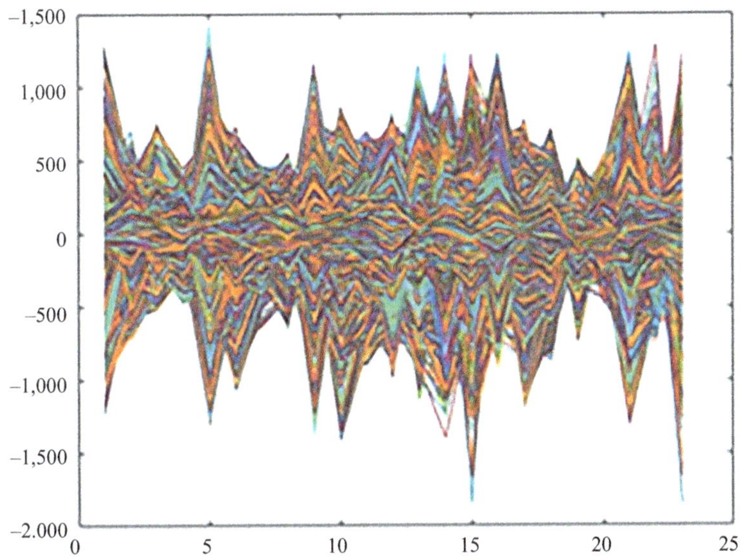

(b)

Figure 7.2 (a) Transient analysis for chb01_03.edf, and (b) transient analysis for chb11_82.edf

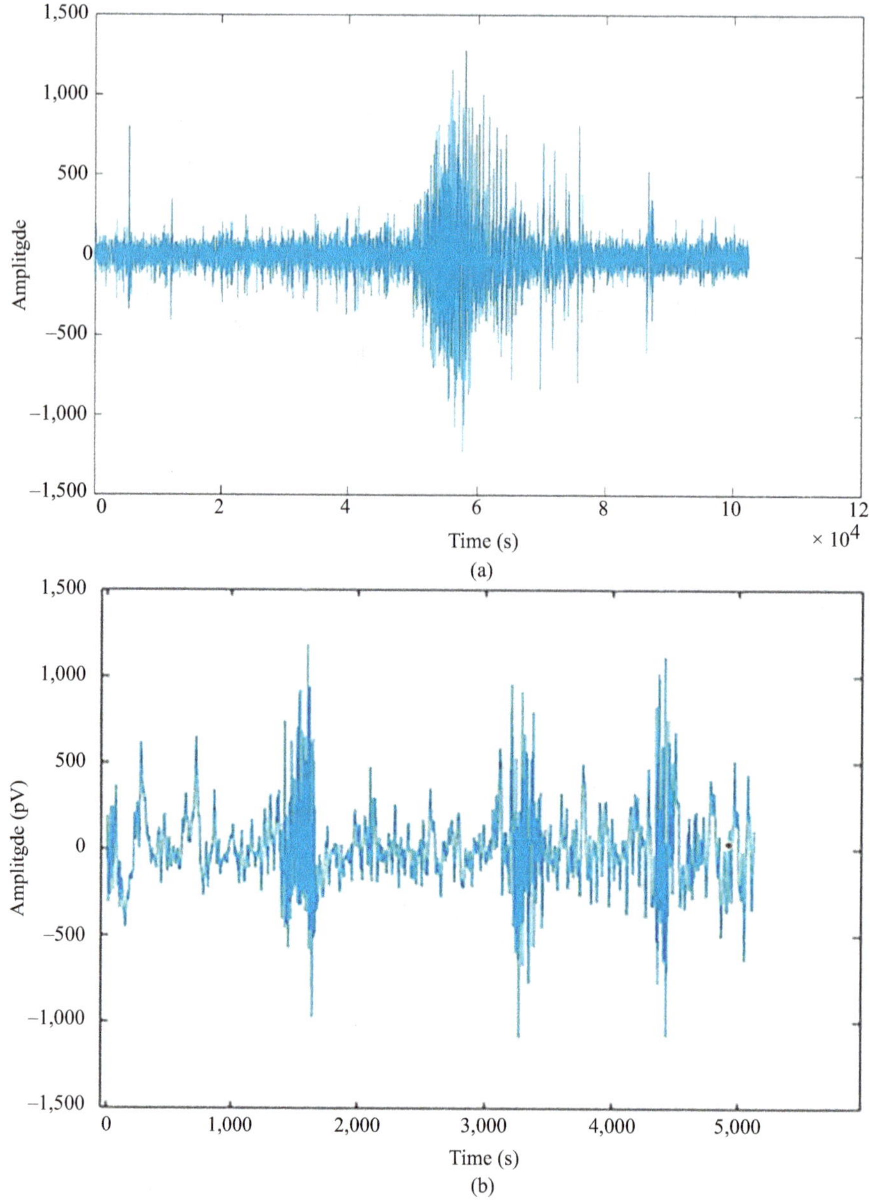

Figure 7.3 (a) Input EEG signal of 2800-3200 seconds from chb01_03.edf, and (b) input EEG signal of 290–310 s of chb11_82.edf

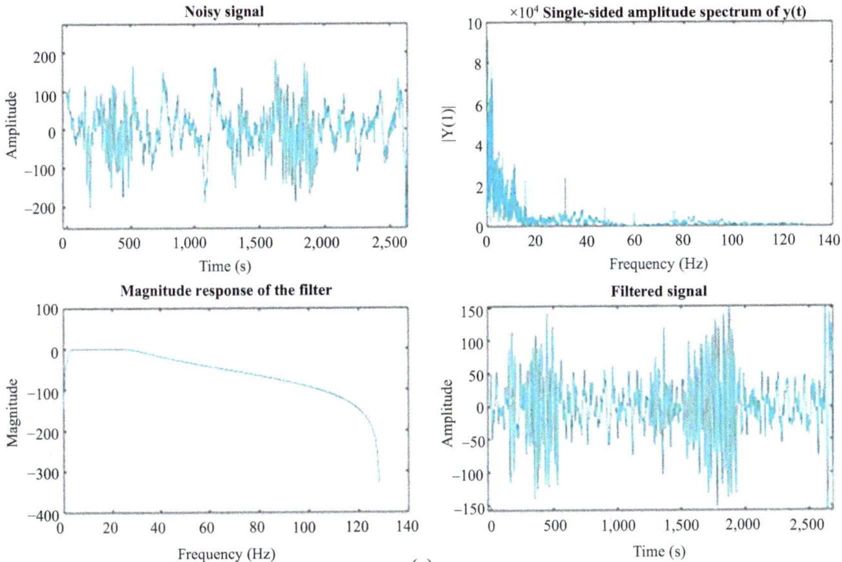

(a)

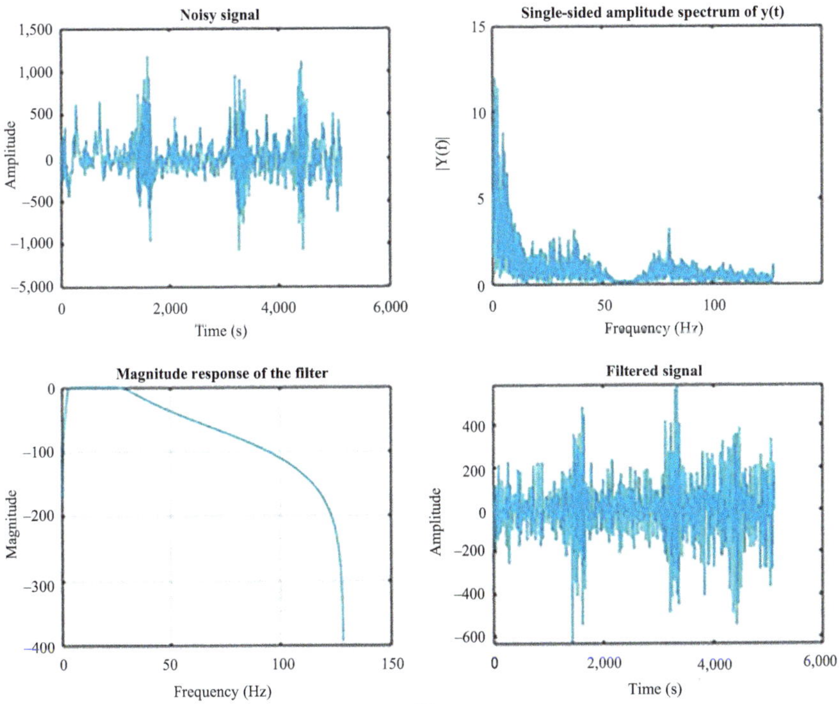

(b)

Figure 7.4 (a) Filtering using bandpass filter after signal extraction chb01_03.edf, and (b) filtering using bandpass filter after signal extraction of chb11_82.edf

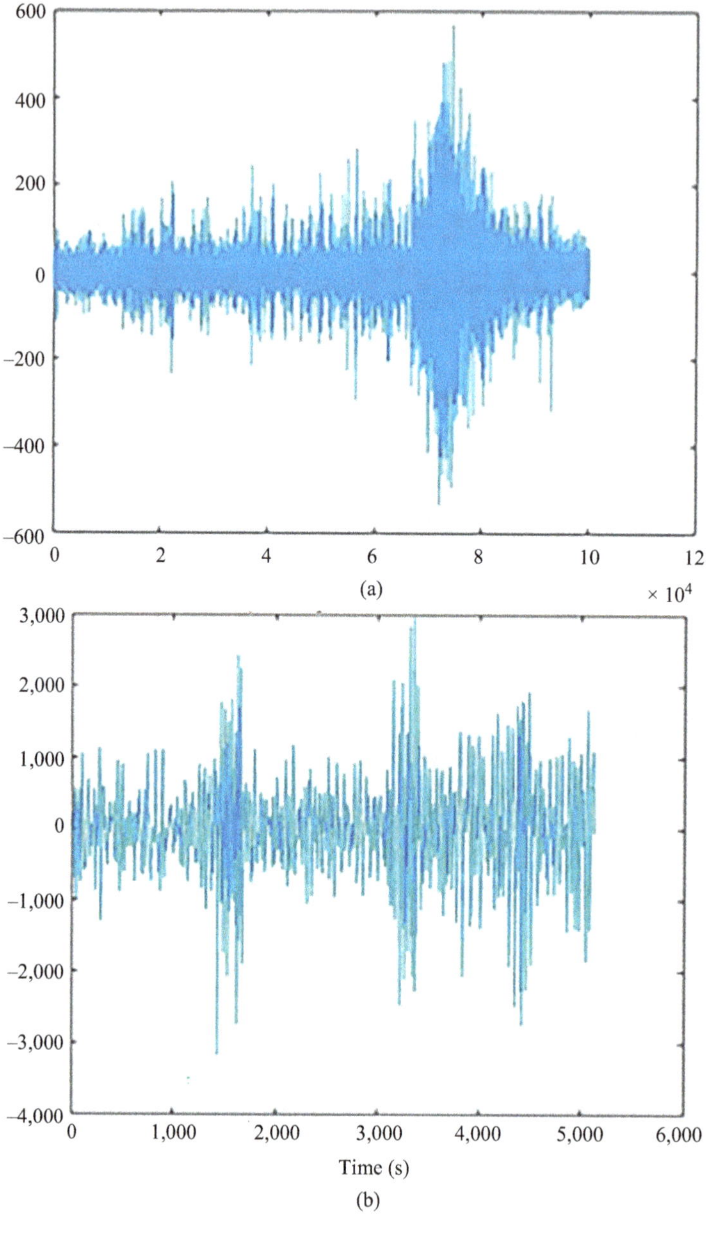

Figure 7.5 Amplification using variable gain amplifier of (a) chb01_03.edf and (b) chb11_82.edf

The hypersynchronous pulse generation from the VLD for different EEG segments is illustrated in Figures 7.6(a) and (b).

The seizure onset has been detected for two different EEG segments using the VLD and the SRA based on the adaptive thresholding scheme for two different EEG segments is shown in Figures 7.7(a) and (b).

7.4.3 Performance comparison analysis

The proposed algorithm is compared with SRA and an already existing algorithm called Analog Front End Linear Support Vector Machine. The SRA algorithm achieves sensitivity of 93%, specificity of 96%, and accuracy of 96%, whereas the

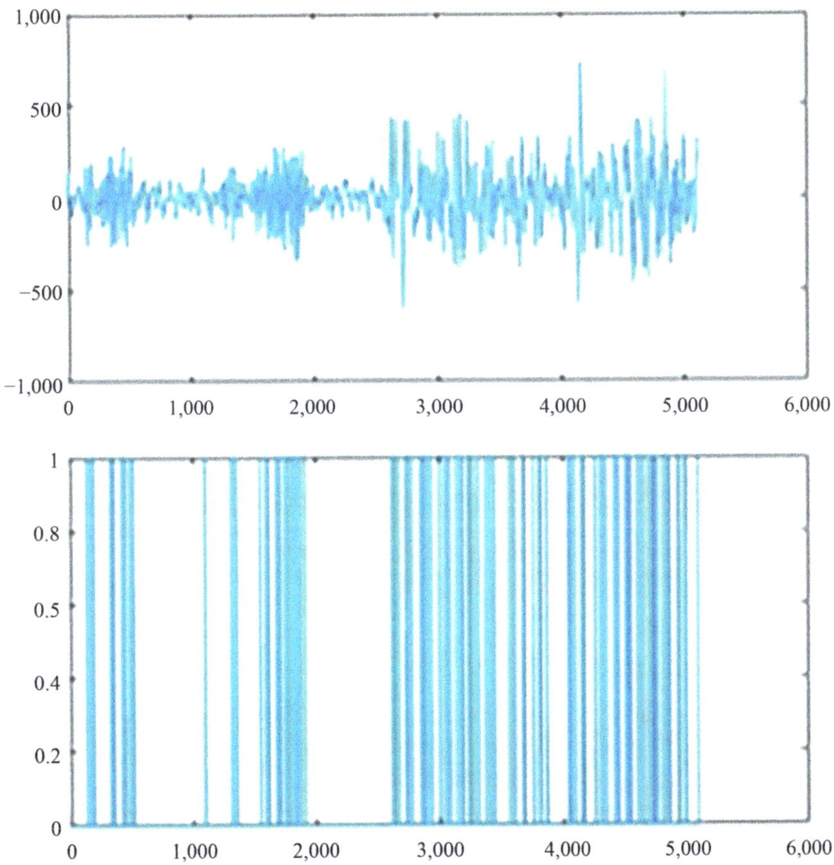

(a)

Figure 7.6 Hypersynchronous pulses generation: (a) chb01_03.edf and (b) chb11_82.edf

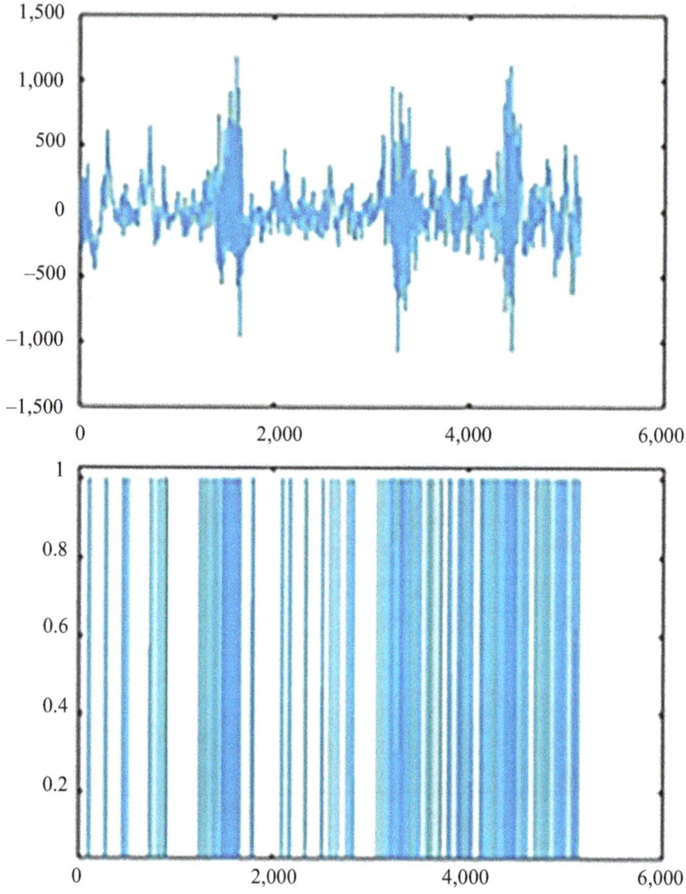

(b)

Figure 7.6 (Continued)

proposed approach achieves an average accuracy of 98.5%, sensitivity of 96%, and specificity of 97%, which is more than other existing algorithms. The same is illustrated in Figure 7.8. The improved performance of the proposed algorithm can be attributed to several factors. First, the algorithm incorporates advanced signal processing techniques and adaptive thresholding based on the Z-score, allowing for more precise detection of seizure onset. Second, the proposed approach leverages heuristic analysis and trial-and-error methods to determine optimal parameter values, enhancing its effectiveness in distinguishing between seizure and non-seizure instances.

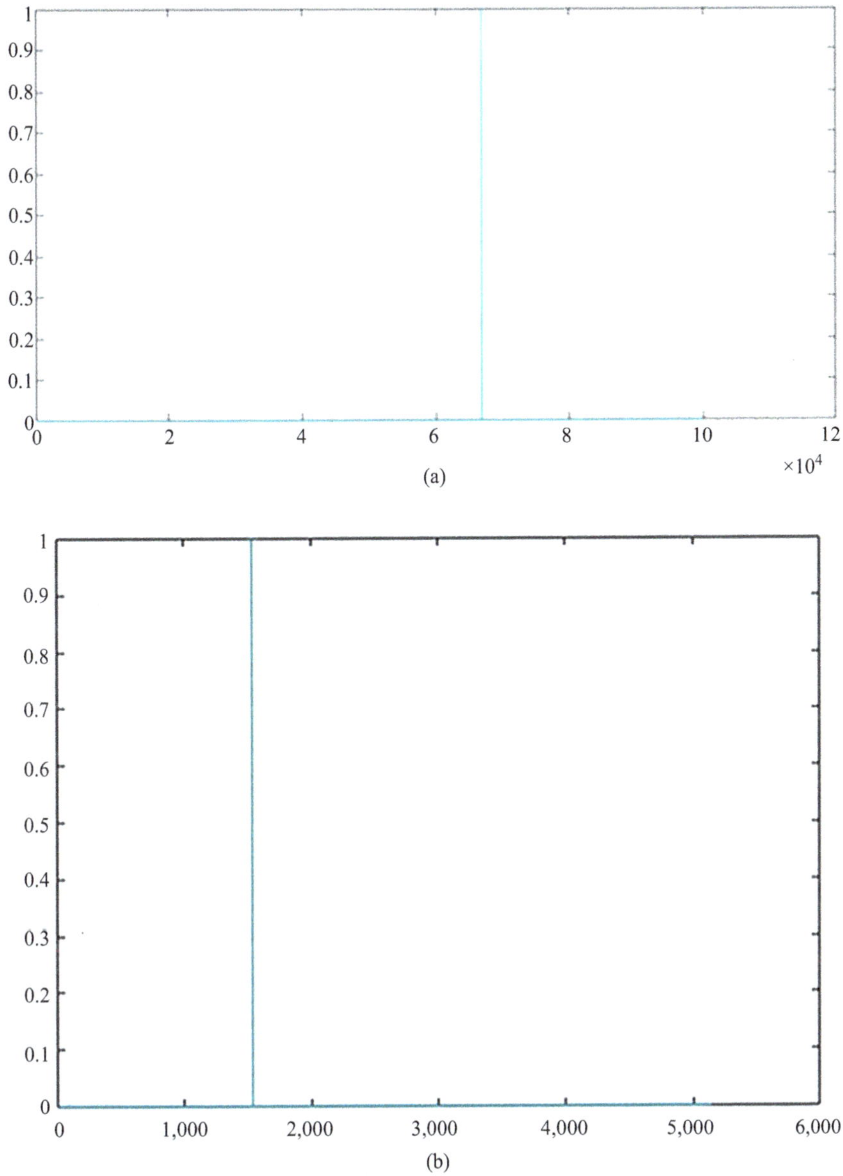

Figure 7.7 (a) Seizure onset detection for chb01_03.edf and (b) seizure onset detection for chb11_82.edf

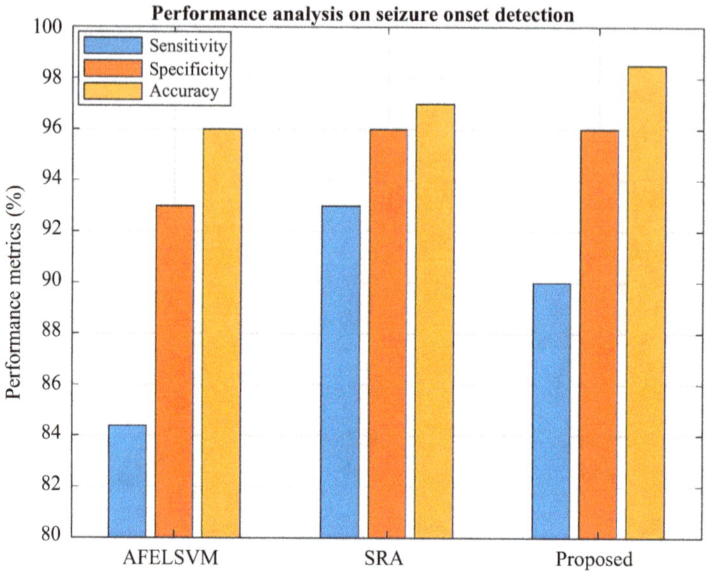

Figure 7.8 Performance analysis

7.5 Conclusion and future work

An automated seizure detection system is proposed to detect seizure onset by continuously monitoring EEG signals and eliminating unwanted hypersynchronous pulses using adaptive thresholding based on the Z-score. This method identifies seizures just before they become serious by first filtering and amplifying the signals, then applying VLD and SRA techniques. Compared with existing algorithms, the proposed system achieves better accuracy, higher sensitivity, and specificity. Consequently, the detection system can recognize and notify medical staff of seizure onset before any serious incident occurs, preventing potential accidents.

In future, the proposed system can be further implemented using the latest integrated circuits and updated technologies. This will enhance its application in wearable devices within the Internet of Healthcare environment, enabling efficient and timely detection of seizures and epilepsy. Conducting extensive clinical trials will validate the system's effectiveness across diverse patient populations, ensuring its reliability and practical applicability in real-world settings.

References

[1] Sun, Y., and Chen, X. "Epileptic EEG signal detection using variational modal decomposition and improved grey wolf algorithm." *Sensors* 23, no. 19 (2023): 8078.

[2] Jaishankar, B., Ashwini, A. M., Vidyabharathi, D., and Raja, L. "A novel epilepsy seizure prediction model using deep learning and classification." *Healthcare Analytics* 4 (2023): 100222.

[3] Yuen, A. W. C., Keezer, M. R., and Sander, J. W. "Epilepsy is a neurological and a systemic disorder." *Epilepsy & Behavior* 78 (2018): 57–61.

[4] Benbadis, R., and Heriaud, L. "Understanding seizures & epilepsy." *Comprehensive Epilepsy Program* Tampa General Hospital, University of South Florida. Available: https://health.usf.edu/medicine/neurology/epilepsy/~/media/Files/Medicine/Neurology/Comprehensive%20Epilepsy%20Program/Epilepsyandseizures.pdf (accessed December 17, 2023).

[5] Lande, T. S., and Butera, R. "IEEE Transactions on Biomedical Circuits and Systems." *IEEE Transactions on Biomedical Circuits and Systems* 1, no. 1 (2007): 1–2.

[6] Shoeb, A., Bourgeois B., Treves S. T., Schachter S. C., and Guttag J. "Impact of patient-specificity on seizure onset detection performance." In *2007 29th Annual International Conference of the IEEE Engineering in Medicine and Biology Society*, pp. 4110–4114. IEEE, 2007.

[7] Sayeed, M. A., Mohanty, S. P., Kougianos, E., and Zaveri, H. P. "eSeiz: An edge-device for accurate seizure detection for smart healthcare." *IEEE Transactions on Consumer Electronics* 65, no. 3 (2019): 379–387.

[8] Salam, M. T., Sawan, M., Hamoui, A., and Nguyen, D. K. "Low-power CMOS-based epileptic seizure onset detector." In *2009 Joint IEEE North-East Workshop on Circuits and Systems and TAISA Conference*, pp. 1–4. IEEE, 2009.

[9] Shoeb, A. H., and Guttag, J. V. "Application of machine learning to epileptic seizure detection." In *Proceedings of the 27th International Conference on Machine Learning (ICML-10)*, pp. 975–982, 2010.

[10] Zhao, W., Wang Y., Sun X., Zhang S., and Li X. "IoMT-based seizure detection system leveraging edge machine learning." *IEEE Sensors Journal* 23, no. 18 (2023): 21474–21483.

[11] Olokodana, I. L., Mohanty, S. P., Kougianos, E., and Olokodana, O. O. "Real-time automatic seizure detection using ordinary kriging method in an edge-IoMT computing paradigm." *SN Computer Science* 1 (2020): 1–15.

[12] Sayeed, M. A., Mohanty, S. P., Kougianos, E., and Zaveri, H. P. "Neuro-detect: A machine learning-based fast and accurate seizure detection system in the IoMT." *IEEE Transactions on Consumer Electronics* 65, no. 3 (2019): 359–368.

[13] Yogarajan, G., Alsubaie, N., Rajasekaran G., *et al.* "EEG-based epileptic seizure detection using binary dragonfly algorithm and deep neural network." *Scientific Reports* 13, no. 1 (2023): 17710.

[14] Hayyolalam, V., Aloqaily, M., Özkasap Ö., and Guizani M. "Edge-assisted solutions for IoT-based connected healthcare systems: A literature review." *IEEE Internet of Things Journal* 9, no. 12 (2021): 9419–9443.

[15] Yogarajan, G., Singh, R. N., Nandhu, S. A., and Rudhran, R. M. "Drowsiness detection system using deep learning based data fusion approach." *Multimedia Tools and Applications* 83 (2024): 36081–36095.

[16] Sharma, G., Joshi, A. M., Yadav, D., and Mohanty, S. P. "A smart healthcare framework for accurate detection of schizophrenia using multichannel EEG." *IEEE Transactions on Instrumentation and Measurement* 72 (2023): 1–9.

[17] Yedurkar, D. P., Metkar, S., Al-Turjman, F., Yardi, N., and Stephan, T. "An IoT based novel hybrid seizure detection approach for epileptic monitoring." *IEEE Transactions on Industrial Informatics* 20, no. 2 (2023): 1420–1431.

[18] Maheshwari, J., Joshi, S. D. and Gandhi, T. K., "Real-time automated epileptic seizure detection by analyzing time-varying high spatial frequency oscillations," *IEEE Transactions on Instrumentation and Measurement* 71 (2022): 1–8, Art no. 4002608

[19] Peterson V., Kokkinos V., Ferrante E., *et al.*, "Deep net detection and onset prediction of electrographic seizure patterns in responsive neurostimulation," *Epilepsia* 64, no. 8 (2023): 2056–2069.

[20] Song. Z., Deng, B., Wang, J., Yi, G., and Yue, W. "Epileptic seizure detection using brain-rhythmic recurrence biomarkers and ONASNet-based transfer learning," *IEEE Transactions on Neural Systems and Rehabilitation Engineering* 30 (2022): 979–989.

[21] Guttag J. *CHB-MIT Scalp EEG Database (version 1.0.0)*. PhysioNet. 2010. Available from: https://doi.org/10.13026/C2K01R (accessed December 17, 2023).

Chapter 8

Prediction models for eye disorders

Ankita Saha[1,2]*, Rajdeep Kabiraj*[3]*, Riddhi Sekhar Dwibedi*[3]*, Shibakali Gupta*[4]*, Chayan Paul*[1]*, Korhan Cengiz*[5] *and Nikola Ivković*[6]

Abstract

Through artificial intelligence and soft computing, assessing the children's eye conditions from the respective mobile photographs can facilitate proper and preliminary detection of eye disorders in a home-like condition. This chapter provides a comprehensive survey on various research projects on prediction models and related marker discovery for retina related disorders. Various age prediction models using features of ocular anterior segment through machine learning methodologies as well as a deep learning-based model for predicting age from the fundus images (i.e., retinal age) that are recently proposed, have been described here. Age-related macular degeneration (ARMD) is a bilateral ocular-disorder that can affect the central part of the retina. Another regression-based biological age clock for the eye (retina) was developed using a transcriptomic dataset of 453 retinal samples, and 167 fibroblast samples. And those 453 retinal samples were categorized into Minnesota Grading System (MGS) levels 1–4, with 105 MGS samples (level 1), 175 MGS samples (level 2), 112 MGS samples (level 3), and 61 MGS samples (level 4). The biological regression models demonstrated effective discrimination among ARMD samples with escalating severity score (MGS1-4). Detection of several prediction model-based marker genes are linked with AMD research. The identified prediction models may work as potential markers for investigating ARMD progression as well as finding supportive therapeutic reagents.

Keywords: Retina disorder; prediction model; regression; gene expression; markers

[1]Department of Computer Science and Engineering, Swami Vivekananda University, India
[2]Department of Science and Management, ABS Academy of Science, Technology and Management, India
[3]Department of Optometry, ABS Academy of Management and Health Science, India
[4]Department of Computer Science and Engineering, University Institute of Technology, Burdwan University, India
[5]Department of Electrical Engineering, Prince Mohammad Bin Fahd University, Saudi Arabia
[6]Faculty of Organization and Informatics, University of Zagreb, Croatia

8.1 Introduction

Determining pediatric eye diseases/disorders at a preliminary stage is a global concern. Traditional screening techniques are depending upon nursing homes/hospitals and optical diagnostic centers that are costly as well as time-consuming. Through artificial intelligence and soft computing, assessing the children's eye conditions from the respective mobile photographs can facilitate proper and preliminary detection of eye disorders in a home-like condition.

In the global scenario, the population that is more than or equal to 60 is evaluated to reach 2.1 billion by 2050. Aging populations generate pressure on healthcare systems. Chronological age is basically an important risk factor for the frailty, age-associated morbidity as well as age-associated mortality. There is huge variability in the health results among individual people with the same chronological age that basically refers to the actual rate of aging at an individual label which is completely variable. In medical science, biological age is a better predictor to the health condition and disease prediction than the traditional chronological age. An accurate prediction of the respective biological age is valuable for risk stratification and interventions.

Age-related macular degeneration (ARMD) is a progressive eye disorder that affects the macula, the retina's central region responsible for central vision and fine detail perception, which is a light-sensitive tissue at the back of the eye. This causes central vision impairment in individuals aged 50 and above [1–3]. In severe cases of ARMD, the macula's impaired function can result in blurred or dark central vision, making everyday activities like reading, writing, or crafting extremely challenging or impossible. Although ARMD can cause central vision loss, it usually does not affect peripheral vision and total blindness is rare. And many people with this condition remain able to live independently. The growing aging population has led to an increase in ARMD cases, making it a major global health issue. Estimates suggest that ARMD affected 196 million individuals worldwide in 2020, with projections indicating a significant increase to 288 million by 2040. ARMD is typically categorized into two distinct forms: "dry" ARMD and "wet" ARMD. The non-exudative and non-neovascular "dry" ARMD affects roughly 85% of the ARMD population. In this case, the macula gradually gets thinner and leads to a progressive loss of function that can occur over an extended period, often years or decades. And the exudative and neovascular or advanced neovascular "wet" ARMD affects around 15% of individuals with ARMD. This is relatively rare and characterized by swift vision deterioration; "wet" ARMD is a less common variant of late ARMD. It is triggered by the growth of irregular blood vessels in the eye's posterior segment, which subsequently damages the macula. While the treatment for neovascular ARMD has shown promising results, non-exudative "dry" ARMD continues to pose a significant challenge. The progression of dry ARMD is characterized by three stages: early and intermediate stages, marked by drusen deposits and sub-retinal pigment epithelium, retinal pigment epithelium (RPE) pigmentary changes leading to mild visual impairment, and an advanced atrophic stage of "dry" ARMD marked by degeneration of critical retinal structures, leading to significant visual loss [4]. ARMD is widely accepted as a multifactorial disease influenced by various factors [5]. Although significant progress has been made in identifying genes and molecular pathways associated with ARMD risk, the precise mechanisms by which particular genetic variants influence disease progression remain unclear (20–27). Genetic analysis has identified more than 35 genetic variants linked to an increased risk of developing ARMD, with significant proportion of these variants concentrated in the complement system. Additionally, genetic variants have implicated importance if lipid biology, extracellular matrix or intercellular matrix remodeling, and inflammation as critical factors that likely influence ARMD pathogenesis. Additional genetic variants linked to ARMD involve genes such as fibroblast growth factor 2, apolipoprotein E, age-related maculopathy susceptibility protein 2, and DNA excision repair protein.

Epigenetic clocks, a type of epigenetic biomarker, have been developed to estimate biological age, based on measurements of various epigenomic features such as DNA methylation, mRNAs, microRNAs. Epigenetic clocks have significant implications for identifying environmental and genetic factors in the aging process, as well as for identifying novel biomarkers for disease diagnosis, and tracking the efficacy of rejuvenation and treatment evaluation. A comprehensive multi-omics approach to biomarker discovery and biological age prediction has been employed, incorporating data from DNA methylation, transcriptomics, frailty assessment, microbiome analysis, proteomics, and neuroimaging. Epigenetic changes, including DNA methylation and RNA expression alterations, exert significant downstream effects and play essential regulatory roles, particularly in the context of aging, where RNA abundance changes are especially pronounced. These age-associated alterations provide a rich source of potential novel biomarkers for aging. For example, DNA methylation age (DNAm age) was first proposed based on a set of age-predictive CpG sites discovered through elastic net penalized regression analysis. An individual's epigenetic age acceleration, whether positive or negative, determines whether they are biologically younger or older than their actual chronological age. By evaluating biological age, scientists have been able to detect individuals with marked differences between their biological and chronological ages, shedding light on the relationships between accelerated biological aging and numerous health issues, including cancer, diabetes, frailty, dementia, and other conditions. Individuals with positive epigenetic age acceleration exhibit epigenetically older (positive acceleration) or younger (negative acceleration) than their chronological age.

Although various models have successfully linked gene expression and methylation patterns to age [6], the application of epigenetic clocks to ARMD and *in vitro* neuronal differentiation remains a relatively unexplored area of research [7]. Research conducted by Hunter *et al.* [8], discovered that the promoters of the GSTM1 (glutathione S-transferase isoform mu1) and GSTM5 (mu5) exhibited hypermethylation in RPE cells derived from donor eyes with ARMD, compared with control samples. Studies conducted by Wei *et al.* [9] and Oliver *et al.* [10] reported contradictory findings regarding the methylation level of IL17RC gene in peripheral blood mononuclear cells (PBMCs) from ARMD patients and healthy controls, highlighting the need for further investigation. A comprehensive genome-wide epigenetic analysis of ARMD identified significant methylation changes, notably hypomethylation of the ARMS2/HTRA1 locus and hypermethylation of the PRSS50 (protease serine 50) locus, distinguishing ARMD from control samples [11,12]. ATAC-Seq analysis performed by Wang *et al.* [13] demonstrated a widespread reduction in chromatin accessibility within RPE cells from ARMD patients, providing valuable insights into the disease's epigenetic landscape. Vallée *et al.* [14] conducted a comprehensive survey highlighted the significance of disrupted circadian rhythm in the pathogenesis of exudative (wet) ARMD with the specific focus on the abnormal activation on the canonical WNT/β-catenin signaling pathway. The study by Ratnapriya *et al.* [15] characterized the genetic architecture of ARMD and developed the EyeGEx (Eye Genotype Expression) database, a valuable resource for interpreting the genetic basis of ocular traits in the post-genome wide association study (GWAS). da Costa *et al.* [16] made a significant contribution to the field by identifying the regulatory signals necessary for the development of retinal organoids, and successfully generated a mature retina *in vitro*, thereby enabling advancement in disease modeling and therapeutic strategies.

In [17], retinal age clocks specific to the retina had been developed and further explored their application in ARMD progression and determining disease severity. Their study demonstrated that the developed age clocks can effectively distinguish between ARMD samples with varying degree of severity, as assessed by MGS scores (1–4), irrespective of the training dataset, whether derived from dermal fibroblast samples, retina samples, or both.

Ma *et al.* [18] introduced an age prediction model that depends upon 276 features of ocular anterior segment through machine learning methodologies. Zhu *et al.* [19] proposed a deep learning-based model that can predict age from the fundus images (i.e., retinal age) and to further conduct research the link in between the retinal age gap (i.e., predicted minus chronological age) and the mortality risk.

In conclusion, this review work provides a novel insight into various prediction models for eye-related disorders and their respective therapeutic markers, providing a thorough guideline to the new researchers in the related visions and issues.

8.2 Materials and methods

8.2.1 Dataset

In [17], at the initial stage, the authors analyzed the human retina gene expression data (NCBI Gene Omnibus Reference ID: GSE115828) available [14,15]. The comprehensive data matrix of gene expression comprised 18,053 distinct genes and 453 samples graded using the Minnesota Grading System (MGS). The 453 MGS samples were categorized into four distinct levels: 105 MGS level 1 (healthy) samples, 175 MGS level 2 (affected), 112 MGS level 3 (affected), and 61 MGS level 4 (affected) samples. The MGS score quantifies ARMD severity. Donor retinas with an MGS1 score were characterized by the absence of ARMD features and were used as controls. In contrast, MGS2-MGS4 samples represented progression of disease severity. Furthermore, their analysis incorporated three additional datasets: neuronal differentiation dataset, primary fibroblast dataset, and dermal fibroblast dataset. The neuronal differentiation dataset (GSE56796) was the largest in terms of gene count, which initially contained 44,562 genes and 24 samples. The primary fibroblast dataset (GSE97265) was smaller, with 6,732 genes and 14 samples. And the dermal fibroblast dataset (GSE113957) had a total of 27,142 genes across 143 samples.

8.2.2 Noise removal step

Using the initial set of gene vectors, the authors applied DBSCAN (Density-Based Clustering of Applications with Noise) algorithm to remove noise and cluster data points (50, 51). Outlier features detected by this method were removed from subsequent analyses.

To determine the optimal epsilon (eps) value, a k-nearest neighbors (KNN) distance plot has been utilized to estimate the knee point, which was then used as the eps-neighborhood value. All other parameters were set to their default values. The density-based clustering algorithm generated clusters, each cluster consisting of core, border, and noisy features. Subsequently, noisy features have been omitted to refine the dataset. The noise-free dataset was then subjected to regression analysis and cross-validation, allowing for the identification of significant patterns and correlations. An assessment of the cluster plot obtained from DBSCAN clustering was performed to evaluate the effectiveness of the clustering technique. To implement the DBSCAN clustering algorithm, two critical user-specified parameters must be required: epsilon (*eps*), which defines the maximum distance between points in a cluster and minimum points (*MinPts*), which determines the minimum number of points required to form a dense region. The epsilon (*eps*) value, representing the radius of the surrounding of each point, defined the epsilon-neighborhood (e-neighborhood), while *MinPts* specified the minimum number of neighboring points within this radius. A point was classified as a core point, if its neighbor count score was equal to or greater than *MinPts* threshold. A point was classified as a border point, if its neighbor count was lesser than *MinPts*, but it was located within the epsilon-neighborhood (e-neighborhood) of a core point. Any point that failed to meet criteria for a core point or a border point was classified as a noisy or outlier point. Our objective was to identify dense regions that can be approximated by analyzing the number of points or objects in close proximity to a specific reference point. Initially, the authors identified the optimal knee-point using a KNN distance plot. The KNN distances were computed, arranged, and analyzed in the ascending order to identify the knee-point value, which marked a significant transition in the data [20]. Subsequently, the values were normalized to a range of 0 to 1, and calculated the derivative to analyze the rate of change. The authors identified the knee point as the first point where the derivative exceeding a predetermined than a certain value (say 1) was treated. The e-neighborhood value was set as the scaled distance score of that identified knee point, serving as a key parameter.

To analyze the retina data with the complete feature set, a kNN distance plot has been generated, where the value of k was determined by adding 1 to the sample size of 453, resulting in k=454. In the plot, the height parameter h has been set to 5,000 to facilitate the determination of the knee point. The identified knee point was subsequently utilized as the eps value in the DBSCAN clustering algorithm to detect and isolate the outlier features within the dataset. By applying DBSCAN clustering algorithm, 75 noisy features that were classified as outliers have been identified and subsequently excluded from further analysis. Following the removal of noisy features, the remaining 17,978 noise-free features to develop a clock model. Using DBSCAN clustering, the authors identified 58 features in the dermal fibroblast dataset that exhibited noisy behavior and removed them from consideration in subsequent analyses. After discarding the noisy features, the authors utilized the existing 14,825 noise-free features to develop a clock model, ensuring a more accurate and reliable outcome.

8.3 Finding predictive models

After the identification and removal of outlier using DBSCAN, the refined datasets were prepared for subsequent analysis. In order to assess the robustness of their model, the authors implemented leave-one-out-cross-validation (LOOCV) on selected datasets, including the samples of retinal tissue, thereby dividing the data into training and test subsets. In the LOOCV approach, a single sample is designated as the test set, whereas the remaining samples comprise the training set. This process is iterated for each remaining sample, ensuring that every sample serves as the test set exactly once. From the expression data, the authors extracted a subset comprising only the training samples and paired it with the age data for those samples obtained from the clinical dataset. Concurrently, the authors selected distinct subsets of the expression data, comprising solely the test samples, and paired it with their corresponding age information from the clinical data.

After dividing the data into training and test sets using LOOCV, the authors employed the "glmnet" package in R to perform regression analysis [21,22]. Specifically, the authors regressed the training expression data against the (response variable) and logarithmically transformed aging data (predicted variable), and obtained an optimal lambda (λ) value by equating the classification error with the least square error (LSE). In the context of the "glmnet" regression technique, which employs a multiple linear model, the target function (T) is typically formulated as a single-objective function, aimed at optimizing a specific criterion. The resulting model yielded a set of optimized coefficients, corresponding to a subset of most informative selected features or genes that collectively minimized the value of the target function T. The target function T was formulated as follows:

$$T = \frac{1}{2m}\sum_{i=1}^{m}\left(y_i - \sum_{j=1}^{n}\beta_i x_{ij} - \beta_0\right)^2 + \lambda\left(\frac{1-\alpha}{2}\sum_{j=1}^{n}\beta_j^2 + \alpha\sum_{j=1}^{n}|\beta_j|\right)$$

Here, the variable m and n signify the total number of samples and features, respectively. In this notation, the variable x_{ij} signifies the gene expression level of jth feature in the ith sample, and y_i denotes the logarithmically transformed chronological age of the ith sample. To incorporate an extra constraint on the coefficients of the predicted variables, a hybrid approach had been utilized for the combined lasso and ridge regularization methods, with equal preference on both. In this context, λ (> 0) served as a tuning parameter that controlled the overall penalty imposed on the specific coefficients, while α (another tuning parameter), ranging from 0 to 1 $(0 < \alpha < 1)$, determined the balance between ridge $(\alpha = 0)$ and lasso $(\alpha = 1)$ techniques. The authors set α to 0.5 and determined the optimal value of λ using LOOCV, which is equivalent to m-fold cross-validation where m represents the total number of samples, and the authors selected the value of λ that resulted in the minimum error plus one-standard-error. Once the optimal λ was determined, the authors proceeded to select the most relevant features and estimated their corresponding coefficients based on the chosen λ value. Subsequently, the features with zero coefficient have been omitted, effectively eliminating non-informative predictors from the model. The authors are then proceeded with the reduced set of non-zero features to develop a predictive model for age.

Utilizing the comprehensive feature set derived from retina data, the authors developed age clock models using MGS level 1 samples, which served as control samples. To evaluate the model's performance, the authors employed LOOCV, to split the specific data into test and training, where a single sample was designated as the test set, while the training set was formed by the remaining 104 samples. This process was iteratively repeated for every individual sample, resulting in 105 iterations, where each sample was used as the test set in one iteration. To train the model, the authors extracted curated subsets of the data, including only the designated training samples, along with their corresponding age values obtained from the clinical data. In the same manner, for testing the model, the authors isolated a subset of particular transcriptomic data, including only the test samples and their associated age information sourced from the clinical dataset. Following the division of data into training and test sets using LOOCV, the authors applied "glmnet" regression to the training methylation data, comprising 104 samples and 17,978 features; the authors used the training methylation data as dependent variable and logarithmically transformed aging data from 104 samples as the independent variable; the authors then identified a lambda (λ) value that achieved a balance between classification error and LSE, where the two errors were equal. The resulting regression model generated coefficients for entire feature set and then isolated the features with non-zero coefficients, which corresponds to the feature selection process. The features with non-zero coefficients were subsequently identified as the selected features, representing the most informative variable for each respective model. Furthermore, the authors calculated the frequency of occurrence rate of each gene across all the developed integrated models. For the restricted retina data feature set, the authors employed the identical methodological pipeline, involving LOOCV followed by "glmnet" regression, using training data that included 104 MGS1 samples and 5,321 features, while test data comprised a solitary MGS1 sample and 5,321 features during each "glmnet" regression iteration.

The authors applied the same methodological pipeline to the restricted feature set of dermal fibroblast data, using 143 samples, where LOOCV was first employed, followed by glmnet regression on the training expression data, comprising 142 samples and 5,321 features as the response variable, and the logarithmically transformed chronological ages of 143 samples as the predictor variable, ultimately determining the optimal λ-score that balanced classification error and LSE. The computed regression model yielded coefficients for the entire gene set, and the authors then isolated the genes with non-zero coefficients, corresponding to the gene selection process. The genes characterized by non-zero coefficients were treated as the selected features, distinctively determined for each of the evolved models.

The authors employed LOOCV to divide the entire restricted feature set of joint data, comprising 248 samples, into training and test sets, facilitating model development and evaluation. In each iteration of LOOCV, one sample served as the test set, and the other 247 samples formed the training set, repeating this process for all 248 samples. After completing LOOCV, the authors utilized "glmnet" regression to the training data, comprising 5,321 features and 247 samples as the response variable; the predictor variable consisted of the logarithmically transformed aging data from 247 training samples, representing chronological ages, ultimately identifying the λ-score that minimized the difference between classification error and LSE. The computed regression model generated coefficients for the entire gene set with non-zero coefficients, effectively selecting the most informative genes. Genes characterized by non-zero coefficients were recognized as the most informative features in each respective model.

8.4 Evaluation metrics

Following the development of age clocks, generated from each data type, the authors predict age in independent data samples using the coefficients derived from the model. To evaluate the performance of the regression model, four major evaluation metrics have been used. Those are (1) age acceleration (AA), (2) median absolute error (MAE), (3) gene occurrence frequency, and (4) correlation coefficient between predicted and original ages [23,24].

8.5 Discussion

The aforementioned study [17] developed and validated retinal age clocks using transcriptomic data from human retina samples. Their results demonstrate that these age clocks can effectively distinguish between ARMD samples with varying degrees of severity. Notably, the application of their age clock model to *in vitro* neuronal differentiation data revealed intriguing results, suggesting potential wider applicability.

The identification of genes with non-zero coefficients in our age clock models provides valuable insights into the underlying biological mechanisms of retinal aging and ARMD. The significant overlap between genes identified in this study and those previously implicated in ARMD and aging research supports the validity of our approach. Their study highlights the potential of epigenetic clocks as biomarkers for monitoring ARMD progression and identifying potential therapeutic targets.

Ma *et al.* [18] introduced an age prediction model that depends upon 276 features of ocular anterior segment through machine learning methodologies. Zhu *et al.* [19] proposed a deep learning-based model that can predict age from the fundus images (i.e., retinal age) and to further conduct research to find the link between the retinal age gap (i.e., predicted minus chronological age) and the mortality risk.

8.6 Conclusion

In conclusion, this review work demonstrates the development and validation of retinal age models using transcriptomic data from human retina samples. The results highlight the potential of these age clocks as biomarkers for monitoring ARMD/other retina-disorder progression and identifying potential therapeutic targets. Future studies can build upon our findings to explore the clinical utility of retinal age clocks in ARMD/other retina-disorder diagnosis and treatment.

Acknowledgments

We want to thank all the researchers of Swami Vivekananda University, Kolkata, India, University Institute of Technology, Burdwan University, Burdwan, India and ABS Academy of Science, Technology and Management, Durgapur, India, and ABS Academy of Management and Health Science, Durgapur, India.

References

[1] Awh CC, Lane AM, Hawken S, Zanke B, and Kim IK. CFH and ARMS2 genetic polymorphisms predict response to antioxidants and zinc in patients with age-related macular degeneration. *Ophthalmology* (2013) 120:2317–23.

[2] Hernandez-Zimbron LF, Zamora-Alvarado R, Ochoa-De la Paz L, *et al.* Age-related macular degeneration: new paradigms for treatment and management of AMD. *Oxid Med Cell Longev* (2018) 2018:8374647. doi:10.1155/2018/8374647.

[3] Miller JW, D'Anieri LL, Husain D, Miller JB, and Vavvas DG. Age-related macular degeneration (AMD): a view to the future. *J Clin Med* (2021) 10:1124. doi:10.3390/jcm10051124.

[4] Fleckenstein M, Mitchell P, Freund KB, *et al.* The progression of geographic atrophy secondary to age-related macular degeneration. *Ophthalmology*. (2018) 125:369–90.

[5] Schaal KB, Rosenfeld PJ, Gregori G, *et al.* Anatomic clinical trial endpoints for nonexudative age-related macular degeneration. *Ophthalmology* (2016) 123:1060–79.

[6] Horvath S. DNA methylation age of human tissues and cell types. *Genome Biol* (2013) 14:R115.

[7] Schick T, Lores-Motta L, Altay L, Fritsche LG, den Hollander AI, and Fauser S. The effect of genetic variants associated with age-related macular degeneration varies with age. *Invest Ophthalmol Vis Sci* (2020) 61:17. doi:10.1167/iovs.61.14.17.

[8] Hunter A, Spechler PA, Cwanger A, *et al.* DNA methylation is associated with altered gene expression in AMD. *Invest Ophthalmol Vis Sci* (2012) 53:2089–105. doi:10.1167/iovs.11-8449.

[9] Wei L, Liu B, Tuo J, *et al.* Hypomethylation of the IL17RC promoter associates with age-related macular degeneration. *Cell Rep* (2012) 2:1151–8. doi:10.1016/j.celrep.2012.10.013.

[10] Oliver VF, Franchina M, Jaffe AE, *et al.* Hypomethylation of the IL17RC promoter in peripheral blood leukocytes is not a hallmark of age-related macular degeneration. *Cell Rep* (2013) 5:1527–35. doi: 10.1016/j.celrep.2013.11.042.

[11] Sobrin L, and Seddon, JM. Nature and nurture-genes and environment predict onset and progression of macular degeneration. *Prog Retin Eye Res* (2014)40:1–15.

[12] Muthiah MN, Keane PA, Zhong J, *et al.* Adaptive optics imaging shows rescue of macula cone photoreceptors. *Ophthalmol Times* (2014) 121 (1): 430–31.

[13] Wang J, Zibetti C, Shang P, *et al.* ATAC-Seq analysis reveals a widespread decrease of chromatin accessibility in age-related macular degeneration. *Nat Commun* (2018) 9:1364. doi: 10.1038/s41467-018-03856-y.

[14] Vallée A, Lecarpentier Y, Vallee R, Guillevin R, and Vallee JN. Circadian rhythms in exudative age-related macular degeneration: the key role of the canonical WNT/beta-catenin pathway. *Int J Mol Sci* (2020) 21:820. doi: 10.3390/ijms21030820.

[15] Ratnapriya R, Sosina OA, Starostik MR, *et al.* Author correction: retinal transcriptome and eQTL analyses identify genes associated with age-related macular degeneration. *Nat Genet* (2019) 51:1067. doi: 10.1038/s41588-019-0430-y.

[16] da Costa JP, Vitorino R, Silva, GM, Vogel C, Duarte AC, Rocha-Santos, T. A synopsis on aging: theories, mechanisms and future prospects *Ageing Res Rev* (2016) 29:90–112.

[17] Mallik S, Grodstein F, Bennett DA, *et al.* Novel epigenetic clock biomarkers of age-related macular degeneration. *Front Med (Lausanne)* (2022) 9:856853. doi:10.3389/fmed.2022.856853. PMID: 35783640; PMCID: PMC9244395.

[18] Ma J, Xu X., Li M, *et al.* Predictive models of aging of the human eye based on ocular anterior segment morphology. *J Biomed Inform* (2021) 120:103855, https://doi.org/10.1016/j.jbi.2021.103855.

[19] Zhu Z, Shi D, Guankai P, *et al.* Retinal age gap as a predictive biomarker for mortality risk. *Br J Ophthalmol* (2023) 107:547–554, https://doi.org/10.1136/bjophthalmol-2021-319807

[20] Bandyopadhyay S, Mallik S, and Mukhopadhyay A. A survey and comparative study of statistical tests for identifying differential expression from microarray data. *IEEE/ACM Trans Comput Biol Bioinform* (2014) 11:95–115. doi:10.1109/TCBB.2013.147.

[21] Hahsler M, Piekenbrock M, and Doran D. DBSCAN: fast density-based clustering with R. *J Stat Softw* (2019) 91:1–30.

[22] Sharma NK, Sharma SK, Gupta A, Prabhakar S, Singh R, and Anand A. Predictive model for earlier diagnosis of suspected age-related macular degeneration patients. *DNA Cell Biol* (2013) 32:549–55. doi: 10.1089/dna.2013.2072.

[23] Davis, MD, Gangnon RE, Lee LY, *et al.* The age-related eye disease study severity scale for age-related macular degeneration: AREDS report*Arch Ophthalmol* (2005) 123(17):1484–98.

[24] Reich NG, Lessler J, Sakrejda K, Lauer SA, Iamsirithaworn S, and Cummings DA. Case study in evaluating time series prediction models using the relative mean absolute error. *Am Stat* (2016) 70:285–92. doi:10.1080/00031305.2016.1148631.

Chapter 9

Artificial intelligence in cataract diagnosis and management with its future directions

N. Ramya[1] *and D. Hemavathi*[1]

Abstract

The development of artificial intelligence (AI) technology brought improvement in various areas of healthcare. The application based on AI offered significant enhancement in the efficiency and quality of patient care in multiple sectors including cataract management. In the worldwide, cataract is the leading cause of the visual impairment in a person. This visual impairment is expected to increase substantially in the aging population. In the present eye care system, it is difficult to develop the tandem that causes shortfall, which becomes highly complex to solve. In general, many factors maximize the frequency of cataract surgeries such as high frequency of eye surgery, population aging, and early intervention. The procedure is limited with more time and fewer resources. It is a major complex for the countries, which depends mainly on the public medical system in terms of cataract management. Therefore, with the general population aging and the advancement in developing countries, AI systems become more significant in the treatment, staging, and screening of the eye condition. This means more population converges in the minimum period in diagnosing and treating the seriousness of the cataract. The gaps in the early and efficient treatment are filled by AI technology by gaining interest in terms of cataract management. AI system offers effective results in detecting age-related eye disease automatically. The advanced technique based on AI can extract high-level features and attain better performance in cataract diagnosis. Due to these advantages, the medical system globally has begun to develop more advanced models with the assistance of AI for cataract management.

Keywords: Cataract; deep learning; retina; CNN; early detection; SVM; ophthalmology

9.1 Cataract and its main causes

In general, the lenses are placed behind the colored part of the eye and it is named as iris. The lens in the eyes helps to focus the light that passes through the eyes. The

[1]Department of Data Science and Business Systems, SRM Institute of Science and Technology, India

light is processed through the eyes and generates a sharp and clear image beside the eyes and it is known as retina. After the person reaches some age, the lens present in the eye becomes poor in flexibility and the vision starts to become unclear. This causes the clouding in the lens and tends to lose the vision completely [1]. There are different reasons for the persons to cause the contract. Some of the major reasons are defined as follows.

- A cataract is formed by making the lens of the eye cloudy, which occurs by clumping the protein in the lens together. In general, the lens of the eye is clear and normal, which allows the light to pass on it. The lens supports to focus of the light onto your retina. This makes the eye to see the object.
- Cataracts make the visibility poor because the light cannot easily pass on to the clump of the protein in your lens. Lately, the clump become thicker and bigger making it complex to see through the eyes. In some cases, the lens changes its color to brown or yellow, which can change how you see colors.
- After it reaches 40 ages, the protein that is present in the lens of the eyes starts to break down naturally. In most cases, the cataract is formed by these natural changes. Also in some cases, injury or accident changes the tissue and makes the eye lens poor. Here the fiber and protein present in the lens start to break down and cause cloudy things in the vision.
- Other health problems that are transferred from the genes also have the chance to affect the lens and improve the risk factor of the cataract. Similarly, other medical conditions like diabetes can also lead to cataract issues. Using steroid medicines for the long term also leads to the development of cataracts.

While the cataract started growing, the cloudiness became worse and the vision became worse. The cataract blocks and scatters the light to pass on the lens and the vision becomes more blurred [2]. It usually happens in both the eyes of the person, but not generally at the same rate. In some cases, the one eye becomes worse than the other one. This leads to differences in the vision between each eye. Some of the common factors that cause cataracts other than the aging factor include:

- Diabetes
- Eye treatment or surgery
- Radiation treatment for cancer and other disease
- Drinking too much alcohol
- The family gene of cataract
- Serious injury in eyes
- Taking steroids medicine for other health problems
- Smoking and
- Facing too much time in the sun without any glasses.

9.2 Types of cataracts

The cataract is a cloudy region in the lens of the eye. The majority of the cataracts [3] are related to the age. These happen due to the normal variations in the subject's eyes, as get older. However, cataracts may happen for some other reasons. There are five primary kinds of cataracts and they are explained as follows.

- **Age-related cataracts**: This kind of cataract occurs when the subject becomes older. In this kind, cataract develops due to the natural variations in the lens of the subject's eye. This kind of cataract is the common kind. Some of the causes of age-related cataracts are
    - Diabetes
    - Too much alcohol consumption
    - Smoking
    - Have a family history of cataracts
    - Usage of steroids-medicines employed to cure several health issues such as allergies or arthritis
- **Traumatic cataracts**: Severe eye damage can affect the subject's lens and cause the cataract. The cataract can form immediately after the eye damage or it can form some years later.
- **Pediatric cataracts**: This type of cataract can happen to children too. This pediatric cataract is rare and genetic. This kind of cataract can also occur due to serious issues during pregnancy or due to diseases during childhood such as uveitis or eye tumors. Children can get cataracts for reasons such as steroids, radiation, or eye injuries.
- **Secondary cataracts**: After performing the surgery for cataracts, it is possible to grow the scar tissues in the eye. This makes the vision cloudy. This is called secondary cataracts.

9.3 Modalities in cataract treatment

There are numerous modalities [4] for treating cataracts such as imaging, surgical, and imaging modalities.

- **Medical modality**: If the visual acuity is 6/24 or better, the pupillary dilatation or refractive glasses with 2.5% phenylephrine can be sufficient to allow for periodic activities without the surgery. The cataracts drop also there in trail for dissolving the cataracts.
- **Surgical modality**: It is needed if the visual acuity is lower than 6/24 or if the cataract is troubling the eye's health because of the medical situation. Some of the surgical strategies are
    - *Phacoemulsification:* To emulsify the cataract, a 2.8 mm incision is employed.
    - *Micro-incision cataract surgery:* To eliminate the affected lens, a small incision is formed in the cornea.
    - *Small incision suture less cataract surgery:* A huge incision is formed in the sclera instead of the cornea.

- **Imaging modality**: The imaging modalities are employed in the surgery of cataracts to validate the cataracts pre-operatively, offer a real-time review to the specialist during the surgery, and instruct the trainee surgeons. Some of the imaging modalities are
    - *Intra-operative optical coherence tomography (iOCT):* It is a mechanism that enables ophthalmic specialists to determine the impacts of surgical manipulations in real time.
    - *Anterior segment optical coherence tomography (ASOCT):* It supports achieving an optical biopsy of numerous neoplastic, degenerative, and dystrophic ocular surfaces. It enables the specialists to detect the eye disorders.
    - *Premium intraocular lenses:* It is employed in cataract surgery to cure the eye's natural lens clouding and offer clear vision at distant and near local points without the requirement for the glasses.
    - *Femtosecond lasers:* It is employed in cataract surgery, where the computer-guided laser is employed to perform accurate cuts in the eye to eliminate the cloudy lens and exchange it with an artificial lens.
- **Other modalities**: No stitch surgery/no injection, advanced extraction methods, ultrasonic and laser measuring approaches for lens implants, adjunctive approaches such as capsular staining and capsular tension rings.

9.4 Limitations in the current model

Many people are affected by cataracts. After reaching a specific age, there are also other reasons for the presence of cataracts. Testing by using the developed model helps to determine the cataract present in the person [5]. Still, some challenges are unclear in the cataract detection research. Some of the complexities in the earlier models are defined as follows.

- There are many automated models developed for detecting cataracts. Some of the models are limited to the use of nuclear cataracts and perform only for the slit lamp. This type of grading model for cortical cataracts is still not possible because it is difficult to determine the cloud formation maturity state.
- In the present times, smartphone-based slit lamp image screening is developed for cataract identifications. However, the methods included pupil dilation, which is a complex process in the final investigation.
- In the machine learning technique, only high-quality imaging is used to train the model. However, the algorithm is not better for the biased or not representative of the population. This makes the parameter selection worse and causes the detection with less accuracy.
- There is a lack of transparency in the machine learning model, which makes it trickier to comprehend the diagnosis procedure. Also, it is more demanding in the substantial amount of processing power and storage. This may result in complexity in determining the fundus image.

- For the diagnosis of the cataract, the combination method is utilized like the convolutional neural network (CNN) and the support vector machine (SVM) based on the slit lamp imaging. This process requires high-quality medical images for the detection; otherwise, it leads to false detection.
- The deep learning model like a Residual network with three-step sequence is utilized in the diagnosis process for identifying the cataract. Here, the model identifies the multiple images from the slit-lamp photography and determines the pupil as non-mydriatic or mydriatic. Further, the severity grading is performed to evaluate the disposition. This method needed the preprocessing procedure before the detection to get accurate results.
- CNN-based ensemble optimization is utilized to grade the cataract based on the input fundus images, which results in accurate cataract diagnosis. Here, the anterior segmenting imaging is replaced with a fundal photo to achieve detection. Moreover, this model has interpretability issues that still exist.

9.5 Screening and diagnosis in cataracts

Cataract screening is the detection process that functions by the eye care specialist to diagnose the presence of cataracts. The cataract is the clouding present in the natural lens in the eyes that leads to minimizing the vision [6]. Screening is the standard procedure that investigates the eye, which includes dilated eye determination, slit-lamp investigation, and acuity testing, which are defined in Figure 9.1.

- **Visual acuity**: This test determines the vision quality at a specific distance. Here, the professionals asked the patient to go through the letters of different sizes from the chart that was placed in the distance. The eyes of the persons are tested separately and combined form to calculate the eyesight accuracy at different distances. This approach is the easiest and painless method to diagnose the cataract.
- **Contrast sensitivity**: It is more similar to the visual acuity. Here, the contrast colors are placed to determine the eyesight. Here, the glare that is formed by the cataract and the decrease in the image contract because of the light scattering are determined.
- **Slit lamp**: It is an advanced type of microscope that evaluates the eyes. Here, the lenses are examined by the doctor to detect the severity and presence of cataracts. The professional asks the patient to place their chin on the slit lamp chin rest. The light in the machine passed to the eyes. By seeing through the slit lamp, the doctor investigates the lens to find the degree of cloud obtained.
- **Pupil dilation**: It is the most generally used method in diagnosing the cataract. While the dilation occurs, the pupil develops the size and provides the view of the whole lens. The specialist examines the lens to identify whether it is affected by cataracts or not based on the quality of vision.

There are different diagnosis models that are developed to detect cataracts in the person. There are various reasons for developing cataracts in the lens of the eye.

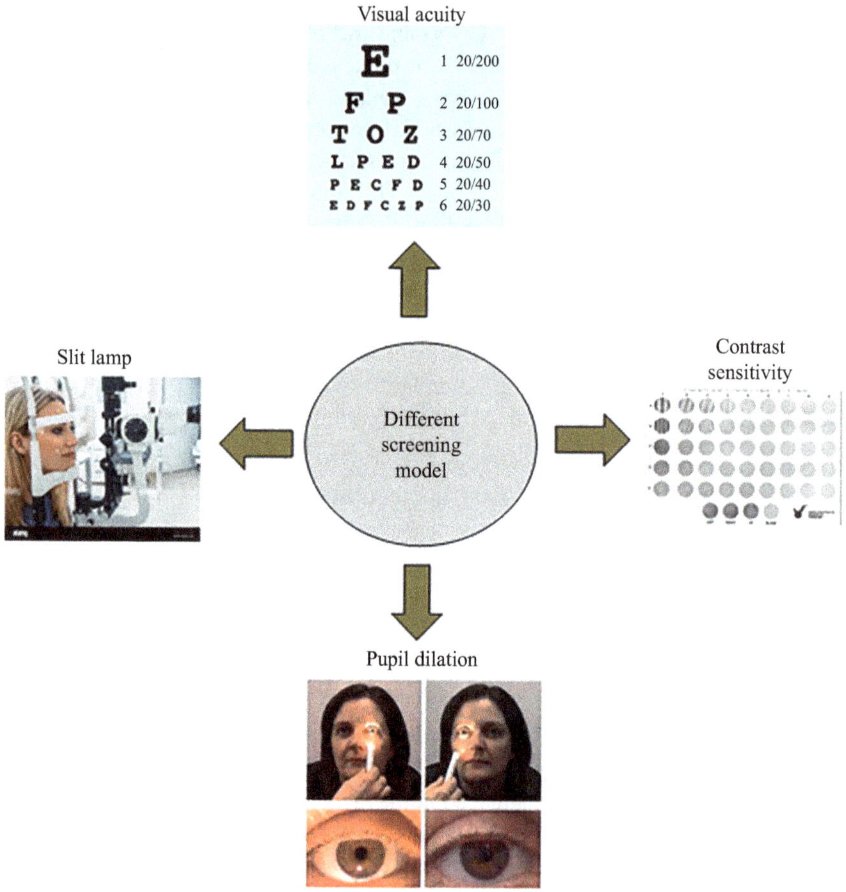

Figure 9.1 Specification of different screening methods for detecting cataracts

Early diagnosis can help to treat the cataract in the right manner and save the vision of the person. Untreated cataract tends to complete vision loss for the person [7]. Therefore, it is necessary to implement the proper diagnosis model.

- In the diagnosis process, the required image detail about the eye is necessary in the initial phase. Some range of tools are used to record the medical images. Further, the images are preprocessed when they are acquired to ensure the perfect alignment and improve the analysis.
- After the preprocessing, the regions of interest are found and recognized. The region of eyes is correlated by the region of interest like the lens or cornea, where the cataract gets developed.
- It is important to take out the feature from the region of interest after it is discovered. This feature includes the details of shape, intensity, color, and

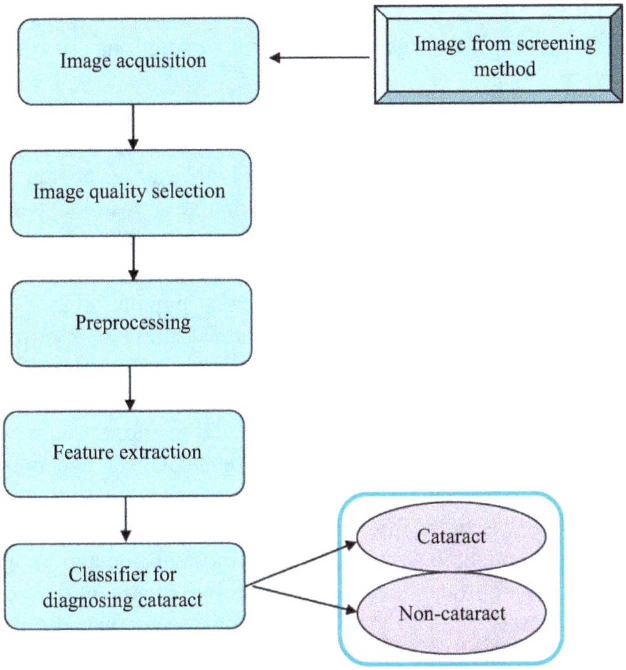

Figure 9.2 Flow diagram representation of the diagnosis procedure of cataract

texture. The improvement of the algorithm in detecting the cataract is based on the feature extraction process.

- Once the feature gets retrieved, the machine learning algorithm gets trained. To achieve the labeled imaging, every presented image evaluates whether it has cataracts or not. The characteristic gets spotted by the program gain by avoiding the non-cataract one.
- In the clinical setting, the approach is implemented to help the detection of early cataracts to validate and test the outcome that occurred from it. Figure 9.2 defines the diagnosis procedure in cataract detection.

9.6 Role of AI in cataracts

Nowadays, AI has provided a significant impact on ophthalmology. The sector has developed from the automation of manual works including executing the ophthalmic images, to deep learning and machine learning. Machine learning [8] is a division of AI that enables the automated device to learn from the existing data by finding out the best weights and parameters within the normal technique such as random forests or SVM. However, deep learning is a division of machine learning that contains the deep neural network (DNN). Numerous layers of network neurons conduct the feature extraction, allowing the technique to learn the fine features in an incremental way. This capacity has resulted in an important breakthrough in the functionality of multiple image classification operations in the sector of ophthalmology.

In the ophthalmology sector, the cataract is the primary cause of curable blindness, resulting in severe or average vision damage. This difficulty is predicted to enhance sustainability, as an outcome of the highly aging population. The services of eye care, nevertheless, have been not able to enlarge in tandem, leading to a problem that is becoming highly complex to recognize. The cataract surgery has been shown to improve the cognitive function between dementia patients and Alzheimer's patients. Some of the factors including population aging, higher frequency of the eye surgery, and timely identification result in an improvement in the cataract surgery frequencies. High time and the limited resources led to primary problems in the cataract management.

Some of the AI models have been implemented to support numerous cataract management aspects. The AI-aided models have displayed high power in some disciplines regarding to the patient care. With the support of the digital revolution, the technology of AI has been applied in all modules of society. Thus, with the enhancement in the developing regions and the normal population aging, AI [9] models can become a significant section of screening, treatment of eye conditions, and the staging. AI seems to be promising in this sector because of its special capacity to internalize the huge data and evaluate huge parameters. To date, AI techniques have been employed in screening and cataract screening operations, risk prediction, forecasting of surgical procedure timings to tune the workflows of operating theaters, and the cataract surgery categorization stages. These roles of AI in cataracts are explained as follows.

- **Early detection**: AI can recognize the disease symptoms earlier than the conventional approaches. Early identification can result in timely treatment and highly enhance patient solutions. AI contributes to the early cataract identification processing standard models on the basis of either machine learning or deep learning.
- **Disease prevention**: AI can support the prevention of cataract disease in some ways, including implementing preventive measures, implementing personalized treatment plans, timely disease recognition, and disease tracking. AI also helps to minimize healthcare errors, enhancing the healthcare sectors.
- **Screening and diagnosis of cataracts**: AI [10] models have offered promising solutions in the screening and the diagnosis of cataracts. AI-trained techniques employed some algorithms and the video recordings achieved employing the recordable and portable device provided high specificity and sensitivity for the diagnosis of cataracts. Through this automatic process, the manual tasks have been minimized and also reduced the requirement of the experienced operator.
- **Intraocular lens power estimation and biometry**: With deep learning, the formulas of intraocular lens (IOL) are taking the merit of AI to improve the prediction solutions. The new formulas that have been implemented in this process are either AI-aided or employ AI-incorporated factors. These AI

formulas [11] have a significant future, as numerous have displayed better prediction accuracy. Apart from the IOL, with the biometry extremes, the patients who have performed the surgery are at enhanced refractive risk as well. AI is highly supported in estimating corneal power and the solutions accurately for the surgery patients.

- **Screening and diagnosis of pediatric cataracts**: To prevent the irreversible amblyopia, the pediatric cataracts must be cured correctly. The diagnosis is most of the time delayed for the sick persons who do not have simple access to the clinicians. The pediatric cataract treatment's time-sensitive nature contributes to the significance of timely identification and prevention. The recent improvements in AI have displayed outstanding solutions and support rectifies these problems.
- **Inter-operative**: AI helps to augment the training of cataract surgery and intra-operative decision-making and offers postsurgical evaluation to improve the surgical mechanisms. AI is utilized to forecast the intra-operative complication's risks and tune the surgical workflows [12]. Virtual reality and AI are employed in tandem to implement smart teaching models for training the cataract surgery.

9.7 Advent of AI with its application in ophthalmology

In an effort to idealize the works in numerous organizations, the majority of them have focused on AI systems, especially in the sector of deep learning and machine learning in recent times. Deep learning is an enhancing AI sector that is a division of machine learning. It includes the utilization of an artificial neural network (ANN) that includes numerous artificial neuron layers to simulate the human brain's physiological functions. The deep learning device is trained to draw out and execute the data in the texts and images, and recognize the speech. Nowadays, AI model's applications in the medical sector have displayed satisfactory outcomes in narrow operations including lung cancer detection, and the colonoscopy polyp's real-time identification. In ophthalmology, where a huge amount of patient data and images are available, the systems of AI have displayed satisfactory solutions in the automated identification of age-based eye disorders including glaucoma, age-oriented macular degeneration, and diabetic retinopathy. The development in computing power and infrastructure has incorporated the quick enhancement of deep learning models in AI implementation [13]. Because of its promising capacity to draw out the unrecognized patterns and the high-level attributes within enormous amounts of data, deep learning models now attain better functionality than the clinicians and human graders in feature-aided diagnosis.

The utilization of AI devices in the clinical sector can highly improve productivity in the workplace and help in patient communication operations and clinical decision-making. Utilizing AI for healthcare diagnosis validations enables the automatic evaluation of imaging. In the sector of ophthalmology, the majority of the AI sectors are being employed for powerful utilization in the treatment, surveillance, and detection of multiple ocular disorders. However, the majority of the experimental phase and the validation should be performed to validate if these models are appropriate for the clinical tasks. Several primary applications of AI in ophthalmology are explained as follows.

- **Glaucoma**: Numerous deep learning models have offered high specificity and sensitivity in detecting the variations of glaucomatous optic nerve. These AI experiments are on the basis of diagnosis features or else normally validated by the experts with numerous image findings and measurements.
- **Ocular oncology**: Mimicking the decision tree technique, a machine learning model was implemented to anticipate the periocular reconstruction course during the basal cell carcinoma's surgical treatment. In addition, some of the machine learning models, through the utilization of ANN, has been implemented to forecast the disease solutions for choroidal melanoma by validating the oncologic history and demographic data.
- **Cataracts**: Machine learning approaches have been implemented to recognize and grade cataracts. The authors experimented with the technique employing the AI technique named ResNet to recognize referable cataracts. Deep learning approaches for the validation of congenital cataracts have also been implemented in the past years.
- **Pediatric ophthalmology**: In the population of pediatric, proper ocular management is complex for vision preservation. The utilization of AI in the treatment and screening practices can support attaining optimal care in ophthalmology. Machine learning models have the power to support screening for high myopia between other refractive faults and categorize the susceptible children to observe the disabilities.
- **Retina**: AI has been employed in recognizing retinopathy-related diseases. AU models have higher sensitivity than the traditional models.
- **Oculoplastics**: In this sector, AI has the power to support the automated measurement and the processing of facial dimensions to support post and pre-op estimations. The semantic segmentation models have been trained to perform the automatic estimates with comparable functionality to the experts.

9.8 AI used in cataract detection and severity classification

Based on the promising solutions of the AI models in numerous eye disorders, there have also been some AI models implemented for the automated grading and detection of cataracts, on the basis of either machine learning or deep learning techniques [14]. Apart from the distinct frameworks of AI, these improved models from conventional experiments also varied based on the kinds of input images employed.

Machine learning models: Machine learning models utilize numerous approaches to forecast the solutions. The classification is one mechanism that is constructed upon semi-supervised and supervised learning. This kind of model enables for concrete classification of solutions. When labels or classes for the training data are not offered, the unsupervised machine learning is still capable to cluster similar inputs, even if it is not capable to categorize the independent clusters. The machine learning architectures can utilize the neural networks for other approaches including genetic programming, tree-aided classification, statistical regression, random forests, and so on. Some of the machine learning approaches employed for performing the cataract classification and segmentation are explained as follows.

- **Support vector machine (SVM)**: It is a conventional supervised machine learning approach that has been highly employed for classification and segmentation purposes. This model is largely employed in distinct cataract ophthalmic images for classifying the cataract. It provides good solutions for any kind of ophthalmic image.
- **Linear regression**: It is one of the famous machine learning models employed to rectify distinct learning operations. The idea of linear regression is still fundamental for other developed models. This model helps to segment and diagnose the cataract and also perform the cataract grading.
- **K-nearest neighbor (KNN)**: It is a simple approach and also developing this technique is very easy. It employs the similarity factors to categorize new cases on the basis of stored instances. It is a non-parametric approach that employs proximity to make predictions regarding the specific data point.
- **Ensemble learning approach**: It employs numerous machine learning strategies to rectify the same issue and normally achieves better classification and segmentation solutions. Some of the ensemble models employed for the classification process are voting, bagging, stacking, and so on.

Advantages of machine learning in the healthcare sector are as follows:

- **Enhancing the diagnosis**: It is employed by healthcare professionals to implement better diagnostic components to validate the diagnostic images.
- **Minimizing the costs**: The models in machine learning employed by the institutions increase the efficacy of the medical sector and also minimize the cost usage. Moreover, the machine learning models can help minimize the resources and time that are wasted on the continuous operations in the healthcare model.
- **Data privacy and security**: With the enhanced health record digitalization, safeguarding patient data is a complex issue. The machine learning model can improve data security by recognizing and responding to cybersecurity attacks in real time.

Deep learning: Deep learning models are neural models with an enlarged amount of communicating layers among the output and input layers. This kind of machine learning includes supervised learning with labeled data sources. The CNN starts with a huge matrix of inputs in the initial layer that further leads to a solution that serves as the input for the subsequent layers in the sequence. The connections among the layers explain the convolution propagating local data. The evaluation from each layer is given to the overall network until the last layer generates the solution. Some of the deep learning approaches employed for performing the cataract classification and segmentation are explained as follows.

- **Convolutional neural network**: It is highly utilized in the sector of ophthalmic image processing and attained a superior performance rate. It progresses via convolutional and pooling layers to draw out the significant features from the given images. The designed model preserves higher accuracy rates.
- **Recurrent neural network (RNN)**: It is a normal feed-forward neural network, where the connections generate an undirected or directed graph. This technique is capable to process the sequential data efficiently for numerous learning approaches. It minimized the error rates of the model and also performed well for both small and large-sized data sources.
- **Long short-term memory (LSTM)**: It is a kind of RNN that employs the gates to process and obtain the data over numerous time sequences. The LSTMs are developed to learn the long-term dependencies among the time sequence data and are employed to classify and process the given images and data. This model produces highly accurate solutions.
- **Hybrid neural network**: It describes that the neural network is integrated into more than two deep networks. The experts are employing hybrid neural networks to rectify distinct learning operations. This model improves the feature extraction, segmentation, and classification processes very effectively.

Advantages of deep learning in the healthcare sector are as follows:

- **Innovating discovery of the drug**: Deep learning supports in medicine discovery and its development.
- **Evaluating the medical imaging**: The medical imaging approaches including CT, MRI, and EEG are employed to recognize dangerous disorders including brain tumors and cancer heart disorders. Thus, deep learning supports doctors to validate the disorders better and offer patients with better treatment.
- **Discovering solutions to the disease**: Deep learning techniques automatically recognize the solutions by extracting the significant features from the input data or image in real time.
- **Enabling the comprehensive interpretation**: Deep learning models are employed to understand the genome and assist the patients obtain an idea regarding the disorder that might trouble them.

9.9 Challenges and future directions

AI has the power to be a supportive component for managing the cataract. However, some concerns and difficulties require to be rectified for the promising translation. Data sources utilized must be heterogeneous to attain an accurate generalizability. Nowadays, medical data is often a target for intruders. Adversarial threats can also perform in these techniques, either by changing the input images or injecting the data during the training process to result in large-scale wrong classification of the AI technique. In addition, trust of end-users such as patients and also physicians in these techniques are significant to obtain the promising translation of clinics. This demands enhancements to the AI technique's explainability such as a detailed demonstration of the decision-making task.

Numerous techniques have been developed to rectify the mentioned problems. Initially, federated learning is highly utilized to enable cross-border or cross-institution AI training without sharing the data. This is a privacy-preserving mechanism that reveals the technique to heterogeneous non-independent [15]. An improvement is referred to as swarm learning. It enables the AI technique parameters to be tuned, thus helping in the enhancement of the generalizable technique. Further, these data sources are enlarged by employing generative networks, especially for rare disorders. In order to improve the AI model's explainability, the techniques contain the heatmaps representing the regions of interest with validation of the included uncertainty and also the feature extractions or the predefined techniques.

Apart from the specific cataract and biometry models, the majority of the AI techniques have not attained a high accuracy level that is acceptable in clinically, and then enhancement is required. In addition, the majority of the intra-operative AI works have concentrated on the growing technologies without proper clinical application, and most of its utilization is still hypothetical. Nevertheless, it is still possible that emerging enhancements can provide practical applications that are not apparent immediately. Moreover, there are still important problems in the development of AI, specifically in implementing the countries because of the lack of data availability, poor infrastructure, technical expertise, and funding. There is also a requirement for accurate training applications to be initially developed to confirm the large catchment for the subjects. Some of the primary complexities of AI encountered in ophthalmology are pictorially shown in Figure 9.3.

Most of the conventional works displayed effective solutions, but it must be considered that none of the traditional works further validated their corresponding models in the real-time test cases. Moreover, the actual utilization of these models in real-world settings remains to be estimated. Further developments are needed in the formula selection and IOL power estimation on the basis of AI for eyes. The patients with a refractive surgery history and the estimation of IOL power employing formulas also give complexities. This is due to the fact that conventional formulas are not developed effectively. This complexity is entirely related in the Asian countries compared with the Western countries. Because of the aging trend in the community of Asia, the number of patients requiring cataract surgery yet with previous refractive surgery is predicted to increase in the upcoming years. Finally, the implementation of a new model or the conventional AI model's refinement, and curation of well-annotated, robust, and large data sources remain a complexity. Moreover, the new technology demands compliance with transparent standards to confirm completeness and transparency.

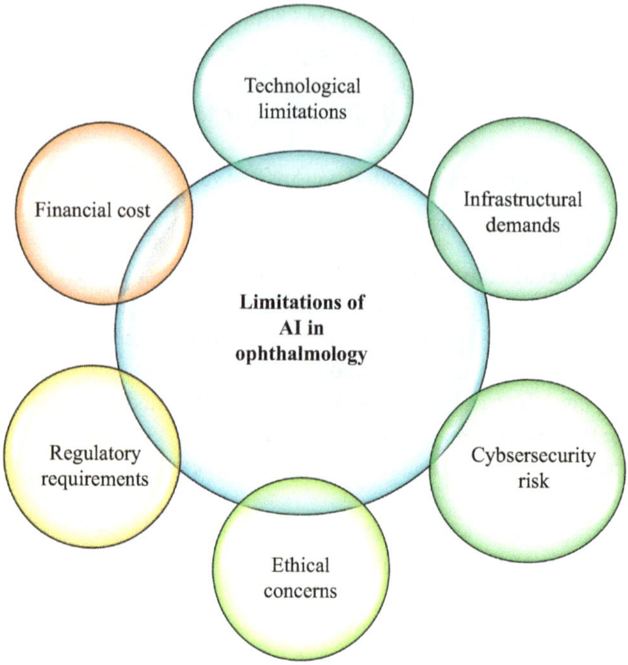

Figure 9.3 Limitations of AI in ophthalmology

9.10 Conclusion

This book chapter has offered a detailed explanation of AI and its significance in cataract diagnosis. Initially, the book chapter explained the cataract and its causes. Then, the chapter analyzed the present model's limitations in the cataract diagnosis. Furthermore, AI models and its advent have been explained with its research gaps and future directions.

References

[1] Lu, Q., Wei, L., He, W., *et al.*, "Lens Opacities Classification System III–based artificial intelligence program for automatic cataract grading", *Journal of Cataract & Refractive Surgery*, vol. 48, pp. 528–534, 2022.

[2] Lindegger, D.J., Wawrzynski, J. and Saleh, G.M., "Evolution and applications of artificial intelligence to cataract surgery", *Ophthalmology Science*, vol. 2, p. 100164, 2022.

[3] Keenan, T.D., Chen, Q., Agrón, E., *et al.*, "DeepLensNet: deep learning automated diagnosis and quantitative classification of cataract type and severity", *Ophthalmology*, vol. 129, pp. 571–584, 2022.

[4] Abbas, Q., Qureshi, I., Yan, J. and Shaheed, K., "Machine learning methods for diagnosis of eye-related diseases: a systematic review study based on ophthalmic imaging modalities", *Archives of Computational Methods in Engineering*, vol. 29, pp. 3861–3918, 2022.

[5] Long, E., Chen, J., Wu, X., *et al.*, "Artificial intelligence manages congenital cataract with individualized prediction and telehealth computing", *NPJ Digital Medicine*, vol. 3, p.112, 2020.

[6] Shimizu, E., Yazu, H., Aketa, N., *et al.*, "Innovative artificial intelligence-based cataract diagnostic method uses a slit-lamp video recording device and multiple machine-learning", *Investigative Ophthalmology & Visual Science*, vol. 62, p. 1031, 2021.

[7] Xue, W., Zhang, J., Ma, Y., *et al.*, "Deep learning-based analysis of infrared fundus photography for automated diagnosis of diabetic retinopathy with cataracts", *Journal of Cataract & Refractive Surgery*, vol. 49, pp. 1043–1048, 2023.

[8] Ueno, Y., Oda, M., Yamaguchi, T., *et al.*, "Deep learning model for extensive smartphone-based diagnosis and triage of cataracts and multiple corneal diseases", *British Journal of Ophthalmology*, vol. 108, pp. 1406–1413, 2024.

[9] Rampat, R., Deshmukh, R., Chen, X., *et al.*, "Artificial intelligence in cornea, refractive surgery, and cataract: basic principles, clinical applications, and future directions", *The Asia-Pacific Journal of Ophthalmology*, vol. 10, pp. 268–281, 2021.

[10] Gutierrez, L., Lim, J.S., Foo, L.L., *et al.*, "Application of artificial intelligence in cataract management: current and future directions", *Eye and Vision*, vol. 9, p. 3, 2022.

[11] Ting, D.S.J., Foo, V.H., Yang, L.W.Y., *et al.*, "Artificial intelligence for anterior segment diseases: emerging applications in ophthalmology", *British Journal of Ophthalmology*, vol. 105, pp. 158–168, 2021.

[12] Muchibwa, C., Eldaw, M.H.S., and Agyeman, M.O., "An assessment of contemporary methods and data-enabled approaches for early cataract detection", *F1000 Research*, vol. 12, p. 998, 2023.

[13] Sung, J., Inomata, T., Akasaki, Y., *et al.*, "Development and validation of artificial intelligence-driven cataract detection model using smartphone-captured images", *Investigative Ophthalmology & Visual Science*, vol. 65, p. 3715, 2024.

[14] Tham, Y.C., Goh, J.H.L., Anees, A., *et al.*, "Detecting visually significant cataract using retinal photograph-based deep learning", *Nature Aging*, vol. 2, pp. 264–271, 2022.

[15] Junayed, M.S., Islam, M.B., Sadeghzadeh, A. and Rahman, S., "CataractNet: an automated cataract detection system using deep learning for fundus images", *IEEE Access*, vol. 9, pp. 128799–128808, 2021.

Chapter 10

Machine learning and blockchain technology in healthcare

M. Arun Anoop[1], P. Karthikeyan[2], Ben Othman Soufiane[3] and S. Poonkuntran[4]

Abstract

Combining machine learning with blockchain encryption provides a promising approach for image forgery detection, ensuring data integrity. The utilization of blockchain technology allows image data to be securely stored on a decentralized and tamper-resistant ledger. Each image is assigned a hash, which is stored on the blockchain, making it difficult for unauthorized modifications to go undetected. Machine learning techniques are employed to extract relevant features from the images, capturing unique characteristics that aid in detecting forgeries. Trained machine learning models, such as extreme learning machine (ELM) or deep learning models, learn patterns and characteristics from a labeled dataset of authentic and forged images. The extracted features and hash values and timestamp of both images on the blockchain ensure transparency and immutability. When a new image is presented, its features are compared with those of authentic images, and a classification algorithm determines its authenticity based on learned patterns. This hybrid approach can be integrated with blockchain technology with extracted feature values along with generated timestamp and hash values from each image used to provide an immutable and decentralized solution for verifying the authenticity of images. In the context of image forgery detection, a consensus mechanism can provide additional security and reliability when extracting timestamps and hash values from images. This decentralization reduces the risk of malicious tampering. If a node tries to manipulate the extracted values, it will be rejected by the consensus algorithm, ensuring the integrity of the data.

[1]Department of Computer Science and Engineering, AJ Institute of Engineering and Technology, India
[2]Department of Electronics and Communication Engineering, Velammal College of Engineering and Technology, India
[3]Department of Computer Science, Higher Institute of Computer Sciences, Gouvernorat de Médenine, Tunisia
[4]School of Computer Science and Engineering, VIT Bhopal University, India

Keywords: Skin cancer images; extreme learning machine (ELM); supervised classifiers; blockchain methods; timestamping; hashing on images

10.1 Introduction

Public blockchains are accessible to all users and are permissionless. Private blockchains, sometimes referred to as regulated blockchains, are managed by a single person. The authority decides who can connect with the network and what rights they have. Private blockchains are not centralized because of access restrictions. Hybrid models of blockchains combine elements of two networks such as public and private networks. Permission-based private systems can be installed by businesses in addition to public ones. They keep public access to the other data while limiting access to particular data contained in the blockchain. Public conduct checks if private transactions have been completed through the use of smart contracts. All Bitcoin transactions are archived on a public ledger, copies, and they are stored on servers all over the earth. Members mine for bitcoin on the open Bitcoin network by resolving mathematical puzzles to produce new blocks. Every new transaction is shared from node to node across the network by the system, which broadcasts it publicly. The blockchain, which functions as Bitcoin's official ledger, is updated permanently with new blocks that miners compile every 10 min or so [1].

Why block chain ideas are necessary when it comes to medical images: It gives flexibility, connectivity, accountability, and data entry confirmation. Health records must be kept remote and protected for various reasons. Blockchain helps avoid certain threats and supports distributed facts and figures defense in the healthcare industry [2].

Why block chain ideas are essential when dealing with digital photographs: Image distribution, including direct images of patients who owned with it, tracking of medical equipment implanted, investigations, teleradiology, and artificial intelligence are midst the latent use cases for blockchain that are particularly related to medical imaging [3].

The essential principle underlying blockchain technologies is that they are a branch of distributed ledger technology (DLT). By disassembling DLT, we can achieve distributed by establishing a peer-to-peer network of nodes, or computers; these nodes collectively constitute a distributed network. Every node handles client-submitted transaction. These transactions are recorded as committed data on all nodes in a replicated database, which is referred to as the ledger. The records in this ledger are organized into blocks and are not subject to change [4].

What is consensus algorithm and what are all the variants still used in this field: Categories of consensus algorithms are non-byzantine based, byzantine based, DAG based and hybrid and its subcategories [5,6].

Combining machine learning with blockchain encryption provides a talented approach for image forgery detection, ensuring data integrity. The use of blockchain technology allows image data to be securely stored on a decentralized and tamper-resistant ledger. Each image is assigned to a hash, which is stored on the blockchain, making it difficult for unauthorized modifications to go undetected. Machine learning techniques are hired to extract relevant features from the images, capturing unique characteristics that aid in detecting forgeries. Trained machine learning models, such as extreme learning machine (ELM) or deep learning models, learn patterns and characteristics from a labeled dataset of authentic and forged images. The extracted features and hash values and timestamp of both images on the blockchain ensure transparency and immutability. When a new image is presented, its features are related with those of authentic images, and a classification algorithm determines its authenticity based on learned patterns. This hybrid approach can be integrated with blockchain technology and extracted feature values along with generated timestamp and hash values from each image used to provide an immutable and decentralized solution for verifying the authenticity of images. In the context of image forgery detection, a consensus mechanism can provide additional security and reliability when extracting timestamps and hash values from images.

The digital images which we have processed here are skin cancer healthy and unhealthy images for showing the integrity of the same images. The main problem in digital images forgery is to prove image authenticity. To prove this image authenticity, a combination of block chain technology will help us for the effective role here. As you know, blockchain in the sense, give hashing-based security thereby assuring image authenticity.

To prove the network security services, security features such as timestamp, hash, nonce, and proof values have been collected for this proposed work. Skin cancer image forgery has been processed for this research work. A small forgery in skin cancer images may lead patient diagnosis wrong and it may lead the patients in trouble in the initial stage itself. So, there must be an authentication proving system that is required. Here, authentication system is the hybridization of machine learning and blockchain technology. So initially the proposed system will process pre-trained convolutional neural network (CNN) models along with ELM and further the system will continue with blockchain modified version for the generation of security features. This decentralization reduces the risk of malicious tampering. If a node tries to manipulate the extracted values, it will be rejected by the consensus algorithm, ensuring the integrity of the data.

10.2 Problem definition

Image authenticity and correlation between image authenticity is used to detect forged images. Hashing algorithm results show varying results in the case of small changes happened in the image. Image forgery prediction will help you understand the region of forgery that has been done, measures values and values like yes or no. But we could not identify whether the image is the actual one or modified. Therefore, an authenticator framework is required to check (hash, timestamp, nonce, proof values) to ensure data integrity. Data integrity is nothing but to ensure that the data which is with us has not been modified.

10.3 Proposed system and its block diagram

Figure 10.1 clearly explains the proposed system details, which consists of skin cancer dataset and the processing steps. The skin cancer dataset consists of forged and genuine images; further images have been processed for pre-processing to remove the noises from image(s). After that, pre-trained CNN models have been processed for feature extraction. As you know, there is no need for feature extraction if we are using CNN models. During research period, we have processed extraction of features based on CNN models to reduce the overfitting problems. The same has also been applied here and those extracted features are compared based on the extracted features of test image(s).

Compared results will get noted and the features listed file will get input of classifiers to check the classification accuracy to detect whether the image is genuine or manipulated. For genuine and manipulated images, we have used labels as Yes/No. Wireless sensor network (WSN) images have redrawn to show the humans in the form of wireless nodes. Humans are represented here as healthcare professionals and patients. Here, hybridized machine learning and blockchain techniques have been used to provide integrity and authenticity.

Image forgery or manipulation may happen during the transmission of report or images to hospital or insurance agent. They may manipulate image(s) or manipulate extracted features. Classifiers are used to check the authenticity prediction. Additionally, a consensus mechanism is added to check the authentic data to ensure data integrity. There is a module named Authenticator to check each image(s)'s (hash values, timestamp) and (hash values, timestamp, nonce, proof values). Nonce is nothing but a number used only once method to give security. It will be processed both in original and forged images, to get to know whether the image is to be rejected or accepted.

Decentralized consensus refers to the process by which a group of distributed entities or participants in a network reach an agreement or consensus without relying on a central authority. It involves achieving agreement on a shared state or

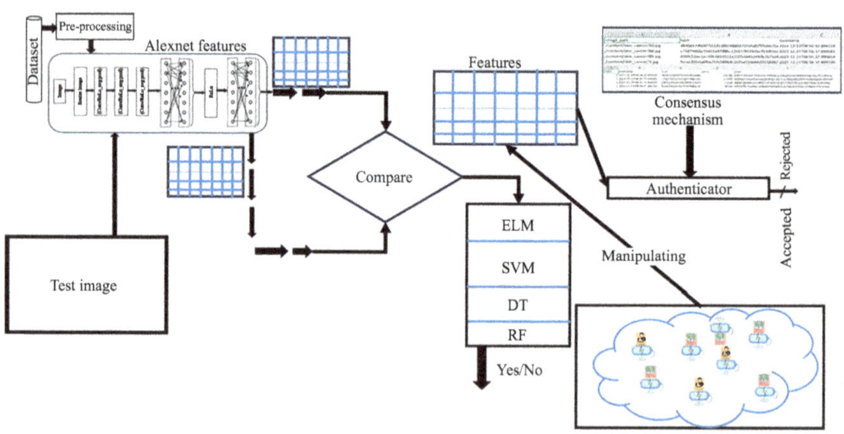

Figure 10.1 Proposed diagram based on wireless sensor network scenario

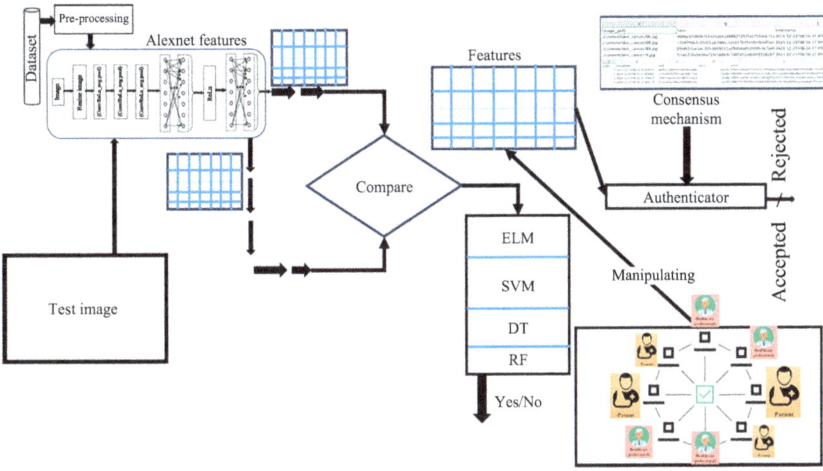

Figure 10.2 Proposed diagram based on de-centralized consensus scenario

decision through the collaboration and coordination of multiple nodes in the network. Decentralized consensus mechanisms, such as blockchain technology, allow participants to validate and agree upon transactions or data without the need for a trusted central entity. Consensus protocols consist of proof of work (PoW)/proof of stake (PoS), and these are used to ensure agreement and security in decentralized networks.

WSNs are a kind of network of small, low-power, and wireless devices called sensors that are deployed to monitor and collect meaningful data from the physical environment. These sensors can be spread across a large area and communicate wirelessly to relay information to a central location/other node in the network. WSNs are commonly used in several applications, including environmental monitoring, industrial automation, smart cities, and healthcare. They enable the collection of real-time data from the physical world, which can be mainly for analysis, decision-making, and control.

In a WSN, people are typically involved in the setup, configuration, and management of the network rather than directly transmitting data.

1. Data transmission in WSN: In WSN, sensors denoted as nodes which can collect data from the surrounding areas and using wireless technology transmission are connected to a central base station or other sensor-nodes in the network.
2. Data collection: Sensor like nodes gather raw facts from their respective sensing capabilities, such as temperature, humidity, or motion sensors, depending on the application. The data collected is stored locally within the nodes.
3. Data aggregation: To conserve energy and reduce network traffic, sensor nodes often perform data aggregation. They combine or summarize the data collected before transmitting it to the base station. Aggregation can involve statistical calculations, filtering, or fusion techniques.
4. Routing: Sensor nodes determine the optimal path to transmit the gathered data to the base station or other target nodes. There are lots of types of router's process protocols that are available and out of that low energy adaptive

clustering hierarchy (LEACH) is commonly used to establish efficient communication routes within the network.

5. Data transmission: Once the route is determined, the sensor nodes transmit the aggregated data wirelessly using radio frequency communication. The base station or target nodes receive and process the transmitted data.

Normal and manipulating data transmission: In a WSN, the data transmitted by sensor nodes is typically assumed to be genuine and accurate. However, it is possible for data transmission to be manipulated or compromised, leading to potential security or integrity issues.

1. Data integrity attacks: Malicious individuals or adversaries may attempt to manipulate the data being transmitted within the WSN. This can involve altering sensor readings, injecting false data, or tampering with the aggregation process. Such attacks can make to inaccurate or misleading information being received by the base station or other nodes.
2. Secure data transmission: To mitigate data manipulation risks, security measures can be implemented. For instance, encryption techniques can be applied to ensure the confidentiality and integrity of the data during transmission. Authentication mechanisms can also be employed to verify the identity of the sensor nodes and ensure that only reliable sensor-nodes can participate in the network.

Above networks can be used to manage digital images and facilitate communication among hospital employees. In healthcare settings, WSNs can be utilized to capture and transmit digital images from medical devices such as X-ray machines, ultrasound scanners, or endoscopy systems. WSNs can be designed to securely transmit these images wirelessly to a central server or storage system for further analysis, diagnosis, or archival purposes. The WSN can consist of sensor nodes equipped with image sensors that capture the images and transmit them to a designated node within this type of network. From there, the images can be processed, stored, and made accessible to healthcare professionals for review and analysis.

WSNs can also facilitate communication and information sharing among hospital employees. For instance, wearable devices equipped with wireless sensors can be used to monitor the health status of hospital staff such as their heart rate, temperature, or location. This data can be transmitted wirelessly within the WSN to a central monitoring system, allowing supervisors or para medico professionals to monitor the well-being and safety of the employees. Cloud platforms offer scalable and secure storage solutions that can manage large volumes of medical images. Additionally, cloud-based processing capabilities can be utilized for tasks like image analysis, feature extraction, or applying machine learning algorithms.

10.4 Types of consensus algorithm

Validation-based consensus algorithms are a type of consensus algorithm where participants in a distributed system validate transactions or blocks based on specific rules and criteria. These algorithms typically rely on the expertise, reputation, or voting power of the participants to reach consensus details outlined in Table 10.1. Here are a few examples:

Table 10.1 Types of consensus algorithm and its survey [7–19]

Sl. no.	Types of consensus algorithm	Types	Survey article's summary
1	Practical Byzantine Fault Tolerance (PBFT)	PBFT is a validation-based consensus algorithm designed to endure Byzantine faults. In PBFT, a leader is chosen among the nodes, and a series of rounds of message exchange and voting takes place to reach consensus on the validity of transactions. Nodes authorize and vote on proposed transactions, and a threshold of votes is compulsory for consensus to be reached. PBFT is a consensus algorithm that can be applied to healthcare blockchain systems to ensure the integrity and consensus of data among multiple participants.	"Secure and Efficient Data Sharing Framework for Healthcare Using Consortium Blockchain" (2019): This survey explores the use of PBFT-based consortium blockchain for secure and efficient data sharing in healthcare. It highlights the advantages of PBFT consensus in terms of scalability, fault tolerance, and resistance to attacks, which are crucial requirements in the healthcare domain. "A Review of Blockchain Consensus Algorithms and Their Application in Healthcare" (2020): This survey provides an overview of various consensus algorithms, including PBFT, and their application in healthcare blockchain systems. It discusses the benefits and challenges of using PBFT in healthcare such as its ability to handle Byzantine faults and maintain data consistency. "Blockchain-based Electronic Health Record System for Healthcare Data Sharing" (2020): This survey examines the use of PBFT consensus in a blockchain-based electronic health record (EHR) system. It discusses the advantages of PBFT, such as fast consensus and fault tolerance, in ensuring data integrity and privacy in healthcare data sharing scenarios. "A Survey on Blockchain Technology for Secure IoT-based Healthcare Applications" (2021): This survey explores the integration of blockchain and IoT in healthcare applications, focusing on PBFT-based consensus algorithms. It discusses the benefits of PBFT in achieving trust, security, and transparency in IoT-enabled healthcare systems.

(Continues)

Table 10.1 (Continued)

Sl. no.	Types of consensus algorithm	Types	Survey article's summary
2	Ripple Consensus Algorithm (RCA)	RCA is the consensus algorithm used in the Ripple payment protocol. It trusts on a network of trusted authorities to validate and agree on the order of transactions. Validators are chosen by the Ripple network, and they join in a voting process to determine the validity and order of transactions. RCA is a consensus algorithm used in the Ripple network, which is a decentralized payment protocol and cryptocurrency.	"Blockchain Technology for Healthcare Data Management: A Review" (2019): This survey provides an overview of blockchain technology in healthcare, including a discussion on the use of RCA in healthcare blockchain systems. It explores the benefits and challenges of applying RCA for secure and efficient healthcare data management. "Blockchain-Based Healthcare Information Exchange: A Survey" (2020): This survey examines the use of blockchain in healthcare information exchange, with a focus on different consensus algorithms, including RCA. It discusses the potential of RCA in ensuring data integrity, interoperability, and privacy in healthcare information exchange scenarios. "A Review of Blockchain Consensus Algorithms and Their Applications in Health Information Exchange" (2020): This survey reviews various consensus algorithms, including RCA, and their applications in health information exchange. It discusses the advantages and limitations of RCA in terms of scalability, security, and transaction speed for healthcare data sharing. "A Comprehensive Study on Blockchain Technology for Healthcare: A Review" (2021): This survey provides a comprehensive study on blockchain technology in healthcare, covering various aspects including consensus algorithms. It explores the potential of RCA in healthcare blockchain systems for secure and interoperable data exchange among healthcare providers.

(Continues)

3	Stellar Consensus Protocol (SCP)	SCP is the consensus algorithm used in the Stellar blockchain network. It is a federated Byzantine agreement protocol that relies on a group of trusted nodes called "quorum slices" to validate and agree on the state of the network. Quorum slices are subsets of the network that include enough trusted nodes to reach consensus.	"Stellar Consensus Protocol: A Review and Potential Applications in Healthcare" (2020): This survey provides an overview of the Stellar Consensus Protocol and its potential applications in healthcare. It discusses the benefits of SCP in terms of scalability, security, and consensus algorithm efficiency for healthcare blockchain systems. "Exploring Stellar Consensus Protocol for Secure and Interoperable Healthcare Data Exchange" (2021): This survey explores the use of Stellar Consensus Protocol for secure and interoperable healthcare data exchange. It discusses the advantages of SCP in ensuring data integrity, privacy, and interoperability in healthcare blockchain networks. "A Comparative Study of Consensus Algorithms in Healthcare Blockchain Networks: Stellar Consensus Protocol vs. Proof of Authority" (2019): This survey compares the Stellar Consensus Protocol with the Proof of Authority consensus algorithm in healthcare blockchain networks. It evaluates their performance, security, and scalability for healthcare data management. "Enhancing Healthcare Data Security with Stellar Consensus Protocol-based Blockchain" (2020): This survey explores how Stellar Consensus Protocol-based blockchain can enhance healthcare data security. It discusses the features of SCP that contribute to secure and tamper-resistant healthcare data storage and sharing.
4	Multi-signature (Multisig): Multisig is a validation-based consensus mechanism used in blockchain systems where multiple parties must sign off on a transaction for it to be considered valid. This mechanism is commonly used in cryptocurrency wallets or smart contracts, where multiple parties must collectively approve a transaction before it can be executed.		

(Continues)

Table 10.1 (Continued)

Sl. no.	Types of consensus algorithm	Types	Survey article's summary
5	Voting-based consensus algorithms	rely on a voting process among the participants in a distributed system to reach consensus on the validity or ordering of transactions.	Proof of Stake (PoS) [20]: While PoS is primarily a validation-based algorithm, it also incorporates voting. In PoS, participants who hold and "stake" a certain amount of cryptocurrency are selected to validate transactions and create new blocks.
6	Authentication-based consensus algorithms	are a type of consensus algorithm that rely on participants proving their identity or authentication credentials to participate in the consensus process. These algorithms ensure that only authenticated and authorized participants can contribute to the consensus mechanism.	Proof of authority (PoA): PoA is an authentication-based consensus algorithm where a predetermined set of approved validators or authorities are responsible for validating transactions and adding blocks to the blockchain [20]. Practical Byzantine Fault Tolerance: PBFT, although primarily a validation-based algorithm, also incorporates authentication. Identity-based Consensus [21]: Identity-based consensus algorithms use participants' unique identities as a basis for authentication. Threshold signature schemes: Threshold signature schemes involve a group of participants jointly creating a signature or cryptographic proof.

10.5 Experiments and results

Work is mainly based on image manipulation forgery detection and further it has utilized consensus algorithm for proving authenticity and integrity network services. ALEXNET pre-trained CNN models connected with the last layer of fully connected layer and all retrieved features for machine learning classifiers to obtain classification accuracy findings and forgery localization. In Chapter 1, a detailed explanation of image manipulation has been provided. Here in this chapter, we have noted some security parameters that have processed to get has, timestamp, nonce and proof values. Figure 10.1 describes image path along with its name, hash values, and timestamp values. All the images are used for testing in the following cases:

A) The consensus algorithm is utilized to extract values of Figures 10.1 and 10.2 before checking the manipulation.
B) The consensus algorithm utilized to extract values of Figure 10.1 and 10.2 after checking the manipulation.

Image manipulation has been processed based on MIASDBv1 dataset. We have used our own manual paint tool for creating and collecting original and forged images (Figures 10.3 and 10.4; Table 10.2).

In Figure 10.5, features are extracted using the CNN ALEXNET pre-trained model and its fully connected last layer considers csv file of all features and gains 96.50% of classification accuracy. Classification accuracy achieved is very less because our manual own paint tool which is not that much effective like photoshop or gimp or other image editing tools. This tool is used for creating only forged images for research level testing process. Out of four models, ALEXNET performed well and it outperforms other pre-trained CNN models (Table 10.3).

A	B	C
image_path	hash	timestamp
/content/skin_cancer/90.jpg	d646e57db097551d1d6f2488027292fab755bb671a	2023-12-23T08:56:37.894319
/content/skin_cancer/88.jpg	c72d74665cf5d11a47d8bc12c037655b16cfb3df3ed	2023-12-23T08:56:37.894481
/content/skin_cancer/89.jpg	03bfc53ae1ec1053859512a25054a452499b767bd4	2023-12-23T08:56:37.894614
/content/skin_cancer/9.jpg	fccac29fa5e06a71fe58864c16f0a510d66d3158287	2023-12-23T08:56:37.894729

Figure 10.3 Extraction of hashes and timestamp

A	B	C	D	E
index	timestamp	hash	nonce	proof
0	2023-12-23T08:56:37.894319	d646e57db097551d1d6f24880	161784	00007e9916d772b54af9c10f9985aa390ab524de266b55553dcdbc255cdfb44d
1	2023-12-23T08:56:37.894481	c72d74665cf5d11a47d8bc12c(	115489	00009d6722ad03e5ea23e2830c3d2713ce84ebd6818692403023bde4f499429f
2	2023-12-23T08:56:37.894614	03bfc53ae1ec1053859512a25(	37286	0000416b099c101d8f2124f18257a14f4114932c6fc31cb9df7e0a12c56f3966
3	2023-12-23T08:56:37.894729	fccac29fa5e06a71fe58864c16	16744	00001f997f5d496cc030d93f99746fe3c5c76e2f023e4ac1ec8da4cff543c0ea

Figure 10.4 Extraction of timestamp, hash, nonce, proof values

Table 10.2 Pre-trained CNN models based on TensorFlow Hub URLs for COVERAGE, CoMoFoD datasets detection based on small and large datasets

Sl. no.	Models	Epochs	Dataset	Val Accuracy
1	RESNET	5	Small	58%
2	ALEXNET	5	Large	**62%**
3	RESNET 50	5	Small	60%
4	VGG16	5	Small	61%

	Model	Accuracy
0	CAT Boosting	96.50
1	Random Forest	94.41
2	Naive bayes	86.01
3	DTC	90.21
4	XGB	97.20
5	Ada Boosting	95.80
6	Bagging C	99.06
7	ET	93.01
8	Bernoullis	92.02
9	Passive Aggressive	97.18
10	SVM	96.71
11	KNN	96.48
12	LR	95.77
13	GB	97.20
14	LDA	94.41
15	ELM	96.50

Figure 10.5 Performance evaluation of the proposed approach using ML methods

Table 10.3 TensorFlow Hub URLs of CNN pre-trained tensor flow models for breast cancer forgery (copy move) detection based on small and large datasets (MIASDBv1 [13])

Sl. no.	Models	Epochs	Val accuracy
1	RESNET	5	60%
2	ALEXNET	5	**67%**
3	RESNET 50	5	66
4	VGG16	5	61

10.6 Conclusion and future work

A possible way for detecting image forgeries and maintaining data integrity is by integrating blockchain encryption with machine learning. Blockchain technology makes it possible to store image data safely on a decentralized, impenetrable ledger. Every image has a unique hash that is verified on the blockchain, making it more difficult for illegal changes to go unnoticed. From a labeled collection of authentic and bogus photographs, trained machine learning machines—like deep learning models or ELM—learn patterns and traits. Blockchain ensures transparency and immutability by keeping the extracted features hash values and timestamp of both photographs. The chance of malicious manipulation is decreased by this decentralization. The consensus mechanism ensures data integrity by refusing attempts by any node to modify the retrieved values. Out of four models, ALEXNET performed well and it outperforms other pre-trained CNN models. Additionally, authenticity has proved based on some security measures and details added in experimentation and results section.

References

[1] What is block chain technology, Available: https://aws.amazon.com/what-is/blockchain/?aws-products-all.sort-by=item.additionalFields.productNameLowercase&aws-products-all.sort-order=asc.

[2] A. Haleem, M. Javaid, R. P. Singh, R. Suman, and S. Rab, "Blockchain technology applications in healthcare: an overview", *International Journal of Intelligent Networks*, 2021;2:130–139. https://doi.org/10.1016/j.ijin.2021.09.005.

[3] M. P. McBee, and C. Wilcox "Blockchain technology: principles and applications in medical imaging", *Journal of Digital Imaging*, 2020;33(3):726–734. https://doi.org/10.1007/s10278-019-00310-3.

[4] M. Zand, X. Wu, and M. A. Morris *Hands-On Smart Contract Development with Hyperledger Fabric V2*, 2021. Sebastopol, CA: O'Reilly Media, Inc.

[5] Z. Hussein, M. A. Salama, and S. A. El-Rahman "Evolution of blockchain consensus algorithms: a review on the latest milestones of blockchain consensus algorithms", *Cybersecurity*, 2023;6:30. https://doi.org/10.1186/s42400-023-00163-y.

[6] J. Yusoff, Z. Mohamad, and M. Anuar "A review: consensus algorithms on blockchain", *Journal of Computer and Communications*, 2022;10:37–50. https://doi.org/10.4236/jcc.2022.109003.

[7] A. Ali, H. Ali, A. Saeed, *et al.*, "Blockchain-powered healthcare systems: Enhancing scalability and security with hybrid deep learning", *Sensors (Basel)*, 2023;23(18):7740. https://doi.org/10.3390/s23187740.

[8] J. Yusoff, Z. Mohamad, and M. Anuar, "A review: Consensus algorithms on blockchain", *Journal of Computer and Communications*, 2022;10(9):37–50. https://doi.org/10.4236/jcc.2022.109003.

[9] A. Dubovitskaya, F. Baig, Z. Xu, *et al.*, "ACTION-EHR: Patient-centric blockchain-based electronic health record data management for cancer care", *Journal of Medical Internet Research*, 2020;22(8):e13598. https://doi.org/10.2196/13598.

[10] M. S. Rahman, M. A. Islam, M. A. Uddin, and G. Stea, "A survey of blockchain-based IoT eHealthcare: Applications, research issues, and challenges", *Journal of Healthcare*, 2022;19:100551. https://doi.org/10.1016/j.iot.2022.100551.

[11] I. Yaqoob, S. Khaled, R. Jayaraman, and Y. Al-Hammadi, "Blockchain for healthcare data management: Opportunities, challenges, and future recommendations", *Neural Computing and Applications*, 2021;33(6):17213–17230. https://doi.org/10.1007/s00521-020-05519-w.

[12] Y. Zhuang, L. R. Sheets, Y. W. Chen, Z. Y. Shae, J. J. P. Tsai, and C. R. Shyu, "A patient-centric health information exchange framework using blockchain technology", *IEEE Journal of Biomedical and Health Informatics*, 2020;24(8):2169–2176. https://doi.org/10.1109/JBHI.2020.2993072.

[13] A. Tandon, A. Dhir, A. K. M. N. Islam, and M. Mäntymäki, "Blockchain in healthcare: A systematic literature review, synthesizing framework and future research agenda", *Computers in Industry*, 2020;122:103290. https://doi.org/10.1016/j.compind.2020.103290.

[14] A. Haleem, M. Javaid, R. P. Singh, R. Suman, and S. Rab, "Blockchain technology applications in healthcare: An overview", *International Journal of Intelligent Networks*, 2021;2:130–139. https://doi.org/10.1016/j.ijin.2021.09.005.

[15] D. Mazieres, "Stellar consensus protocol", *Stellar White Paper*. [Online]. Available: https://stellar.org/papers/stellar-consensus-protocol.

[16] W. Y. Ng, T. E. Tan, P. V. H. Movva, *et al.*, "Blockchain applications in health care for COVID-19 and beyond: A systematic review," *The Lancet Digital Health*, 2021;3(12):e819–e829. https://doi.org/10.1016/S2589-7500(21)00210-7.

[17] X. Zhu, "Consensus algorithms in blockchain: A survey to create decision trees for blockchain applications," *KTH Royal Institute of Technology*.

[Online]. Available: https://kth.diva-portal.org/smash/get/diva2:1772618/FULLTEXT01.pdf.

[18] N. Badri, L. Nasraoui, and L. A. Saïdane, "A comprehensive review of blockchain integration in remote patient monitoring for e-health," *International Journal of Network Management*, 2024;34(2):e2254. https://doi.org/10.1002/nem.2254.

[19] J. Suckling, J. Parker, D. Dance, *et al.* (2015) *Mammographic Image Analysis Society (MIAS) database v1.21. Apollo – University of Cambridge Repository.* https://doi.org/10.17863/CAM.105113.

[20] M. Kaur, M. Z. Khan, S. Gupta, A. Noorwali, C. Chakraborty, and S. K. Pani "MBCP: Performance analysis of large scale mainstream blockchain consensus protocols", *IEEE Access*, 2021;9:80931–80944. https://doi.org/10.1109/ACCESS.2021.3085187.

[21] M. A. Bouras, Q. Lu, S. Dhelim, and H. Ning "A lightweight blockchain-based IoT identity management approach", *Future Internet*, 2021;13(2):24. https://doi.org/10.3390/fi13020024.

Chapter 11

RF energy harvesting system for wearable health monitoring devices

Mounira Ben Yamna[1]*, Nabil Dakhli*[2]*, Hedi Sakli*[1,3] *and Mohamed Aoun*[1]

Abstract

This chapter introduces an efficient rectenna able to gather surrounding radio frequency (RF) power to charge wearable health monitoring (WHM) devices. A rectangular patch antenna is crafted utilizing computer simulation technology (CST), operating at 3.5 GHz, achieving notable specifications such as a gain of 2.49 dB and a reflection coefficient of −40 dB. Moreover, a simple serial harvester developed in Advanced Design System (ADS) with a single Schottky diode SMS7630, operating at 3.5 GHz, attains an impressive power conversion efficiency (PCE) of 63.5% at zero dBm RF input power with 300 Ω of a load resistor. The performance of the rectenna visualized in ADS shows promising results, with a reached PCE of 45% at 0 dBm and a DC (direct current) output voltage of 2.5 V, sufficient to power low-power medical devices. A novel approach to enhancing rectenna performance is outlined, demonstrating the effectiveness of leveraging circuit optimization within both ADS and CST. ADS yields optimal results in achieving impedance matching network and PCE for the rectifier circuit, while CST ensures superior performance in terms of both reflection coefficient and gain for the antenna. This optimization leads to increased output voltage and RF-to-DC conversion efficiency. This research demonstrates promising potentiality for enhancing the trustworthiness and effectiveness of WHM implements, providing consistent power sources and decreasing dependence on traditional batteries.

Keywords: RF energy harvesting; rectenna; antenna; rectifier; wearable health monitoring devices

[1]MACS Laboratory, National Engineering School of Gabes, University of Gabes, Tunisia
[2]Innov'COM Lab, Sup'COM, University of Carthage, Tunisia
[3]EITA Consulting, Montesson, France

11.1 Introduction

During the last years, an exponential increase in the demand for low-energy embedded systems and sensors has occurred [1], finding applications in various domains such as health surveillance, transport, environmental surveillance, and engineering [2]. In the medical healthcare field, sensors play a crucial role in remotely monitoring patients' conditions and transmitting health signals to relevant entities like doctors, nurses, and remote servers.

Wireless body area networks measure and record various healthcare parameters, including body temperature, blood pressure, heartbeat rate, electrocardiograms, sweat rate, and electro-dermal activity. Wearable health monitoring (WHM) and intelligent medical implants (IMI) (Figure 11.1) are anticipated to become integral to daily lives in the upcoming decade, offering unique scanning and sensing features compared to mobile phones and laptops [3]. These devices typically come equipped with communication links, providing users access to online information. This technology extends to doctors who can carry wearable devices, facilitating easy contact and location identification within a hospital setting.

Wireless technologies process and analyze the collected medical system data, which can be stored locally or transmitted to a medical center for further analysis. Wearable sensors play a pivotal role in collecting data subjected to analysis by medical software, triggering alerts for urgent healthcare treatment. wearable monitoring device (WDM) serves as vital tools for monitoring activities and accessories within medical centers, contributing to the efficient operation and monitoring of various companies. Their versatility extends to assisting diverse patients managing conditions like diabetes, asthma, epilepsy, and Alzheimer's disease. Wearable devices also show promise in addressing prevalent health issues such as sleep disorders, obesity, and cardiovascular diseases, playing a significant role in collecting valuable data for clinical research trials and studies [4].

The rapid proliferation of linked sensor systems has led to the emergence of the Internet of Things (IoT) in health maintenance [5]. However, a significant challenge faced by IoT sensors lies in their inherently low power capacity, typically functional only when powered by a battery. Recent research endeavors focus on enhancing energy efficiency and communication reliability of IoT sensor networks to extend their operational lifespan, considering energy harvesting as a promising solution. Energy harvesting involves capturing surrounding power from the air and transforming it into DC power to supply applications. Various methods are employed for energy harvesting, including solar energy, thermic power, and radio frequency (RF) power. Sun power has limitations, such as being unavailable at night, restricting its applicability in indoor settings like hospitals and smart homes [6]. RF energy, crucial for portable and internal medical equipment, significantly contributes to prolonging battery life without frequent replacements or recharging.

RF energy harvesting stands out as among the most appropriate and user-oriented methods for extracting energy from the environment [7]. The key element in RF energy harvesting is the rectenna. The rectenna is an integration of an

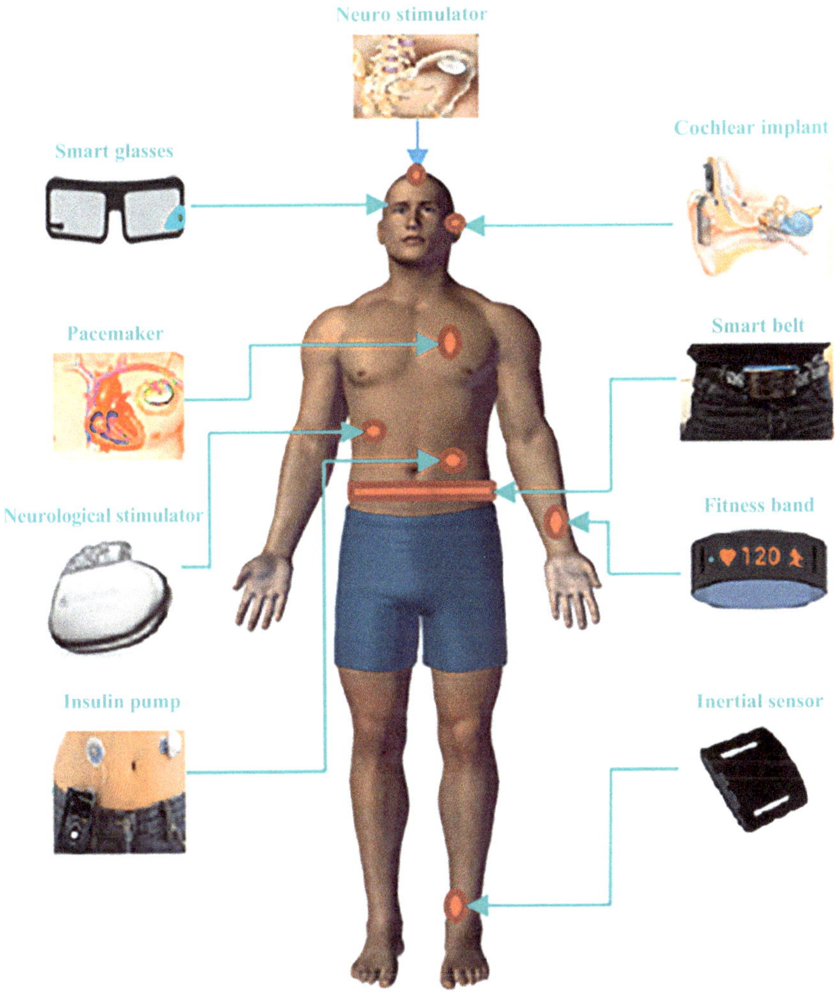

Figure 11.1 Various WHM and IMI devices are commonly employed in precision healthcare

antenna and rectifying circuit [8]. This process involves harvesting RF waves using an antenna and converting them into DC power through a rectifying circuit, commonly employing the patch antenna for optimal energy capture.

11.2 Related works

Alkhalaf *et al.* [9] introduced a concept for a rectenna functioning at 2.45 GHz, designed for wireless power transfer medical sensors. A patch was introduced on the human organism to improve the wireless power transfer (WPT). Hosain *et al.*

[10] created a novel rectenna incorporating a PIFAntenna and a doubler rectifier utilizing two Schottky diodes, operating at 0.915 MHz. The reached peak DC voltage is 7.5 V at a distance of 2 cm between the transmitter and receiver antennas, employing a 20 kΩ resistor. The produced DC energy was utilized to control the pulse generator of a deep brain simulation (DBS). The DBS is a promising treatment for neuropsychiatric disorders, requiring compact, wireless devices. The DBS system comprises three main components: an implantable pulse generator (IPG), conductive terminals, and a program coder. The IPG, implanted in the subclavicular or chest region, serves as the central signal generator for the therapy. Experimental results indicated the successful operation of the designed rectenna. The advantages of the proposed rectenna lie in enabling the DBS system to function not powered by a battery, consequently, this could contribute to reducing the cost of DBS procedures.

A rectenna and a source operating at 2.4 GHz, was developed by DeLong *et al.* [11]. A 47.7% of PCE was obtained. It delivered 1.2 mW over a span of 42 cm spanning from the transmitting to the receiving antennas. This system provides a sturdy replacement for coil-based position sensors commonly utilized in the medical field. Future prospects involve further miniaturization the rectifier, allowing implantation sub-dermally. This rectenna might then be integrated into an insulin pump.

A self-powered wearable sensor network has been developed by Yang *et al.* [12], featuring a triple band rectenna, a storage unit, a microcontroller, and communication unit. A triple-band rectifying circuit transforms harvested RF power into DC power, the reached peak of PCE is 59% at −10 dBm input power. This compact wearable sensor is well-suited for monitoring human body health. Lin *et al.* [13] proposed a wearable rectenna to power a medical device. Cordura fabric is selected as the textile material for wearable application. The system is crafted to operate at 2.45 GHz. The rectenna achieves a 2.2 V of output DC voltage over 40dB of input power (−40 dBm to 0 dBm), demonstrating excellent performance when utilized to supply wearable human sensor. Bouchoucha *et al.* [14] developed rectenna operating at 2.45 GHz. The rectifying circuit demonstrated a PCE of 72% at 0 dBm. This rectenna is particularly designed for supply implanted and connected medical devices. Abdi and Aliakbarian (2019) [15] developed a novel rectenna featuring a PIFAntenna and a single Schottky diode. For −20 dBm, the reached DC output voltage is 0.2V with 10 kΩ load resistor. The realized PCE is approximately 40%. This rectenna designed for power implanted healthcare devices.

11.3 Rectenna performance in health monitoring devices

As the utilization of IoT devices in healthcare continues to rise, including WHM and implanted medical sensors, the demand for a consistent power supply grows. Nevertheless, RF energy harvesting provides a potential answer to bridge the energy gap by capturing electromagnetic (EM) waves and converting it into RF signal then into DC signal.

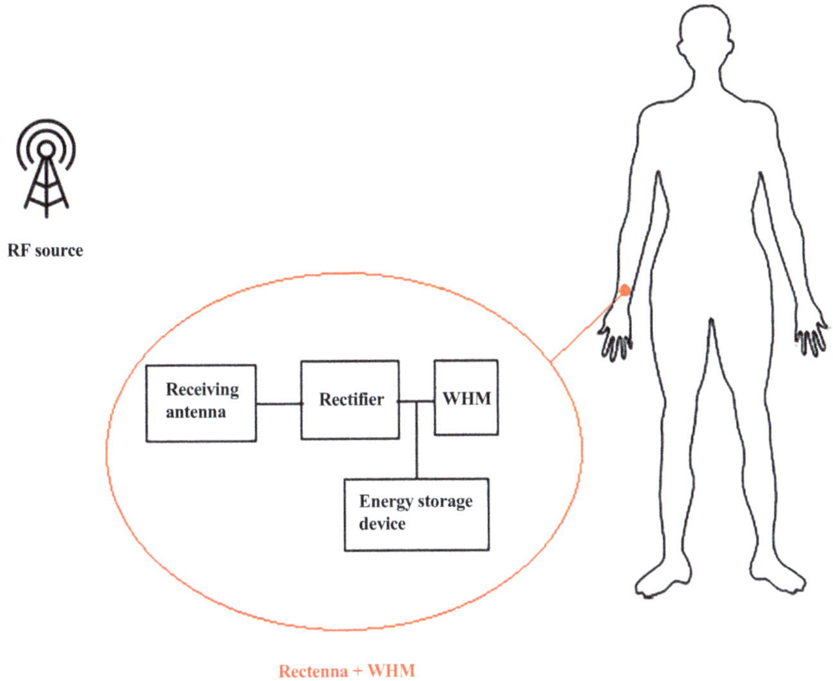

Figure 11.2 Suggested RF harvester for health monitoring devices

Figure 11.2 outlines the core elements of the suggested RF energy harvester. Typically, it comprises an antenna, a rectifying circuit, an energy control module, and a storage device. The primary functionalities of this system for IoT applications in health monitoring are:

- *Telemonitoring:* Through IoT-enabled wearable, streaming data regarding the status of a patient, tool effectiveness, and battery state is transmitted to medical professionals. It enables telemonitoring and facilitates prompt medical procedure in case of any anomalies or problem.
- *Alerting and emergency response:* During emergency circumstances, like the detection of anomalies or tool problems, IoT-enabled wearable transmit warning to medical personnel. This capability enables rapid reaction and possibly life-preserving measures.
- *Automatic data transmission:* Wearable medical devices with IoT abilities autonomously send data to medical personnel with no patient involvement. This streamlined procedure assures uninterrupted data gathering, leading to improved evaluation and intervention decisions.
- *Patient comfort:* IoT-enabled WMH devices improve patient comfort by excluding the necessity of repeated on-site consultations, enhancing convenience for patients while ensuring efficient control of their health status.

- *Preventive maintenance:* IoT technology enables the ongoing monitoring of connected medical wearables and their performance. Through the analysis of data patterns, medical professionals can preventively detect foreseen problems or anticipate the necessity for device repair or substitution. This approach minimizes the possibility of unforeseen deficiencies.

These advancements are geared toward enhancing patient safety, diminishing medical consultation, and finally enhancing the patient's well-being. Nonetheless, IoT abilities in WHM devices necessitate additional power for tasks such as Information analysis, wireless transmission, and detection. Stabilizing the energy demands of these attributes with the imperative to optimize battery lifespan can present a complex challenge. The RF energy harvesting offers a cost-effective solution and ensures uninterrupted power supply for IoT healthcare devices, rendering it a performant energy source. Typically, to transfer data to external peripherals, IoT wearable medical devices utilize low energy wireless communication protocols like Zigbee or Bluetooth Low Energy.

11.4 Rectenna design

The rectenna is the primary component in the electric field (E-field) and magnetic field (H-field) (EH) system. It includes an antenna and a rectifying circuit. The antenna captures the EM waves present in the ambient air and converts them into an RF signal. The rectifier transforms this RF signal into a DC signal. The rectifier comprises a high-frequency (HF) filter, a rectifying circuit such as a Schottky diode or transistor, a DC filter, and a storage/load element (Figure 11.3) [16].

11.4.1 Antenna design

The suggested antenna is depicted in the Figure 11.4, featuring dimensions of 40×40 mm^2, operating at 3.5 GHz. It is positioned on the FR4 substrate. The antenna comprises a radiating element with a square configuration, a complete ground plane, and a 50 Ω microstrip feed line. The patch dimension is 26×20 mm^2. The inset length is 5 mm, and the inset gap is 1.5 mm. The length and the width of the feed are 10 and 3 mm, respectively.

The antenna was simulated using computer simulation technology (CST) simulator, which uses the Finite Integral Technique. The reflection coefficient S_{11} is illustrated in Figure 11.5, enabling us to visualize the matching level and bandwidth at the desired resonance frequency for this antenna. In CST, the optimization tool, commonly referred to as the "Optimizer," utilizes numerical optimization algorithms to adjust model parameters to minimize or maximize a predefined

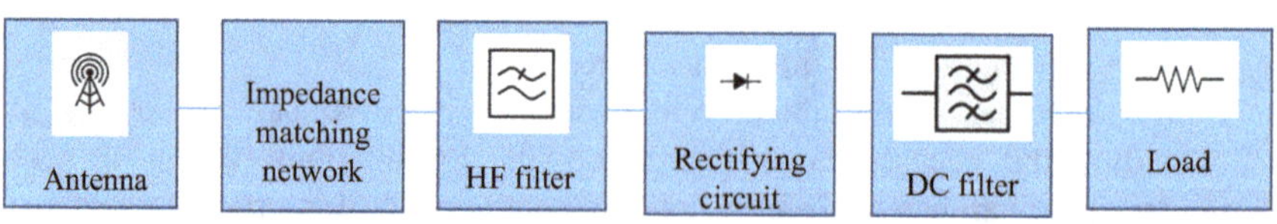

Figure 11.3 The rectenna structure

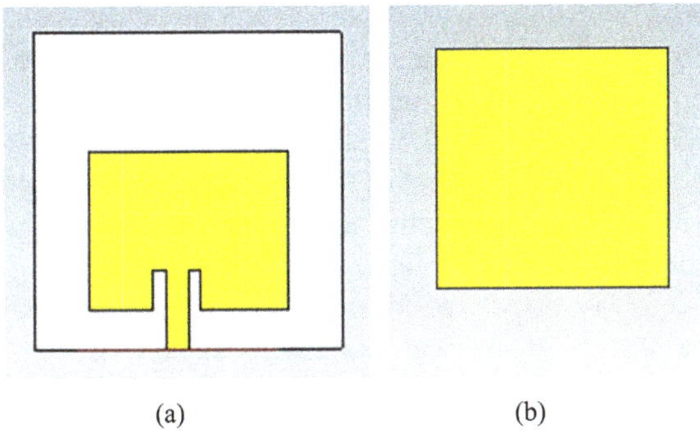

Figure 11.4 Structure of the proposed antenna in (a) frontal perspective, and (b) back perspective.

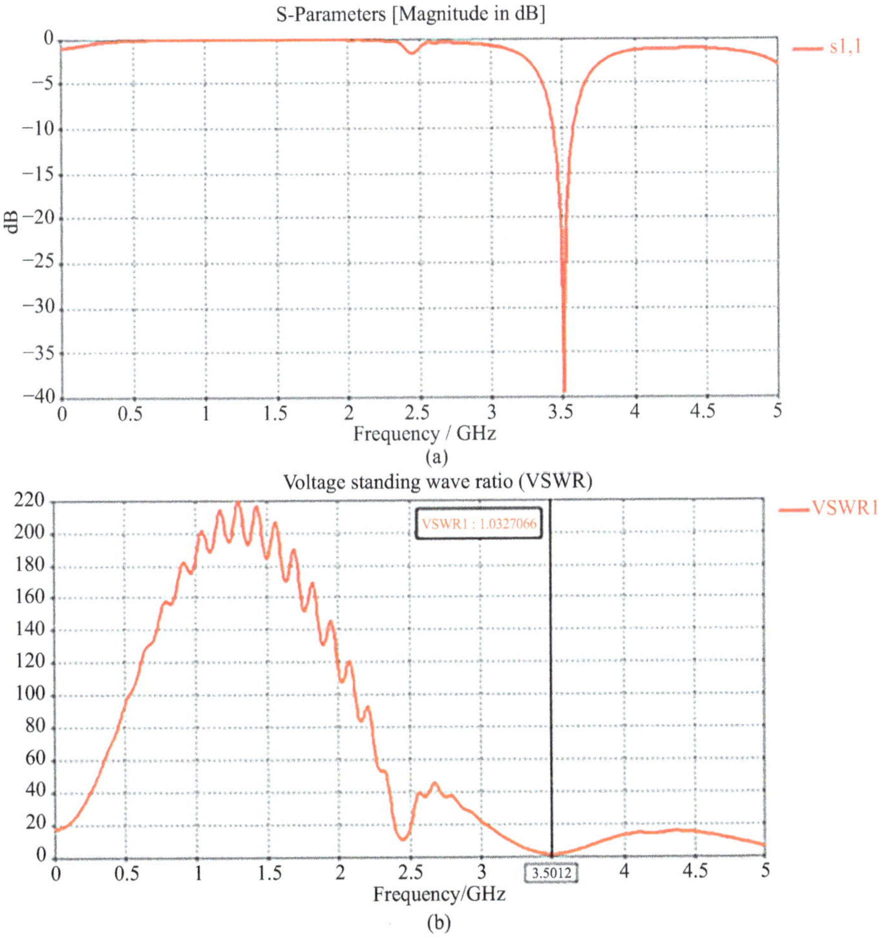

Figure 11.5 (a) S_{11} coefficient and (b) VSWR.

objective function. CST offers several optimization methods, each tailored to specific types of problems. Gradient descent is the method used in this case; it enables us to select the optimal dimensions mentioned previously.

The reflection coefficient at 3.5 GHz is equal to −40 dB, as shown in Figure 11.5, indicating a good match. The VSWR quantifies the efficiency of power transmission between the antenna and transmission line (TL), with an optimal value being below 1.2, in our antenna case VSWR is equal to 1.03. According to Figure 11.6, we observe that the 3D radiation patterns and polar plots of the simulated antenna using CST are nearly omnidirectional.

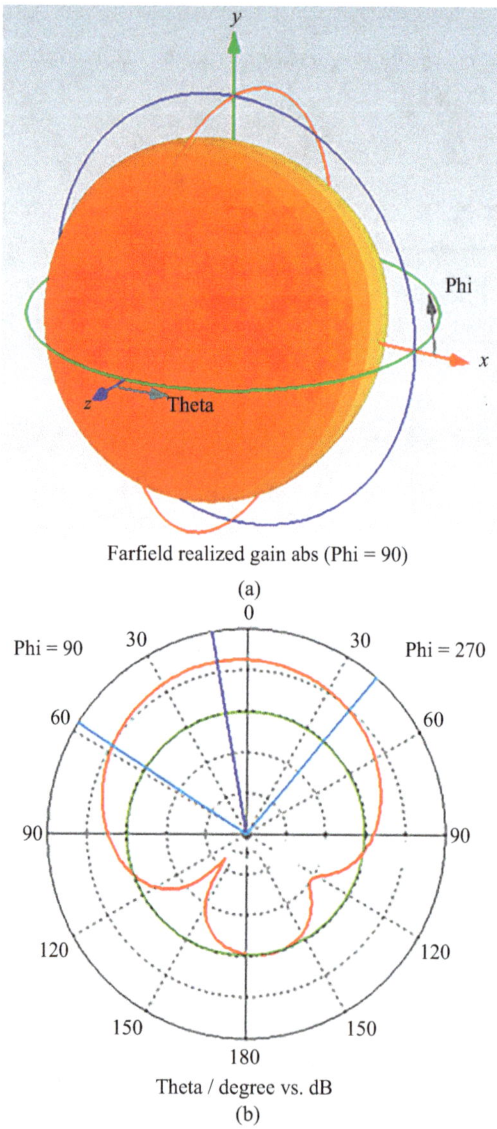

Figure 11.6 The pattern radiation of the antenna: (a) 3D representation and (b) polar representation

11.4.2 Rectifier design

A simple serial rectifier is proposed, implemented through microstrip technology, operates at 3.5 GHz (Figure 11.7). The selected substrate is composed of FR4 substrate characterized by a thickness (h) measuring 1.6 mm and dielectric properties $\varepsilon = 4.4$, $\tan\delta = 0.02$. The structure includes an impedance matching network (IMN), a HF filter, a Schottky diode, a DC filter, and a storage/load component. The design and simulation of the harvesting circuit's performance, specifically in terms of PCE and reflection coefficient concerning RF input power and frequency, are carried out using Keysight Advanced Design System (ADS) software.

In this section, we illustrate the importance of leveraging the optimization functionalities within ADS to achieve optimal impedance and performance for the circuit. This process entails conducting iterative simulations, employing trial and error methods to fulfill predefined performance criteria. These objectives encompass achieving optimal impedance matching, maximizing DC voltage, and enhancing PCE. By systematically refining our circuit design through iterative optimization, we aim to attain superior performance characteristics, ensuring our system operates at its full potential.

The ADS software employs an iterative optimization algorithm method to fine-tune the lengths of TL1, TL2, and TL3, systematically varying them and evaluating the resulting DC output power. This iterative process adjusts the parameters until it converges to the optimal combination that yields the best reflection coefficient. By continuously refining the values through successive iterations, the algorithm effectively navigates through the parameter space to identify the configuration that maximizes the system's performance. This approach ensures that the rectenna circuit achieves optimal efficiency and output power, enhancing its overall effectiveness in wireless power harvesting applications.

For the rectifier design, the optimal dimensions of the TLs are outlined in Table 11.1.

A rectifier performance is evaluated by its PCE as defined below.

$$PCE\ (\%) = \frac{P_{DC}}{P_{in}} \times 100 \tag{11.1}$$

where P_{DC} and P_{in} are the DC power and the RF power, respectively.

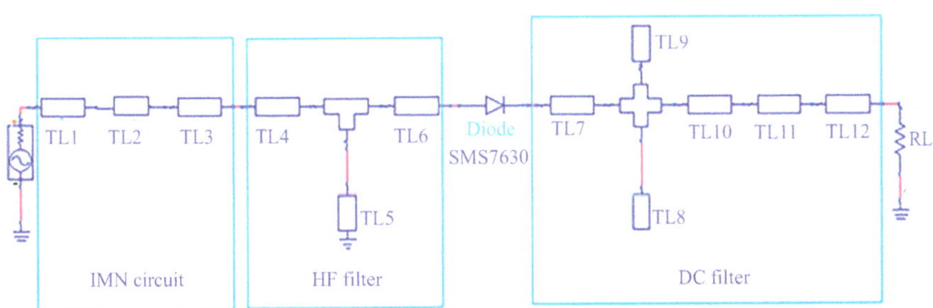

Figure 11.7 Rectifier design

Table 11.1 Dimensions of the transmission lines (TL) used in the proposed rectifier

TL	Width (mm)	Length (mm)
TL1	1	9.4
TL2	4.7	8.5
TL3	4.7	2.4
TL4	3	3
TL5	3	11
TL6	3	5.1
TL7	3	2.9
TL8	3	12
TL9	3	5
TL10	3	3
TL11	17.3	11.5
TL12	3	3

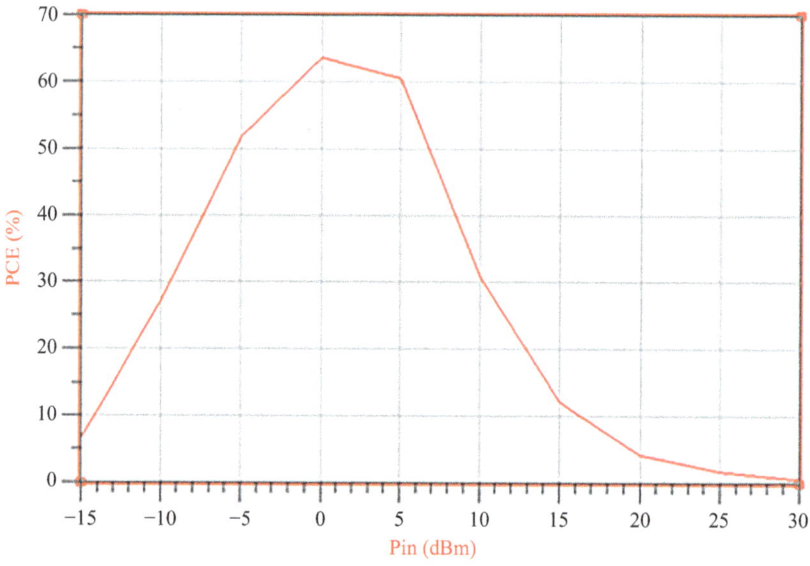

Figure 11.8 The efficiency of the suggested rectifying system vs. P_{in}

As shown in Figure 11.8, the maximum PCE reaches 63.5% for 0 dBm of RF input power and it is greater than 50% over 13dB (from −5 dBm to 8 dBm).

11.4.2.1 HF input filter

It operates as a bandpass filter strategically located before the diode to inhibit the reflection of harmonics produced by the rectifier back toward the antenna. The HF input filter consists of a $\frac{\lambda}{4}$line short-circuited to the ground (TL5).

11.4.2.2 DC output filter

It functions as a low-pass filter, designed to block both the fundamental and harmonics, to grant passage just the DC component. The structure of the proposed DC filter is depicted in Figure 11.2a, incorporating four micro-strip lines (TL8, TL9, TL10, TL11). TL8 is configured as a $\lambda/4$ TL, terminated with an open circuit. This design ensures effective short-circuit behavior at both the fundamental and odd-order harmonics. TL9 is configured as an $\lambda/8$ line, open-circuited. It can be inferred that TL9 effectively blocks the 2nd, 6th, 10th, and 14th harmonics, and so forth. TL10, being a $\lambda/16$ line, is designed to block the 4th harmonic. Meanwhile, TL11, as a quarter-wavelength line and with a wide structure, functions akin to a parallel capacitor, contributing to ripple smoothing on the output voltage.

11.4.2.3 The choice of the Schottky diode

The Schottky diode plays a crucial role in the rectifier, serving a critical function in the RF-DC conversion circuit and overall rectifier performance. Consequently, the careful selection of an appropriate diode is imperative. The corresponding representation of the Schottky diode is illustrated in Figure 11.9.

A Schottky diode is characterized by its breakdown voltage B_v, series resistance R_s, zero bias junction capacitance C_{j0}, and resistance junction R_j. Considering the power level of the rectifier, the maximum power capacity of the diode ($\mathrm{Pc}_{\max}$) is influenced by the breakdown voltage $\mathrm{B_v}$ and the load R_L.

$$\text{which:} \quad \mathrm{Pc}_{\max} = \frac{Bv^2}{4 \times RL} \tag{11.2}$$

Beyond this power level ($\mathrm{Pc}_{\max}$), the rectifier becomes vulnerable to breakdown, resulting in a substantial degradation in performance. Based on the observations in Figure 11.10, it can be deduced that a Schottky diode with a low breakdown voltage and a high load is preferable for low-power rectifiers. Conversely, for high-power rectifiers, a Schottky diode with a higher breakdown voltage and a low load resistance proves to be more suitable [17]. In our case, the SMS-7630 Schottky diode from Skyworks is opted, which boasts a breakdown voltage of 2V. This diode is paired with a resistor load of 1,600 Ω to meet the low power requirements.

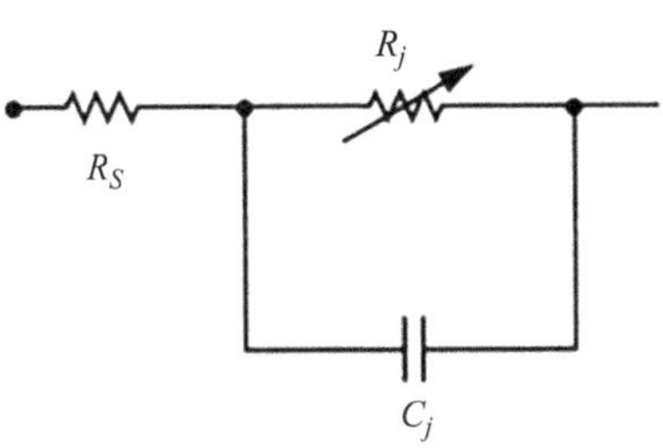

Figure 11.9 Equivalent electrical model of Schottky diode

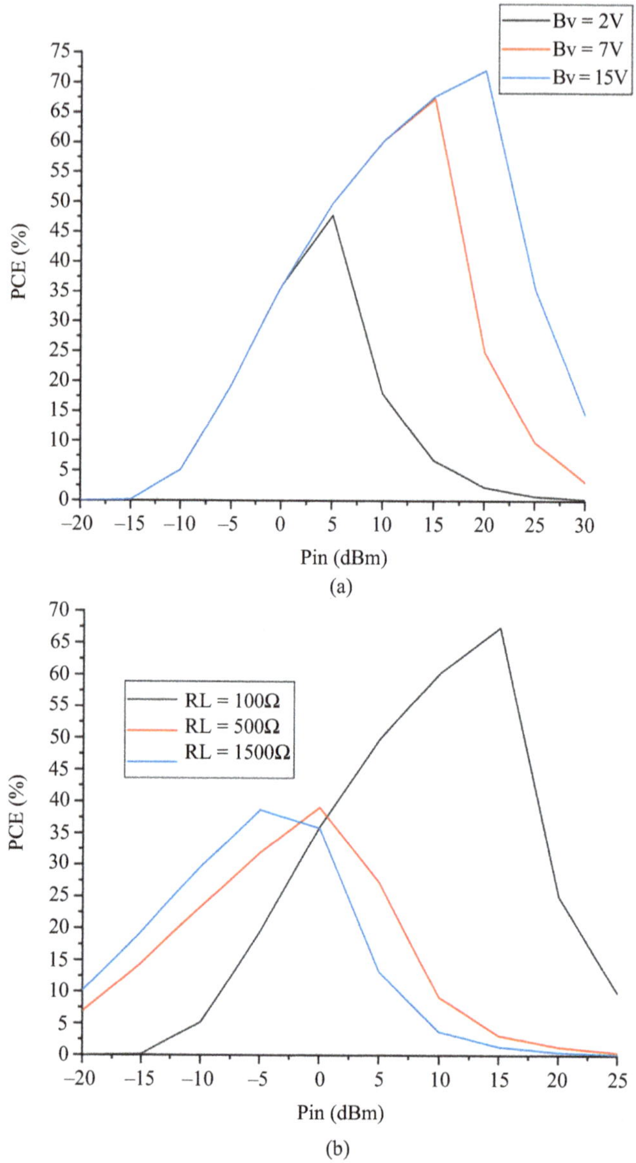

Figure 11.10 Variation of (a) B_v and (b) the load on rectifier efficiency

11.4.2.4 Impedance matching network

The IMN is composed of three transmission lines (TL1, TL2, TL3), to ensure an optimal energy transfer from the antenna to rectifying system. Visualizing the reflection coefficient S_{11} of the rectifier at 3.5 GHz in Figure 11.11, it is clear that the circuit is good matched, which $S_{11} < -35$ dB.

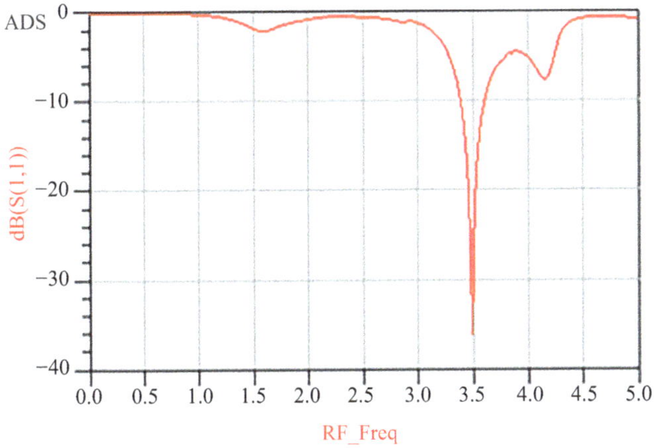

Figure 11.11 Reflection coefficient of the proposed rectifier

11.5 Rectenna performances

This section is dedicated to examining the competence of the rectenna with regard to efficiency and DC voltage. Understanding the efficiency of the rectenna is crucial in evaluating its performance in transforming received EM waves into utilizable electrical power. The efficiency of the rectenna is the proportion between the DC power by the RF power acquired by the receiving antenna.

With the goal of assess the received power at the receiving antenna, we employed the Friis transmission equation. Utilizing the same antenna configuration as previously described for both transmission and reception scenarios, with a fixed separation distance of 100 mm between them, enabled consistent comparative analysis. By applying the Friis equation as shown in Figure 11.12, which accounts for factors such as distance and antenna characteristics, we obtained a comprehensive understanding of the power transfer efficiency across the system. This approach ensures a rigorous evaluation of the rectenna's performance under standardized conditions, facilitating accurate assessment and meaningful comparisons.

The Friss transmission equation is written as follows:

$$P_r = \frac{P_T \times G_T \times G_r \times {\lambda_0}^2}{(4 \times \pi \times R)^2}, \tag{11.3}$$

where (P_T, G_T) are the transmitting power and the gain of the transmitter antenna, respectively; and (P_r, G_r) are the receiving power and the gain of the receiver antenna; R is the distance between the two antennas and λ_0 is the wavelength.

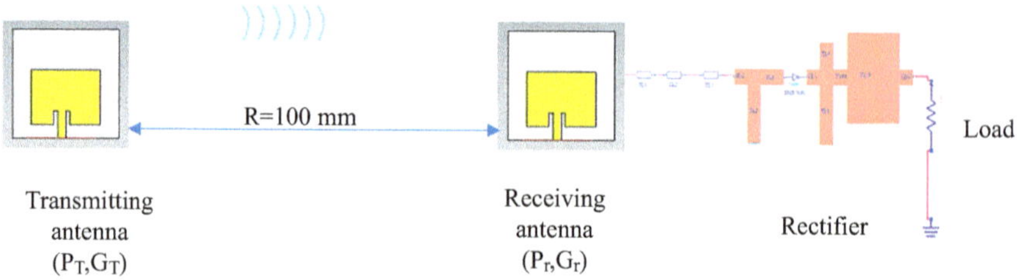

Figure 11.12 Friis transmission model

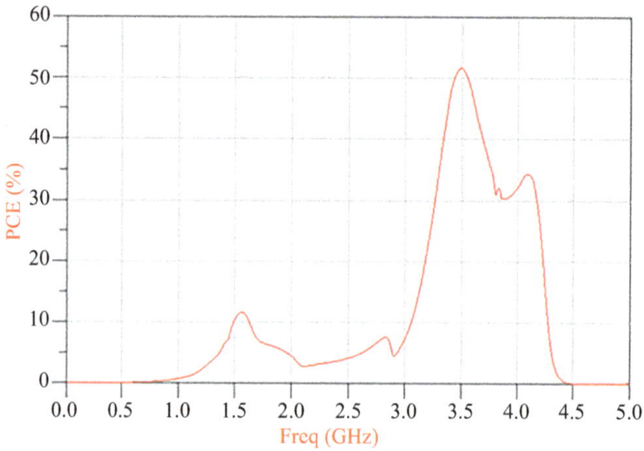

Figure 11.13 PCE of the proposed rectenna vs. frequency

The simulation results in ADS show that the maximum PCE reached at 0 dBm is 51.7% at 3.5 GHz (Figure 11.13) and the DC voltage is 2.4 V sufficient to supply low power WHM devices.

11.6 Conclusion

This study introduces a better version rectenna that is made for wearable health monitors that don't need batteries. Using rectennas to gather energy seems like a good way to power these devices without batteries, which could help patients in the long run. The new design of the rectenna performs very well at 3.5 GHz. It includes a rectangular patch antenna with 2.49 dB of gain and return loss (S_{11}) equal to −40 dB. The rectifier circuit that converts harvested energy also works impressively, with an efficiency of 63.5% at 0 dB at 3.5 GHz. The overall rectenna able to convert 51.7% of RF power into DC power at 0 dBm, and it manages to generate a steady DC voltage of 2.4 V at 3.5 GHz. It seems very suitable for harvesting energy for WHM devices.

References

[1] Jameel, F., Ristaniemi, T., Khan, I., and Lee, B. M., "Simultaneous harvest-and-transmit ambient backscatter communications under Rayleigh fading", *EURASIP Journal on Wireless Communications and Networking*, 166(1), 1–9, 2019.

[2] Ovsthus, K., and Kristensen, L. M. "An industrial perspective on wireless sensor networks—A survey of requirements, protocols, and challenges", *IEEE Communications Surveys & Tutorials*, 16(3), 1391–1412, 2014.

[3] Khan, R. A., and Pathan, A. S. K., "The state-of-the-art wireless body area sensor networks: A survey", *International Journal of Distributed Sensor Networks*, 14(4), 1550147718768994, 2018.

[4] Sabban, A. "Green wearable sensors for medical, energy harvesting, communication, and IoT systems". Sabban, A. (ed.), In *Advances in Green Electronics Technologies*. IntechOpen; 2023. http://dx.doi.org/10.5772/intechopen.112352.

[5] Jameel, F., Duan, R., Chang, Z., Liljemark, A., Ristaniemi, T., and Jantti, R., "*Applications of backscatter communications for healthcare networks*", *IEEE Network*, 33(6), 50–57. https://doi.org/10.1109/MNET.001.1900109.

[6] Perera, T. D. P., Jayakody, D. N. K., Sharma, S. K., Chatzinotas, S., and Li, J., "Simultaneous wireless information and power transfer (SWIPT): Recent advances and future challenges", *IEEE Communications Surveys & Tutorials*, 20(1), 264–302, 2017.

[7] Alkhalaf, H. Y., Ahmad, M. Y., and Ramiah, H., "Design of rectifier circuit to harvest the RF energy for wearable medical devices". In *Kuala Lumpur International Conference on Biomedical Engineering* (pp. 381–388). Cham: Springer International Publishing, 2021.

[8] Surender, D., Halimi, M. A., Khan, T., Talukdar, F. A., Koul, S. K., and Antar, Y. M., "2.45 GHz Wi-Fi band operated circularly polarized rectenna for RF energy harvesting in smart city applications", *Journal of Electromagnetic Waves and Applications*, 36(3), 407–423, 2022.

[9] Yahya Alkhalaf, H., Yazed Ahmad, M., and Ramiah, H., "Self-sustainable biomedical devices powered by RF energy: A review". *Sensors*, 22(17), 6371, 2022.

[10] Hosain, M. K., Kouzani, A. Z., Tye, S. J., Abulseoud, O. A., Amiet, A., Galehdar, A., Berk, M, "Development of a compact rectenna for wireless powering of a head-mountable deep brain stimulation device", *IEEE Journal of Translational Engineering in Health and Medicine*, 2, 1–13, 2014.

[11] DeLong, B. J., Kiourti, A., and Volakis, J. L., "A radiating near-field patch rectenna for wireless power transfer to medical implants at 2.4 GHz", *IEEE Journal of Electromagnetics, RF and Microwaves in Medicine and Biology*, 2(1), 64–69, 2018.

[12] Yang, L., Zhou, Y. J., Zhang, C., Yang, X. M., Yang, X. X., and Tan, C., "Compact multiband wireless energy harvesting based battery-free body area

networks sensor for mobile healthcare", *IEEE Journal of Electromagnetics, RF and Microwaves in Medicine and Biology*, 2(2), 109–115, 2018.

[13] Lin, C. H., Chiu, C. W., and Gong, J. Y., "A wearable rectenna to harvest low-power RF energy for wireless healthcare applications". In *2018 11th International Congress on Image and Signal Processing, BioMedical Engineering and Informatics (CISP-BMEI)* (pp. 1–5). IEEE, 2018.

[14] Bouchoucha, Y., Omri, D., and Aguili, T., "Efficient ISM band rectenna for RF energy harvesting in IoT healthcare devices". In *2023 International Conference on Communications, Computing, Cybersecurity, and Informatics (CCCI)* (pp. 1–8). IEEE, 2023.

[15] Abdi, A., and Aliakbarian, H., "A miniaturized UHF-band rectenna for power transmission to deep-body implantable devices", *IEEE Journal of Translational Engineering in Health and Medicine*, 7, 1–11, 2019.

[16] Wu, P., Chen, Y. D., Zhou, W., Ren, Z. H., and Huang, S. Y., "A wide dynamic range rectifier array based on automatic input power distribution technique", *IEEE Microwave and Wireless Components Letters*, 30(4), 437–440, 2020.

[17] Halimi, M. A., Khan, T., Kishk, A. A., and Antar, Y. M., "Rectifier circuits for RF energy harvesting and wireless power transfer applications: A comprehensive review based on operating conditions", *IEEE Microwave Magazine*, 24(1), 46–61, 2022.

Chapter 12

Telemedicine and remote patient monitoring with AI

Anita Mohanty[1], Ambarish G. Mohapatra[1] and Subrat Kumar Mohanty[2]

Abstract

In this chapter, we take a thorough look at the impact that artificial intelligence (AI) has had on the fields of telemedicine and remote patient monitoring, catalyzing a revolutionary transformation in the healthcare delivery landscape. Using a thorough analysis, we expose the various ways in which the integration of AI technologies in these domains has the potential to improve patient outcomes, increase healthcare accessibility, and maximize the effective use of resources. We also explore a wide range of AI-driven applications, including smart teleconsultations, advanced remote health monitoring, and predictive analytics. As change agents, these applications are essential for improving the precision of diagnoses, creating individualized treatment regimens, and increasing the general effectiveness of medical procedures. This chapter aims to highlight the transformational potential that technology brings to the forefront of patient care and well-being by clarifying the complex interactions between AI and healthcare.

Keywords: Telemedicine; remote patient monitoring; artificial intelligence; predictive analytics; virtual consultations; wearable devices; patient engagement

12.1 Introduction

Recently, the healthcare landscape has undergone a transformative shift, fueled by technological advancements and the incorporation of artificial intelligence (AI) into medical practices [1]. Telemedicine, a cornerstone of this evolution, enables

[1]Electronics Engineering, Silicon University, India
[2]IQAC & Academics, Templecity Institute of Technology and Engineering, India

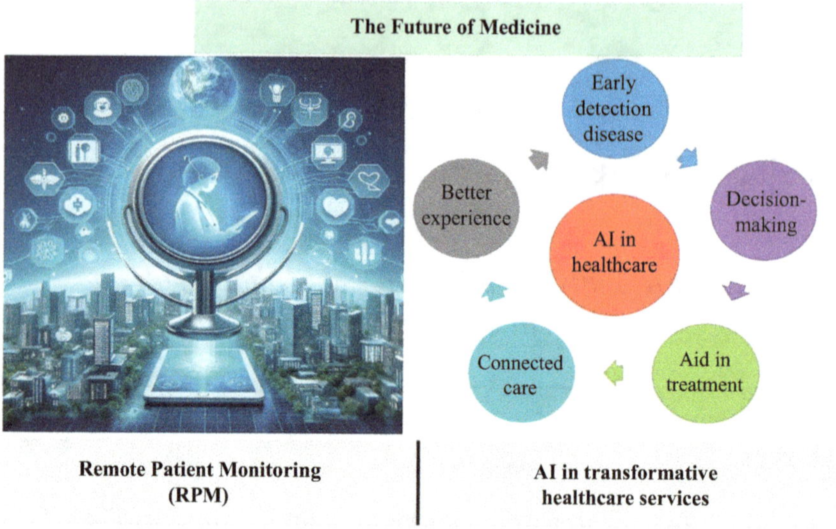

Figure 12.1 AI-driven healthcare and the future of medicine

the remote delivery of healthcare services, transcending geographical barriers and facilitating timely access to medical expertise. Leveraging communication technologies such as video conferencing and secure messaging platforms, telemedicine not only enhances patient convenience but has also proven invaluable in situations where immediate, in-person care is challenging. This newfound accessibility empowers patients to connect with healthcare experts seamlessly, minimizing the necessity for physical appointments and ensuring continuous care delivery. Figure 12.1 highlights the insight of future medicine and the role of AI in delivering smart healthcare-based solutions. The major aspects such as early detection of diseases, decision-making, connected healthcare practices, and better user experience are portrayed in Figure 12.1.

(a) Remote patient monitoring: real-time insights into patient health

Remote patient monitoring (RPM) complements telemedicine by using wearables and sensors for real-time health data (e.g., vital signs). RPM enables early anomaly detection, prompt intervention, and personalized treatment, enhanced by AI for advanced analysis and predictive modeling. This integration fosters holistic healthcare and patient participation [2].

(b) The synergy of AI in transformative healthcare services

The convergence of telemedicine, RPM, and AI revolutionizes healthcare delivery. AI algorithms analyze extensive remote monitoring datasets for early disease detection and treatment optimization [3]. AI-powered virtual assistants streamline administrative tasks and enhance patient-provider communication. While offering improved outcomes and cost savings, these technologies pose data security and regulatory challenges, requiring balanced innovation and privacy safeguards.

12.1.1 Overview of telemedicine and remote patient monitoring

In recent years, the healthcare industry has significantly transformed with telemedicine and RPM. These technologies have become essential for providing accessible, efficient, and patient-centered healthcare services. Telemedicine [4] breaks geographical barriers by using telecommunications for remote healthcare delivery, including virtual consultations and secure messaging, which is crucial during pandemics or in rural areas with limited healthcare infrastructure. RPM complements telemedicine through connected devices and sensors that provide real-time health insights, enabling remote monitoring, early anomaly detection, and tailored treatments [5]. This is particularly valuable for managing chronic conditions, post-surgery recovery, and preventive care. Additionally, AI enhances these technologies by analyzing vast datasets, predicting disease progression, and aiding personalized treatments. AI-powered virtual assistants streamline tasks and improve patient communication [6]. Despite the promise of enhanced efficiency and patient-centric care, challenges such as data security, regulatory compliance, and equitable access must be addressed.

12.1.2 Growing significance in the era of digital healthcare

The rise of digital healthcare is prominently marked by the expanding roles of telemedicine and RPM, which are reshaping conventional healthcare models and introducing a new era of patient-centric, accessible, and efficient healthcare services. Telemedicine is pivotal in improving healthcare accessibility by overcoming geographical barriers, facilitating virtual consultations, and ensuring real-time communication through secure platforms [7]. It plays a crucial role in remote diagnosis and treatment, particularly valuable during public health crises, while enhancing patient care through convenient access to medical advice and follow-ups from home.

RPM empowers patients in managing their health by leveraging wearable devices and sensors to continuously track vital signs and chronic conditions, providing real-time health data insights [8]. This capability supports early detection and personalized interventions, enabling healthcare providers to optimize treatment strategies and deliver proactive, preventive care measures.

The integration of AI and data analytics amplifies the impact of telemedicine and RPM, enabling AI to analyze monitoring data for predictive insights and personalized treatments. AI-driven virtual assistants further enhance patient engagement and operational efficiency, promising a technologically advanced and patient-centered digital healthcare environment [9].

12.2 AI in teleconsultations

AI is revolutionizing teleconsultations, enhancing healthcare delivery with increased efficiency, accuracy, and personalized patient care. AI's integration in teleconsultations empowers healthcare professionals and enhances patient experiences, creating a more sophisticated and accessible healthcare ecosystem. AI improves diagnostics and decision support by analyzing medical data and symptoms, reducing errors, ensuring timely interventions, and providing precise diagnoses and treatment recommendations during remote sessions. AI-powered virtual health assistants offer personalized support, answering questions, providing medication details, and offering post-consultation guidance. These assistants use natural language processing (NLP) and machine learning (ML) to boost patient engagement and treatment adherence [10,11]. Additionally, AI employs predictive analytics to analyze patient data and lifestyle factors, predicting health risks and enabling proactive, personalized preventive strategies. However, challenges such as data privacy, ethics, and regulatory compliance must be addressed. Secure data handling, addressing AI biases, and transparent patient communication are critical to balancing innovation with ethical responsibilities.

12.2.1 Intelligent virtual consultations powered by AI

AI's integration into healthcare marks a new era of intelligent virtual consultations, promising personalized, efficient, and globally accessible healthcare [12,13]. This transformative approach harnesses advanced AI to provide unrestricted online medical access, facilitating expert advice at users' fingertips. Intelligent virtual consultations leverage AI to enhance diagnostic accuracy and decision-making for healthcare providers. By swiftly analyzing symptoms and diagnostic data, AI reduces errors and enhances treatment effectiveness during virtual sessions, thereby improving overall patient care.

AI-powered virtual health assistants engage directly with patients using NLP and ML, offering medical information, answering queries, and supporting post-consultation care tailored to individual needs. This interaction enhances patient engagement, treatment adherence, and health outcomes in virtual settings. Integrating predictive analytics into these consultations enables proactive healthcare management, predicting risks and recommending preventive actions based on patient data and lifestyle factors.

However, addressing ethical challenges such as patient privacy protection, minimizing AI biases, and ensuring transparent communication is crucial. Balancing innovation with ethical considerations ensures these advancements enhance healthcare accessibility while fostering patient trust. AI-driven developments promise a future of personalized, efficient, and patient-centered healthcare, transcending geographical barriers to expert medical guidance.

12.2.2 Natural language processing for enhancing communication

In the ever-evolving landscape of technology, NLP has emerged as a revolutionary tool, transforming how we communicate and interact with information. NLP, a subfield of AI, focuses on enabling machines to understand, interpret, and generate human-like language [14]. Its applications span various domains, profoundly enhancing communication between humans and machines.

NLP facilitates seamless human-machine interaction by enabling computers to understand and respond to natural language. Advancements in speech recognition and language understanding have led to intelligent virtual assistants, chatbots, and voice-activated systems, making technology more accessible through intuitive communication methods. Additionally, NLP improves accessibility and inclusivity by breaking communication barriers and understanding diverse linguistic expressions, accents, and speech patterns. In fields like healthcare, education, and customer service, NLP-driven applications create user-centric tools that cater to diverse language proficiencies and disabilities. By grasping context, sentiment, and user intent, NLP enhances personalization in recommendation systems, targeted advertising, and content customization. Despite significant strides, NLP faces challenges such as biases, privacy issues, and contextual nuances, necessitating responsible development for a future with refined language models and advanced conversational agents.

12.2.3 Case studies demonstrating successful AI-driven telehealth platforms

1. Babylon Health: AI-enhanced symptom checker and virtual consultations

Overview: Babylon Health, a UK-based telehealth platform, incorporates AI into its services to offer a sophisticated symptom checker and virtual consultations. The AI-powered symptom checker utilizes NLP to understand user inputs about their symptoms. It provides personalized health assessments, recommends appropriate actions, and guides users to relevant resources. Additionally, Babylon Health offers virtual consultations with healthcare professionals, where AI assists clinicians by analyzing patient data, medical history, and relevant literature to support decision-making [15].

Success factors:

- **Enhanced triage**: Babylon's AI-driven symptom checker assists users in determining the urgency of their health concerns, facilitating efficient triage and reducing unnecessary emergency room visits.
- **Time-efficient consultations**: The integration of AI in virtual consultations aids healthcare providers in quickly accessing relevant patient information, leading to more focused and time-efficient interactions.
- **Improved access to care**: By offering 24/7 access to AI-driven symptom checking and virtual consultations, Babylon Health enhances healthcare accessibility, particularly for users in remote or underserved areas.

2. Ada Health: AI-powered health assessment and navigation

Overview: Ada Health, a Berlin-based telehealth platform, employs AI to provide users with personalized health assessments and navigation. The platform's AI-driven chatbot interacts with users, asking detailed questions about their symptoms, medical history, and lifestyle. The AI analyzes this information to generate a comprehensive health assessment and offers guidance on potential conditions, preventive measures, and appropriate next steps [16]. Ada Health's emphasis on user education and empowerment distinguishes it as an AI-driven platform for informed decision-making.

Success factors:

- **User empowerment**: Ada Health's AI-driven chatbot empowers users by providing detailed insights into their health, fostering a sense of autonomy and informed decision-making.
- **Preventive health guidance**: The AI system not only assists in identifying existing health concerns but also offers personalized recommendations for preventive measures and lifestyle modifications.
- **Seamless integration with healthcare providers**: Ada Health's platform facilitates seamless communication between users and healthcare providers, streamlining the process of sharing health assessments and facilitating virtual consultations when necessary.

3. TytoCare: AI-integrated remote examination tools

Overview: TytoCare is a telehealth platform that combines hardware and AI to enable remote medical examinations. The platform provides users with a handheld examination device equipped with various sensors, allowing them to capture vital signs, and images of the throat, ears, skin, and lung sounds. The collected data is then analyzed by AI algorithms to assist healthcare providers in diagnosing and treating various conditions remotely [17]. TytoCare's AI integration enhances the precision and reliability of remote examinations.

Success factors:

- **Comprehensive remote examinations**: TytoCare's AI-powered examination device enables users to conduct thorough remote examinations, contributing to a more comprehensive virtual healthcare experience.
- **Enhanced diagnostics**: The integration of AI aids healthcare providers in interpreting examination data, leading to more accurate diagnostics and informed treatment decisions.
- **Reduced need for in-person visits**: TytoCare's platform reduces the necessity for in-person visits, particularly for routine examinations, allowing for timely interventions and improved healthcare accessibility.

12.3 Predictive analytics for remote health monitoring

Remote health monitoring integrates predictive analytics, revolutionizing healthcare with proactive, personalized care. Platforms such as Philips eCareCoordinator and Current Health leverage analytics from wearables and health records to detect trends and early signs of deterioration, enabling timely interventions. These models enhance personalized care and optimize resource allocation, improving outcomes across chronic disease management, hospitalization prevention, and mental health support [18].

12.3.1 Applications of AI in predicting health outcomes

AI has emerged as a powerful tool in predicting health outcomes [19] across various domains of healthcare. Some notable applications include:

1. **Disease diagnosis and prognosis:** AI algorithms can analyze medical data, including symptoms, medical history, and test results, to accurately diagnose diseases and predict their progression. For example, AI-powered systems can assist in diagnosing conditions like cancer, cardiovascular diseases, and neurological disorders, enabling early intervention and better management.
2. **Personalized treatment planning:** By analyzing patient data, such as genetic information, lifestyle factors, and treatment history, AI can recommend personalized treatment plans tailored to individual patients. This approach enhances treatment efficacy, reduces adverse effects, and improves patient outcomes.
3. **Risk stratification and prevention:** AI models can assess an individual's risk of developing certain health conditions based on their demographics, lifestyle factors, and medical history. By identifying high-risk individuals, healthcare providers can implement targeted preventive measures and interventions to mitigate the risk of disease onset.
4. **Remote patient monitoring:** AI-enabled remote monitoring devices can continuously collect and analyze patient data, such as vital signs, activity levels, and medication adherence, in real time. This allows healthcare providers to remotely monitor patients' health status, detect early warning signs of deterioration, and intervene promptly to prevent adverse outcomes.
5. **Drug discovery and development:** AI algorithms can analyze vast amounts of biological data to identify potential drug targets, predict drug efficacy, and optimize drug design. This accelerates the drug discovery and development process, leading to the development of novel therapeutics for various diseases.
6. **Predictive analytics in healthcare operations:** AI-driven predictive analytics can optimize healthcare operations by forecasting patient admissions, resource utilization, and staffing needs. By anticipating demand and optimizing workflows, healthcare facilities can improve efficiency, reduce wait times, and enhance patient satisfaction.
7. **Public health surveillance:** AI-based models can analyze epidemiological data, social media trends, and other sources of information to detect disease outbreaks, monitor the spread of infectious diseases, and forecast healthcare needs. This enables timely public health interventions and resource allocation to mitigate the impact of outbreaks.

Figure 12.2 depicts AI's diverse applications in predictive analytics for remote health monitoring. Centered on "Applications of AI in predicting health outcomes," it integrates disease diagnosis, personalized treatment planning, risk stratification, remote monitoring, drug discovery, healthcare operations, and public health surveillance. These functionalities advance proactive, personalized care by optimizing resource allocation and providing real-time insights. While AI promises significant

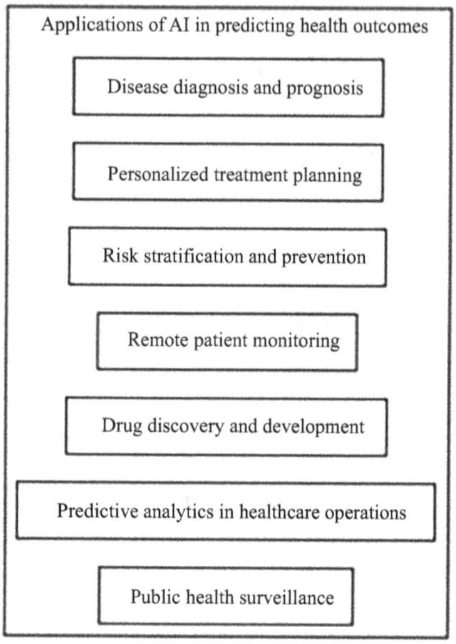

Figure 12.2 Interconnection of various applications of AI in predicting health outcomes within the context of predictive analytics for remote health monitoring

benefits in healthcare, ongoing research, validation, and integration are crucial for maximizing its potential and improving patient and population health outcomes.

12.3.2 Monitoring chronic conditions through predictive modeling

Monitoring chronic conditions through predictive modeling involves leveraging data and algorithms to anticipate disease progression, exacerbations, and related outcomes [20]. Figure 12.3 shows the different steps through which predictive modeling works to monitor chronic conditions. Here's how it typically works:

1. **Data collection:** Relevant data sources include electronic health records (EHRs), medical imaging, lab results, patient-reported outcomes, wearable devices, and environmental factors. These data points provide a comprehensive view of the patient's health status and risk factors.
2. **Feature selection:** Data pre-processing techniques are applied to clean and pre-process the data. Feature selection methods are then used to identify the most relevant variables or features that contribute to predicting outcomes related to chronic conditions. These features may include demographic information, medical history, biomarkers, and lifestyle factors.
3. **Model development:** Predictive models, such as ML algorithms, are trained using historical data to learn patterns and relationships between the selected

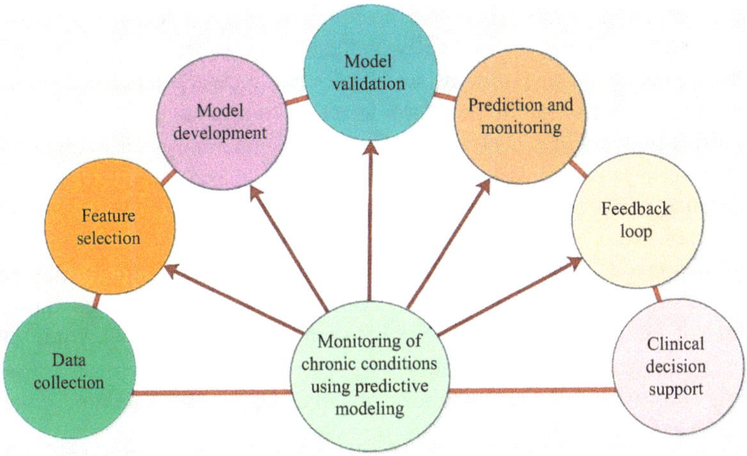

Figure 12.3 Different steps predictive modeling follows to monitor chronic conditions

features and health outcomes. Common algorithms include logistic regression, random forests, support vector machines, and neural networks. The choice of algorithm depends on the specific characteristics of the data and the prediction task.

4. **Model validation:** The performance of the predictive model is evaluated using validation techniques such as cross-validation, holdout validation, or bootstrapping. This ensures that the model generalizes well to new, unseen data and is not overfitting or underfitting the training data.
5. **Prediction and monitoring:** Once the predictive model is validated, it can be deployed to predict future health outcomes for individual patients. Patients at higher risk of disease progression or adverse events can be identified early, allowing healthcare providers to intervene proactively with personalized interventions, such as medication adjustments, lifestyle modifications, or care management programs.
6. **Feedback loop:** Continuous monitoring and evaluation of the predictive model's performance are essential to ensure its accuracy and relevance over time. Feedback from healthcare providers and patients can be used to refine the model and improve its predictive capabilities.
7. **Clinical decision support:** Predictive modeling results can be integrated into clinical workflows to provide decision support to healthcare providers. Alerts and notifications can be generated to flag patients at high risk, prompting timely interventions and preventive measures.

12.3.3 Real-time data analysis for early intervention

Real-time data analysis is essential for early intervention in chronic conditions, utilizing advanced predictive modeling techniques to continuously monitor patient data. This approach allows healthcare providers to swiftly detect deviations from baseline health parameters, potentially indicating deterioration [21]. Continuous monitoring of vital signs, symptoms, and health parameters offers a comprehensive view of patient health, enabling prompt detection of anomalies. Predictive modeling identifies patterns in real-time data that signal potential health complications, prompting timely intervention. Customized alerts based on individual patient characteristics notify providers when specific thresholds are exceeded, facilitating proactive care. ML algorithms forecast future health outcomes, identifying at-risk patients to prevent adverse events. Integration with remote monitoring devices enables real-time data capture and analysis, supporting continuous monitoring regardless of location. Actionable insights from real-time analysis guide clinical decisions, such as medication adjustments, lifestyle changes, or specialist referrals, ultimately improving patient outcomes, reducing hospitalizations, and enhancing quality of life for patients with chronic illnesses.

12.4 Wearable devices and AI

The integration of wearable devices and AI is revolutionizing telemedicine and RPM by enhancing personalized care through continuous, real-time health data. Wearables like smartwatches track vital signs, activity, and sleep patterns, sending this data to AI algorithms for health status analysis [22]. AI detects subtle changes in metrics, enabling early identification of health issues such as irregular heart rates or disrupted sleep, prompting timely intervention. By analyzing wearable data, AI offers personalized health advice, suggesting lifestyle changes, medication adjustments, or referrals based on individual health profiles. AI-powered telemedicine platforms facilitate remote diagnostics and consultations, aiding in diagnosing conditions, interpreting medical images, and recommending treatments. For chronic disease management, wearables and AI monitor conditions like diabetes and hypertension, analyzing data for trends, predicting complications, and optimizing treatment plans. These technologies also enhance patient engagement and adherence with personalized feedback, reminders, and incentives, improving involvement in their care. Additionally, AI-driven telemedicine expands healthcare access, especially in remote areas, reducing reliance on in-person visits and improving healthcare equity.

12.4.1 Integration of AI algorithms in wearable health tech

The integration of AI algorithms in wearable health technology revolutionizes RPM and healthcare delivery. AI-equipped wearables analyze physiological data like heart rate and activity levels in real time, detecting abnormalities early and providing timely health insights [23]. These algorithms offer personalized recommendations based on user data, considering factors such as age, medical history, and lifestyle. By analyzing trends in wearable data, AI predicts health outcomes and identifies risks before they manifest. AI-equipped wearables act as virtual health coaches, offering personalized guidance on staying active, improving sleep, and adhering to medications. They enable remote monitoring for chronic conditions, alerting healthcare providers to changes and facilitating timely interventions without in-person visits. AI learns from user data to enhance accuracy over time, refining predictions and recommendations, while ensuring data security with encryption and anonymization.

12.4.2 Remote monitoring of vital signs and health metrics

Remote monitoring of vital signs and health metrics is crucial for managing chronic conditions and enabling timely interventions [24]. Devices like wearable sensors continuously collect vital signs such as heart rate, blood pressure, blood glucose, oxygen saturation, and activity levels. This data is wirelessly transmitted to secure cloud platforms or healthcare systems, allowing remote access for healthcare providers. Specialized software and AI algorithms analyze patient data, detecting trends or anomalies and alerting providers to potential health issues. This enables timely responses, such as medication adjustments or telehealth consultations. Patients engage by accessing their health data, tracking progress, and making informed decisions with their providers. Research shows remote monitoring improves health outcomes, reduces hospitalizations, and optimizes chronic disease management, proving cost-effective by reducing office visits and hospital stays.

12.4.3 The role of AI in interpreting and contextualizing wearable data

The role of AI in interpreting and contextualizing wearable data is paramount for extracting meaningful insights and facilitating personalized healthcare interventions [25]. Table 12.1 shows how AI enhances this process:

Table 12.1 Role of AI in interpreting and contextualizing wearable data

Steps	Role	Explanation
1	Data analysis	AI algorithms analyze data from wearables—vital signs, activity, sleep—to detect patterns and anomalies indicating changes in health status.
2	Pattern recognition	AI algorithms excel at identifying patterns in wearable data, distinguishing normal physiological variations from abnormal trends, such as irregular heart rate indicating cardiac issues.
3	Contextual understanding	AI algorithms contextualize wearable data by incorporating wearer demographics, medical history, environment, and lifestyle. This enhances personalized health insights and accuracy.
4	Predictive analytics	AI-powered wearables use predictive analytics to forecast health outcomes, like predicting migraines from sleep patterns, activity levels, and environmental factors.

(Continues)

Table 12.1 (Continued)

Steps	Role	Explanation
5	Personalized recommendations	AI algorithms analyze wearable data to offer personalized recommendations like lifestyle changes, medication adjustments, or behavioral interventions tailored to the wearer's health needs and goals.
6	Continuous learning	AI algorithms evolve with continuous learning from new data, refining interpretations and predictions. This iterative process enhances understanding of health patterns and improves accuracy in forecasting outcomes.
7	Clinical decision support	AI-driven wearable data analysis aids clinical decision-making in healthcare, guiding diagnosis, treatment, and chronic condition management for personalized and proactive care delivery.
8	Data integration	AI merges wearable data with EHRs, genetic data, and patient-reported outcomes for a comprehensive view of health, enhancing informed decision-making by healthcare providers.

12.5 Enhancing diagnostic accuracy

AI has transformed diagnostic accuracy across medical fields with advanced algorithms and data analytics. It swiftly interprets medical images like X-rays and MRI scans, aiding precise diagnoses. By detecting subtle patterns in patient data, AI identifies early signs of disease and predicts progression, prompting timely interventions. AI-driven clinical decision support provides evidence-based recommendations tailored to patient data, empowering providers. It stratifies patients by risk for targeted screening and prevention [26]. AI also customizes genetic analysis and treatment based on individual profiles, enhancing precision. NLP extracts insights from clinical data, while wearable data enables remote monitoring. AI integrates datasets, advancing comprehensive diagnostics and improving patient outcomes.

12.5.1 AI-assisted diagnostics in telemedicine

AI-assisted diagnostics in telemedicine are transforming remote healthcare by leveraging advanced algorithms and data analytics to enhance diagnostic accuracy and efficiency [27]. AI algorithms analyze patient data remotely, including medical history, symptoms, and diagnostic tests, providing insights without in-person consultations. They prioritize cases based on severity, ensuring urgent cases receive prompt attention and optimizing resource allocation. AI aids in generating comprehensive differential diagnoses, improving accuracy even in complex cases, and suggests personalized treatment plans based on patient data and evidence-based guidelines. Continuous learning from new data inputs refines AI's diagnostic capabilities, keeping it updated with evolving medical knowledge. This approach expands access to specialized care, particularly in underserved or remote areas, by connecting patients with expert diagnostic support remotely.

12.5.2 Image and signal processing for accurate remote diagnostics

Image and signal processing play pivotal roles in remote diagnostics, particularly in telemedicine [28]. Techniques in image processing enhance the quality of medical images during remote transmission, employing methods such as noise reduction and contrast enhancement for precise interpretation. Algorithms extract key features from images, aiding healthcare providers in focusing on critical areas like lesions or anatomical structures, thereby enhancing assessment accuracy. Pattern recognition algorithms detect subtle anomalies by analyzing pixel intensity and spatial relationships, improving diagnostic precision beyond human capability. Signal-processing techniques filter noise from physiological signals like ECGs and EEGs, ensuring clarity in remote settings where signal quality may degrade. Real-time monitoring capabilities allow prompt identification of abnormalities, supporting timely intervention and better patient outcomes. These technologies integrate seamlessly with AI algorithms to automate diagnostic tasks and enhance decision-making, further bolstering accuracy in remote healthcare scenarios.

12.5.3 Reducing diagnostic errors through machine learning algorithms

ML algorithms play a pivotal role in healthcare by reducing diagnostic errors [29]. They analyze extensive patient data—medical history, symptoms, test results, and imaging—to identify patterns indicating diseases. ML excels in complex pattern recognition, detecting subtle disease indicators missed by humans. It predicts patient risks for diseases or adverse outcomes, guiding preventive interventions. ML-driven clinical decision support systems offer evidence-based recommendations, integrating patient data and guidelines for accurate diagnoses. Continuously learning from new data, ML algorithms refine diagnostic accuracy over time, improving error prevention. This advancement enhances healthcare delivery by providing clinicians with robust tools to make informed and precise diagnostic decisions.

12.6 Personalized treatment plans

Personalized treatment plans are individualized healthcare strategies that cater to each patient's specific needs, preferences, and characteristics [30]. They begin with a thorough patient assessment, considering medical history, current health status, genetic factors, lifestyle, and personal preferences. Advanced data analytics, including ML, analyze patient data to uncover patterns and predictors relevant to the patient's condition. Healthcare providers stratify patients based on risk to guide personalized treatment decisions aligned with evidence-based guidelines. Treatments, from pharmacological options to lifestyle modifications, are selected based on the patient's preferences and medical profile. Continuous monitoring allows adjustments to treatment plans for optimal outcomes, emphasizing patient engagement and education. A collaborative approach involves a team of healthcare professionals to comprehensively address the patient's health needs, ensuring holistic care delivery.

12.6.1 Tailoring treatment strategies based on AI-driven insights

Customizing healthcare interventions for individuals using AI involves employing advanced algorithms and analytics [31]. Table 12.2 details AI-driven treatment strategies for patient monitoring.

Table 12.2 Some treatment strategies based on AI-driven insights for patient monitoring

Sl. no.	Treatment strategies	Explanation
1	Data analysis	AI algorithms process extensive patient data—medical records, genetic info, diagnostic tests, and treatment results—to identify patterns and predictors crucial for treatment response.
2	Predictive analytics	AI predicts treatment outcomes and identifies patients likely to respond favorably to interventions, customizing strategies for optimal efficacy and minimal adverse effects.
3	Precision medicine	AI enables precision medicine by pinpointing biomarkers and genetic variants linked to treatment response, enabling tailored therapies based on individual molecular profiles.
4	Personalized risk stratification	AI categorizes patients by risk for treatment complications, enabling personalized risk mitigation and closer monitoring of high-risk individuals by healthcare providers.
5	Treatment optimization	AI evolves treatment strategies based on patient data, refining plans and optimizing outcomes by analyzing responses and adapting to individual needs over time.
6	Clinical decision support	AI-driven clinical decision support systems offer evidence-based treatment recommendations tailored to patient data and medical literature, aiding healthcare providers in informed treatment decisions, dosing, and monitoring.
7	Integration with electronic health records (EHRs)	AI seamlessly integrates with EHR systems, offering real-time insights and supporting clinical decision-making by delivering personalized treatment recommendations based on patient data and clinical history.

12.6.2 Precision medicine in remote patient care

Precision medicine [32] in remote patient care delivers personalized healthcare interventions tailored to individuals' unique genetic makeup, lifestyle, environment, and health history, irrespective of their location. Genetic profiling identifies disease-related variants influencing treatment responses and prognoses. Integrating genetic, medical, wearable, and patient-reported data provides comprehensive health insights and guides personalized treatment strategies. AI-driven risk assessment categorizes individuals by disease susceptibility, optimizing resource allocation and intervention prioritization. Tailored treatment plans incorporate patient-specific data, recommending targeted therapies, lifestyle adjustments, and preventive measures. Remote monitoring via wearables tracks vital signs and medication adherence in real time, facilitating proactive care. AI-powered clinical decision support systems enhance diagnostic accuracy and treatment outcomes by providing evidence-based recommendations. Patient engagement through remote platforms empowers active participation in care, offering education, self-management tools, and support groups to promote personalized treatment adherence.

12.6.3 Patient-specific recommendations for better outcomes

Patient-specific recommendations are personalized healthcare strategies designed to cater to the unique needs, preferences, and characteristics of individual patients [33], aiming to optimize health outcomes comprehensively. These recommendations entail developing customized treatment plans that take into account the patient's medical history, genetic profile, lifestyle choices, and personal preferences, thereby maximizing treatment effectiveness while minimizing potential adverse effects. By tailoring interventions to address specific patient needs and underlying causes, healthcare providers can significantly improve treatment outcomes and promote long-term health and well-being. Moreover, patient-specific recommendations focus on reducing the risk of complications by addressing individual risk factors such as genetic predispositions, lifestyle habits, and coexisting conditions. They also include tailored patient education and guidance to enhance understanding of the condition and treatment options, empowering patients to actively participate in their care. Regular monitoring and follow-up are integral, ensuring ongoing assessment of progress, adjustment of treatment strategies as needed, and early intervention to prevent complications. Additionally, these recommendations integrate technology such as wearable devices and mobile health apps for remote monitoring, fostering patient engagement, adherence to treatment plans, and ultimately improving outcomes through advanced, integrated healthcare solutions.

12.7 Challenges and considerations

Implementing patient-specific recommendations involves addressing critical challenges and considerations [34] to ensure effectiveness. Data privacy and security are paramount, requiring robust measures to protect sensitive health data and comply with regulations like HIPAA (Health Insurance Portability and Accountability Act) in the United States or GDPR (General Data Protection Regulation) in Europe. Ensuring data quality and interoperability across EHRs, wearable devices, and patient-reported outcomes is essential for accurate insights. Ethical concerns include obtaining informed consent, respecting patient autonomy, and avoiding biases in decision-making processes. Addressing health disparities and promoting equitable access to care are crucial, necessitating efforts to mitigate algorithmic biases that could exacerbate inequalities. Enhancing health literacy and patient engagement is vital for understanding and adherence to personalized treatment plans. Healthcare providers need adequate training and support to seamlessly integrate these recommendations into clinical workflows amidst cost and resource constraints. Continuous evaluation and improvement are necessary to assess impact on outcomes, care quality, and costs, ensuring sustainable delivery of personalized healthcare interventions.

12.7.1 Ethical considerations in AI-powered telemedicine

Ethical considerations are pivotal in AI-powered telemedicine to safeguard patient safety, privacy, and autonomy [35]. Key issues include ensuring informed consent, where patients must be fully aware of AI algorithm usage and consent to their data being used for analysis and treatment recommendations. Transparency and accountability are crucial; healthcare providers need to openly discuss AI algorithms, their functions, limitations, and biases, taking responsibility for decisions based on AI-generated insights. Addressing bias and promoting fairness is essential to prevent care disparities arising from biased AI algorithms. Protecting patient privacy and data security is paramount, requiring robust measures against unauthorized access. Ensuring equity and access involves overcoming barriers like socio-economic disparities and technological literacy. Clinical oversight is vital in interpreting AI-generated recommendations alongside human judgment. Continual evaluation and improvement of AI algorithms are necessary to maintain efficacy, accuracy, and safety standards. Upholding patient autonomy means involving patients in decision-making processes regarding AI recommendations, respecting their rights to choose their care paths.

12.7.2 Addressing privacy and security concerns

Addressing privacy and security concerns in AI-powered telemedicine is essential for maintaining patient confidentiality and trust [36]. Robust measures such as data encryption during transmission, strict access controls, and anonymization or de-identification of patient data before AI analysis are crucial steps. It is also imperative to minimize data collection to necessary elements, store data securely in encrypted databases, and conduct regular audits and monitoring to detect and mitigate vulnerabilities promptly. Obtaining patient consent and ensuring transparency about data use are vital for building trust. Compliance with regulations like HIPAA and GDPR is non-negotiable, requiring ongoing updates and adherence. Comprehensive employee training on privacy and security practices is essential to instill a culture of awareness and accountability. These efforts collectively ensure that patient data remains confidential and secure, fostering confidence in AI-powered telemedicine systems.

12.7.3 Balancing technology with the human touch in remote healthcare

Achieving a balance between technology and human interaction is crucial in remote healthcare to deliver holistic and patient-centered care [37]. Personalized patient-provider communication via telemedicine should harness technology while emphasizing empathy, active listening, and interpersonal skills to address emotional and psychological needs effectively. Integrating tech-driven assessments with human insight ensures thorough patient evaluations that consider social factors and preferences, forming comprehensive treatment plans. While providing patients with educational resources, self-management tools, and remote monitoring capabilities through technology, healthcare providers must offer personalized guidance to empower patients in decision-making and care participation. Human clinical judgment remains essential in interpreting AI recommendations and aligning them with individual patient needs, fostering critical thinking and ethical practice. Cultural sensitivity is paramount, adapting communication approaches to respect diverse backgrounds during remote consultations. Maintaining continuity of care through enduring patient-provider relationships via virtual consultations ensures consistent support, follow-up, and collaborative care while upholding ethical standards, patient privacy, autonomy, and confidentiality.

12.8 Future trends and innovations

Future trends and innovations in remote healthcare are poised to revolutionize healthcare delivery, enhancing accessibility, efficiency, and patient outcomes. Advancements in AI and ML will drive more accurate diagnoses, personalized treatment recommendations, and predictive analytics. AI-powered virtual assistants and chatbots will amplify patient engagement and support healthcare providers in clinical decision-making, ushering in a new era of precision and efficiency.

Telemonitoring and RPM will leverage wearable devices and IoT sensors to monitor vital signs and adherence to treatment plans in real time. This proactive approach enables early detection of health issues, reducing hospital readmissions and improving overall patient health. Virtual reality (VR) and augmented reality (AR) technologies will transform medical education, training, and patient interaction, enhancing teleconsultations with immersive simulations and overlaying medical information onto real-world environments.

Blockchain technology will bolster data security and transparency, ensuring secure sharing of medical records while protecting patient privacy. Integrated telemedicine platforms and ecosystems will streamline communication and data exchange between healthcare providers and patients, supporting a wide array of telehealth services. Genomics and precision medicine will further personalize treatment strategies, while recognizing social determinants of health (SDOH) will drive efforts to address disparities and improve healthcare equity globally.

Regulatory reforms and policy changes will underpin the expansion of remote healthcare services, fostering standards and guidelines to ensure quality, safety, and equitable access. Embracing these innovations promises to transform healthcare delivery, meeting the diverse needs of patients worldwide in the digital age.

12.8.1 Emerging technologies shaping the future of telemedicine

Emerging technologies are shaping the future of telemedicine, revolutionizing healthcare delivery and improving patient access to medical services [38]. Table 12.3 shows some key technologies driving this transformation:

Table 12.3 Some emerging technologies shaping the future of telemedicine

Sl. no.	**Emerging technologies**	**Contribution to telemedicine**
1	Artificial intelligence (AI) and machine learning	AI-powered algorithms in telemedicine analyze patient data for precise diagnoses, personalized treatment plans, and predictive analytics. ML supports virtual assistants, chatbots, and clinical decision tools, enhancing healthcare and outcomes.
2	Remote patient monitoring (RPM)	RPM technologies like wearables, IoT sensors, and mobile health apps monitor vital signs and symptoms, enhancing remote patient management and enabling early health issue detection, thereby improving outcomes.
3	Teleconsultation platforms	Integrated teleconsultation platforms and telemedicine apps enable virtual consultations for remote diagnosis, treatment, and follow-up care, ensuring secure communication and collaboration between patients and providers.
4	Telemedicine-enabled diagnostics	Telemedicine-enabled diagnostic tools like digital stethoscopes and high-resolution cameras allow remote examinations, aiding accurate diagnoses and treatment planning through real-time medical condition visualization.
5	Virtual reality (VR) and augmented reality (AR):	VR and AR technologies are transforming medical education, training, and patient engagement in telemedicine. VR offers immersive learning experiences, while AR overlays medical information during virtual consultations for enhanced visualization.
6	Blockchain technology	Blockchain technology improves security, interoperability, and transparency in telemedicine by decentralizing patient records and transactions, ensuring data privacy, integrity, and secure sharing among stakeholders.
7	5G connectivity	5G technology supports high-speed, low-latency communication networks, improving telemedicine quality with real-time video conferencing, remote monitoring, and data-intensive applications in remote areas.
8	Genomics and precision medicine	Advancements in genomics and precision medicine enable personalized treatment tailored to genetic makeup, lifestyle, and environmental factors, integrated into telemedicine platforms for optimized remote patient care.

12.8.2 AI advancements and their potential impact on remote patient monitoring

AI advancements are poised to revolutionize RPM [39] by significantly enhancing its accuracy, efficiency, and effectiveness. Through predictive analytics, AI algorithms can analyze extensive datasets collected from wearable sensors and mobile apps, identifying patterns and early indicators of health deterioration. This capability enables healthcare providers to intervene pre-emptively, potentially averting adverse outcomes before they escalate. Additionally, AI can stratify patients based on their risk profiles, leveraging data such as vital signs and medical history to pinpoint individuals at higher risk of developing certain conditions or experiencing complications. This allows for targeted interventions and personalized preventive measures, optimizing patient care in remote settings.

Furthermore, AI-driven decision support systems generate personalized alerts and recommendations aligned with clinical guidelines and individual patient data. These alerts prompt timely interventions and adjustments to treatment plans, enhancing responsiveness and efficacy in managing patient health remotely. AI's capability to continuously learn from new data inputs and patient outcomes ensures ongoing improvement in diagnostic accuracy and predictive capabilities. By integrating with EHRs, AI-powered RPM systems provide comprehensive patient data, supporting holistic assessments and informed decision-making by healthcare providers. Ultimately, AI holds immense promise in optimizing resource allocation, streamlining workflows, and delivering proactive, data-driven care that improves patient outcomes and reduces healthcare costs in RPM programs.

12.8.3 Opportunities for further integration and collaboration in digital healthcare

Opportunities in digital healthcare integration and collaboration abound, promising to improve patient care, streamline workflows, and enhance outcomes [40]:

1. **Interoperability standards**: Standardizing healthcare systems, devices, and data exchange ensures seamless integration for comprehensive patient care and provider collaboration.
2. **Integrated health platforms:** Consolidating data from EHRs, telemedicine, remote monitoring, and patient apps supports holistic care coordination and enhances collaboration across care settings.
3. **Telehealth networks**: Connecting providers, specialists, and patients facilitates collaborative care delivery, virtual consultations, and access to specialized services.
4. **Data sharing and analytics**: Using advanced analytics and shared data promotes insights from clinical and operational data, driving continuous improvement in patient care.
5. **Digital health ecosystems**: Bringing together stakeholders fosters innovation, knowledge sharing, and collaborative solutions to healthcare challenges.
6. **Interdisciplinary care teams**: Collaborative teams of healthcare professionals leverage digital tools to deliver patient-centered care across specialties.

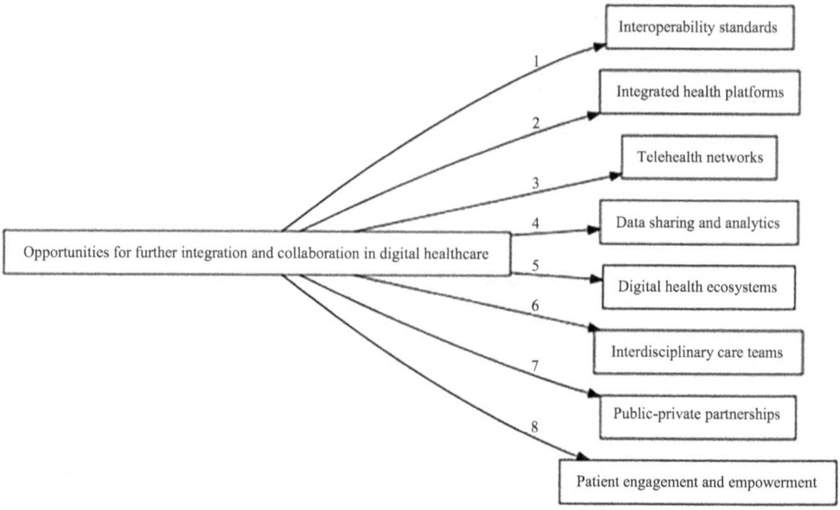

Figure 12.4 Key opportunities for further integration and collaboration in digital healthcare

7. **Public-private partnerships**: Collaborating across sectors enhances resource sharing, technology adoption, and policy development to improve healthcare access and quality.
8. **Patient engagement**: Empowering patients with health information, tools, and decision-making support promotes active participation and improves health outcomes.

These opportunities highlight the potential for digital healthcare to transform care delivery, enhance collaboration, and drive improvements in healthcare access and quality.

Figure 12.4 illustrates key opportunities in digital healthcare integration and collaboration, focusing on enhancing patient care, streamlining workflows, and improving outcomes. These include establishing interoperability standards and fostering patient engagement to optimize healthcare delivery and build a cohesive ecosystem.

12.9 Conclusion

The integration of AI with telemedicine and RPM is revolutionizing healthcare delivery. AI enhances remote monitoring by analyzing patient data for predictive insights and personalized recommendations, extending healthcare access and improving outcomes. This synergy promises to reshape healthcare, driving innovation and advancing patient-centered care through efficient, accurate, and personalized health interventions.

12.9.1 Recap of the transformative role of AI in telemedicine and remote patient monitoring

AI's pivotal role in telemedicine and RPM transforms healthcare delivery. It refines diagnostics, tailors treatments, and predicts outcomes using advanced algorithms.

AI analyzes extensive patient data from wearables and IoT, enabling proactive care and early issue detection. Decision support systems offer personalized alerts, ensuring timely interventions. AI's integration promises to enhance care access, improve outcomes, and revolutionize healthcare models.

12.9.2 Anticipated future developments and the continued evolution of AI-driven healthcare

Looking ahead, the future of AI-driven healthcare promises profound transformations in delivery and patient experiences. Key anticipated developments include advancements in AI algorithms, focusing on enhancing accuracy, scalability, and interpretability. Innovations in deep learning and NLP will empower AI systems to analyze complex medical data, offering precise predictions and actionable insights.

Personalized medicine will expand, with AI tailoring treatments based on genetic profiles, lifestyle factors, and patient preferences. Precision medicine and biomarker discoveries will optimize therapies, balancing efficacy and minimizing side effects. Predictive analytics powered by AI will foresee and prevent health crises, refining risk stratification and early warning systems. This proactive approach will bolster personalized health coaching and population health management, driving better outcomes and reduced costs.

AI's integration into clinical workflows will streamline decision-making, diagnosis, and treatment planning. Clinical decision support systems will furnish evidence-based recommendations, mitigate risks, and streamline administrative tasks, enhancing care delivery efficiency. Remote monitoring and telehealth will expand, aided by wearable tech and IoT sensors, facilitating remote consultations and diagnostics from patients' homes.

Ethical and regulatory frameworks will evolve to safeguard patient privacy, safety, and equity amidst AI's rapid adoption in healthcare. Collaboration across academia, industry, and healthcare will foster novel AI algorithms and validate their real-world applications. This interdisciplinary approach will cater to diverse patient needs and fuel healthcare innovation.

In summary, AI-driven healthcare's anticipated evolution holds transformative potential, promising to redefine care delivery, elevate patient outcomes, and advance medical practice globally. Embracing AI's capabilities and fostering collaborative efforts across the healthcare spectrum will unlock unprecedented opportunities to revolutionize healthcare and enhance patient lives.

References

[1] J. Bajwa, U. Munir, A. Nori, and B. Williams, "Artificial intelligence in healthcare: transforming the practice of medicine", *Future Healthcare Journal*, vol. 8, no. 2, pp. e188–e194, 2021.

[2] L. P. Serrano, K. C. Maita, F. R. Avila, *et al.*, "Benefits and challenges of remote patient monitoring as perceived by health care practitioners: a systematic review", *The Permanente Journal*, vol. 27, no. 4, pp. 100–111, 2023.

[3] S. Yelne, M. Chaudhary, K. Dod, A. Sayyad, and R. Sharma, "Harnessing the power of AI: a comprehensive review of its impact and challenges in nursing science and healthcare", *Cureus*, vol. 15, no. 11, pp. 1–14, 2023.

[4] A. Haleem, M. Javaid, R. P. Singh, and R. Suman, "Telemedicine for healthcare: capabilities, features, barriers, and applications", *Sensors International*, vol. 2, pp. 1–13, 2021.

[5] R. Uddin, and I. Koo, "Real-time remote patient monitoring: a review of biosensors integrated with multi-hop IoT systems via cloud connectivity", *Applied Sciences*, vol. 14, no. 5, pp. 1–36, 2024.

[6] D. Mdhlalose, and G. Mlambo, "Integration of technology in education and its impact on learning and teaching", *Asian Journal of Education and Social Studies*, vol. 47, no. 2, pp. 54–63, 2023.

[7] S. Z. S. Khassim, and B. Al-Haimi, "Telemedicine: transforming healthcare accessibility and quality with sustainable technological advancement", in H. El-Chaarani, I. El Dandachi, S. El Nemar, and Z. El Abiad (eds.), *Navigating the Intersection of Business, Sustainability and Technology*, Springer, Berlin, pp. 137–149, 2024.

[8] E. E. Thomas, M. L. Taylor, A. Banbury, *et al.*, "Factors influencing the effectiveness of remote patient monitoring interventions: a realist review", *BMJ Open*, vol. 11, no. 8, pp. 1–9, 2021.

[9] S. K. Jagatheesaperumal, P. Mishra, N. Moustafa, and R. Chauhan, "A holistic survey on the use of emerging technologies to provision secure healthcare solutions", *Computers and Electrical Engineering*, vol. 99, pp. 1–24, 2022.

[10] D. M. El-Sherif, M. Abouzid, M. T. Elzarif, A. A. Ahmed, A. Albakri, and M. M. Alshehri, "Telehealth and artificial intelligence insights into healthcare during the COVID-19 pandemic", *Healthcare (Basel)*, vol. 10, no. 2, pp. 1–12, 2022.

[11] F. L. Segui, R. Ander Egg Aguilar, G. de Maeztu, *et al.*, "Teleconsultations between patients and healthcare professionals in primary care in Catalonia: the evaluation of text classification algorithms using supervised machine learning", *International Journal of Environmental Research and Public Health*, vol. 17, no. 3, pp. 1–24, 2020.

[12] M. Mekhi, "An artificial intelligence based virtual assistant using conversational agents", *Journal of Software Engineering and Applications*, vol. 14, no. 9, pp. 455–473, 2021.

[13] N. Villafuerte, S. Manzano, P. Ayala, and M. V. Garcia, "Artificial intelligence in virtual telemedicine triage: a respiratory infection diagnosis tool with electronic measuring device", *Future Internet*, vol. 15, no. 7, pp. 1–29, 2023.

[14] D. Khurana, A. Koli, K. Khatter, and S. Singh, "Natural language processing: state of the art, current trends and challenges", *Multimedia Tools and Applications*, vol. 82, pp. 3713–3744, 2023.

[15] A. Baker, Y. Perov, K. Middleton, *et al.*, "A comparison of artificial intelligence and human doctors for the purpose of triage and diagnosis", *Frontiers in Artificial Intelligence*, vol. 3, pp. 1–9, 2020.

[16] G. Sun, and Y. Zhou, "AI in healthcare: navigating opportunities and challenges in digital communication", *Frontiers in Digital Health*, vol. 5, pp. 1–5, 2023.

[17] R. Wagner, T. C. Lima, M. R. T. da Silva, *et al.*, "Assessment of pediatric telemedicine using remote physical examinations with a mobile medical device", *JAMA Network Open*, vol. 6, no. 2, pp. 1–11, 2023.

[18] G. Castelyn, L. Laranjo, G. Schreier, and B. Gallego, "Predictive performance and impact of algorithms in remote monitoring of chronic conditions: a systematic review and meta-analysis", *International Journal of Medical Informatics*, vol. 156, pp. 1–12, 2021.

[19] O. Ali, W. Abdelbaki, A. Shrestha, E. Elbasi, M. A. A. Alryalat, and Y. K. Dwivedi, "A systematic literature review of artificial intelligence in the healthcare sector: Benefits, challenges, methodologies, and functionalities", *Journal of Innovation & Knowledge*, vol. 8, no. 1, pp. 1–19, 2023.

[20] I. Preethi, and K. Dharmarajan, "Diagnosis of chronic disease in a predictive model using machine learning algorithm", *2020 International Conference on Smart Technologies in Computing, Electrical and Electronics (ICSTCEE)*, IEEE, pp. 1–6, Bengaluru, India, 2020.

[21] E. Nimy, M. Mosia, and C. Chibaya, "Identifying at-risk students for early intervention—a probabilistic machine learning approach", *Applied Sciences*, vol. 13, no. 6, pp. 1–13, 2023.

[22] W. Wang, and W. Hsu, "Integrating artificial intelligence and wearable IoT system in long-term care environments", *Sensors*, vol. 23, no. 13, p. 5913, 2023.

[23] S. Shajari, K. Kuruvinashetti, A. Komeili, and U. Sundaraj, "The emergence of AI-based wearable sensors for digital health technology: a review", *Sensors*, vol. 23, no. 23, p. 9498, 2023.

[24] S. T. Creavin, M. Garg, and A. D. Hay, "Impact of remote vital sign monitoring on health outcomes in acute respiratory infection and exacerbation of chronic respiratory conditions: systematic review and meta-analysis", *ERJ Open Research*, vol. 9, no. 2, pp. 1–13, 2023.

[25] A. T. Shumba, T. Montanaro, I. Sergi, *et al.*, "Wearable technologies and AI at the far edge for chronic heart failure prevention and management: a systematic review and prospects", *Sensors*, vol. 23, no. 15, pp. 1–26, 2023.

[26] T. Habuza, A. N. Navaz, F. Hashim, *et al.*, "AI applications in robotics, diagnostic image analysis and precision medicine: current limitations, future trends, guidelines on CAD systems for medicine", *Informatics in Medicine Unlocked*, vol. 24, pp. 1–31, 2021.

[27] J. Choi, S. Woo, and A. Ferrell, "Artificial intelligence assisted telehealth for nursing: a scoping review", *Journal of Telemedicine and Telecare*, vol. 31, no. 1, pp. 140–149, 2023.

[28] X. Yang, "Remote diagnosis and detection technology for electrical control of intelligent manufacturing CNC machine tools", *Scientific Programming and Digital Twins in Medical Informatics*, vol.2022, pp. 1–14, 2022.

[29] I. A. Scott, "Using information technology to reduce diagnostic error: still a bridge too far?", *Internal Medicine Journal*, vol. 52, no. 6, pp. 908–911, 2022.

[30] D. Stefanicka-Wojtas, and D. Kurpas, "Personalised medicine—implementation to the healthcare system in Europe", *Journal of Personalised Medicine*, vol. 13, no. 3, pp. 1–10, 2023.

[31] A. Tiwari, "Custom AI models tailored to business-specific content needs", *Jurnal Komputer, Informasi dan Teknologi*, vol. 4, no. 2, pp. 1–21, 2024.

[32] K. B. Johnson, W. Wei, D. Weeraratne, *et al.*, "Precision medicine, AI, and the future of personalized health care", *Clinical and Translational Science*, vol. 14, no.1, pp. 86–93, 2021.

[33] D. Bhati, M. S. Deogade, and D. Kanyal, "Improving patient outcomes through effective hospital administration: a comprehensive review", *Cureus*, vol. 15, no.10, pp. 1–12, 2023.

[34] A. H. Krist, S. T. Tong, R. A. Aycock, D. R. Longo, "Engaging patients in decision-making and behavior change to promote prevention", *Studies in Health Technology and Informatics*, vol. 240, pp. 284–302, 2017.

[35] H. Siala, and Y. Wang, "SHIFTing artificial intelligence to be responsible in healthcare: A systematic review", *Social Science & Medicine*, vol. 296, pp. 1–15, 2022.

[36] B. Murdoch, "Privacy and artificial intelligence: challenges for protecting health information in a new era", *BMC Medical Ethics*, vol. 22, no. 122, pp. 1–5, 2021.

[37] A. Halbig, S. K. Babu, S. Gatter, M. E. Latoschik, K. Brukamp, S. von Mammen, "Opportunities and challenges of virtual reality in healthcare—a domain experts inquiry", *Frontiers Virtual Reality in Medicine*, vol. 3, pp. 1–20, 2022.

[38] R. Damasevicius and O. O. Abayomi-Alli, "The future of telemedicine: emerging technologies, challenges, and opportunities", in L. Gaur and N. Z. Jhanjhi (eds.), *Metaverse Applications for Intelligent Healthcare*, IGI Global Scientific Publishing, pp. 306–338, 2024. https://doi.org/10.4018/978-1-6684-9823-1.ch010.

[39] T. Shaik, X. Tao, N. Higgins, *et al.*, "Remote patient monitoring using artificial intelligence: current state, applications, and challenges", *WIREs Data Mining and Knowledge Discovery*, vol. 13, no. 2, pp. 1–31, 2023.

[40] Q. Wang, M. Su, M. Zhang, and R. Li, "Integrating digital technologies and public health to fight Covid-19 pandemic: Key technologies, applications, challenges and outlook of digital healthcare", *International Journal of Environmental Research and Public Health*, vol. 18, no. 11, pp. 1–12, 2021.

Chapter 13

Smart in-home health monitoring system using IoT: architecture and enhancements

K. Priyadarsini[1], S. Karthik[2], J. Jeba Sonia[1], P.C. Karthik[1], U.V. Anbhazhagu[3] and V.G. Saranya[2]

Abstract

This chapter discusses how Internet of Things (IoT) technologies can be used for the creation of home-based monitoring systems for healthcare. The feasibility of IoT for remote, real-time monitoring of patients' basic health parameters with ease offers overwhelming benefits, especially in pandemic contexts like COVID-19. Healthcare systems are overburdened, and care providers are minimally exposed while adopting this paradigm. A variety of available technologies and platforms are reviewed to emphasize the recent trends, issues, and possible directions in developing cost-effective, scalable, and user-friendly IoT-based health monitoring systems. A case study is also provided to demonstrate a real-world implementation of such a system, with an emphasis on the integration of hardware, server-side infrastructure, and web interfaces for remote patient monitoring.

Keywords: IoT; internet of medical devices; patient monitoring; smart health system

13.1 Introduction

The COVID-19 pandemic has established an urgent need for physical distancing wherever it is possible. Such a precaution is particularly important for those who have symptoms of infectious diseases. Medical professionals—such as doctors, nurses, and support staff—are frequently in close proximity to patients and thus are one of the most at-risk groups. While there are protocols like the employment of

[1]Department of Data Science and Business Systems, School of Computing, College of Engineering and Technology, SRM Institute of Science and Technology, India
[2]Department of ECE, College of Engineering and Technology, SRM Institute of Science and Technology, Vadapalani Campus, India
[3]Department of Computing Technologies, School of Computing, College of Engineering and Technology, SRM Institute of Science and Technology, India

personal protective equipment, constant hand hygiene, and daily symptom checks in place, patients still report to hospitals and stay for prolonged durations, risking the transmission of disease.

A pragmatic solution to reducing the risk of transmission is to reduce unnecessary contact between medical staff and possibly infected patients. In most instances, a first-time examination in the hospital will suffice, following which the patient can stay at home while doctors keep track of their vital signs remotely. This approach provides several advantages: healthcare workers are exposed to fewer infections, patients have fewer contact points with others, and hospital resources beds are available for more severe cases.

Recent developments in the Internet of Things (IoT) have proven its efficacy in different industries such as finance, commerce, education, and governance. Nevertheless, its use in real-time medical health monitoring is still limited. The majority of current patient monitoring systems either send data to only a smartphone or locally store it on the device. Although some systems send data to centralized servers, many are not scalable or are hardware constrained.

A portable and trustworthy health monitoring device that would allow patients to stay at home but give healthcare providers instant access to the vital signs of those patients could greatly enhance medical monitoring, especially with the COVID-19 pandemic. Most patients are hospitalized just for observation, which could be done remotely if suitable monitoring systems existed. Through the use of IoT and combined databases, clinicians will have remote access to patient information and enable patients to recover from home. Additionally, the system can provide automated alerts or intervention suggestions, e.g., medication titration or surgical referral-based on trending vital signs, e.g., abnormal heart rhythms. If the patient's status does not improve, they can be referred quickly for in-person assessment.

This chapter is dedicated to the development and deployment of IoT-based health monitoring systems that are intended to benefit the healthcare sector. These systems must be cost-effective, lightweight, easy to use, and available to the masses, while having a level of accuracy that enables healthcare professionals to make informed decisions based on the data.

The IoT enables the interconnection of physical objects through the internet, allowing them to collect and exchange data. The concept of IoT has evolved through advancements in embedded systems, sensor technology, algorithm development, and data analytics. In healthcare, this vision has given rise to the idea of the "smart hospital," where devices are connected via wired or wireless networks to automate and enhance medical operations. These devices are capable of collecting health-related data and transmitting it in real time for further analysis. While IoT is already being used across domains such as transportation, automotive systems, smart devices, and infotainment [1], its role in healthcare focuses on a wide range of applications involving sensors, diagnostic equipment, smart monitoring systems, and advanced imaging tools. These technologies contribute significantly to improving living standards and productivity across both developed and developing nations.

The IoT makes data exchange worldwide by linking digital, mechanical, and computing devices through automation without involving humans. Through the COVID-19 pandemic period, IoT technology has been particularly useful in remote monitoring of patients' health. Many deaths have been attributed to delayed or flawed health information [2,3]. With the help of sensors, IoT devices are able to send real-time alarms about abnormalities such as nutritional deprivation or variations in vital signs. Cloud platforms hosting COVID-19 patient information have made it simpler to access care, even for children living in remote or underprivileged areas. The devices also have the capacity to track routine activities and find potential health dangers [4,5]. In contemporary healthcare environments, innovative technology becomes an integral part of successful treatment. IoT has immense potential not just in diagnosis and ongoing monitoring but also in aid of surgical interventions and assessment of therapeutic advancements [6]. The use of IoT technologies amid the COVID-19 pandemic has helped improve and responsive patient management. IoT technology is especially beneficial for real-time diagnosis and early intervention, and it can save lives from causes such as diabetes, cardiac arrest, respiratory diseases, and hypertension. These systems leverage networked health devices that send important information—oxygen levels, pulse rate, weight, and glucose levels—to doctors through mobile phones or cloud-based applications. Through granting real-time access to such crucial health data, IoT supports timely medical interventions and improves patient outcomes [7,8].

13.1.1 Scope

IoT can project real-time patient health monitors in realistic, day-to-day applications. During the treatment of COVID-19 inpatients, IoT can overcome current issues by allowing electronic storage and remote retrieval of all related medical information. Convergence of available technologies can drastically enhance the delivery of healthcare, leading to more widespread usage by doctors and medical practitioners. Being a fast-developing and widespread technology, IoT promises to enable assistance in surgery, improve the accuracy of diagnostics, and facilitate better access to essential medical information and data. In addition, IoT systems can streamline medical supply chains by guaranteeing timely availability of necessary equipment and drugs. Smart IoT devices can function autonomously, promoting sustainable diagnostic work. By merging data storage with public and institutional computing, and using sophisticated software tools, IoT can facilitate the creation of smart medical infrastructures that are in line with the objectives of the Healthcare 4.0 ecosystem.

13.1.2 Need for the study

One of the major issues confronting the healthcare industry during the COVID-19 pandemic has been the absence of a strong digital infrastructure that can process and analyze large amounts of patient data in real time. The use of IoT has proven to be a potential solution, providing scalable and efficient data acquisition and monitoring functions. In order to reach greater heights of performance and reliability, healthcare professionals need to embrace systems integrating IoT hardware and smart software platforms. Because it is flexible and can simplify the process of collecting and analyzing data, IoT has the ability to tackle a multitude of healthcare challenges, particularly those exacerbated during pandemics such as the COVID-19 pandemic.

13.2 Literature review

13.2.1 Summary of literature assessment

The hypothesis presented to be explored in this have a look at is set the layout of this type of affected person tracking machine that ="hide">uses="tips Box"> the strategies of IoT technologies to bring result in the scientific subject. This device should be low priced, without problems transportable, simple enough that a not unusual character can use it and safely correct so that medical provider can rely on it.

We start by looking at the two encyclopedic records sources offered. Then, we expand our studies into a way to use literature to evaluate to track advances in science and medical studies. I usually move over scholarly techniques for comparing the level of partnership. The relevance of scientific reports is then evaluated and use more than a few approaches. The topic of common local cooperative power may be subsequently evaluated.

Aggarwal *et al.* [9] proposed a scalable and flexible personalized wireless device for wireless local area network (WLAN), emphasizing its potential to combine different networking approaches for healthcare monitoring systems. Another work focused on the emergence of synergistic systems facilitated by IoT across industries such as semiconductors, electronics, and telecommunications. Likewise, Senthamilarasi *et al.* [10] discussed mobile-based heart monitoring systems based on IoT frameworks, illustrating how smartphones can enable wireless health monitoring solutions.

Shin and Mao delved into data sharing in IoT contexts aimed at promoting information exchange among connected devices. Tartarisco and Paniclo advanced techniques for the preservation of sensor coverage and connectivity within massive wireless sensor networks (WSNs), critical for the optimal deployment of healthcare IoT systems. Their technique entails combining scientific modeling, parallel computing, mobile communication, and distributed data mining within organizational setups [10].

Eileen Elena Turcua analyzed the use of IoT for increasing access to healthcare services and enhancing social welfare delivery. She conducted research on the combination of power identities, multi-agent systems, and IoT technologies to construct an overall framework for enabling patient-centered care. Senthamilarasi *et al.* [10] also highlighted the necessity of IoT-based systems in the provision of efficient health services using enhanced connectivity and automation. Haleem and Javaid [11] gave a visionary overview of the IoT, highlighting crucial considerations for developing and deploying IoT in different sectors such as healthcare. The authors stressed that infrastructure supporting pervasive placement and embedding of sensing and surveillance devices should be developed. Mobile devices would be used as main access points for data reading and transmission according to the authors, serving as the foundation for coordinated communication across distributed IoT setups. J. L. Kalju helped design systems that can detect multiple biological markers, which play a crucial role in applications like rate-responsive cardiac pacing. In biomedical engineering, Schwiebert, Sharma, and Weinmann tested portable health monitoring systems composed of hybrid semiconductor-based sensors. Their results substantiate the utility of using different sensor materials to acquire real-time health information [12].

Gentili G.B. suggested a laser-based method for the detection of wireless ventricular activity by using changes in modulated wave amplitudes that are passed through the body. The approach illustrates how future sensing technologies can be used in non-invasive health monitoring. WLAN and micro-sensor networks have also been investigated to continuously monitor and record physiological states in real-time care environments [10].

Dilmaghani proposed a sensor-based network for remotely monitoring severe health conditions, especially from the home setup of a patient. The system could monitor parameters such as pulse rate and ambient temperature using embedded sensors in a compact, portable gadget. Senthamilarasi *et al.* [10] also highlighted the convenience of wearable IoT devices, which can continuously monitor physiological data and send it directly to centralized health systems for real-time analysis and clinical decision-making.

Kietzmann and Prpic presented the term "Internet of Everything", which is an extension of the underlying concept of the IoT that brings not just devices but also people, data, and processes into a collective digital environment. The authors highlighted the integration of mundane things such as smartphones, tablets, and wearable sensors into wise systems that are able to exchange and transform information. This convergence has created the "Internet of People," where machines and people coexist in harmony. The framework indicates a future scenario where people, organizations, and physical objects are connected to facilitate communication, automation, and delivery of services across industries, including health.

Gupta suggested a mobile healthcare monitoring system based on microcontrollers, namely a small CPU board based on the Raspbian operating system, which is a Linux-based desktop environment. The system was implemented on a Raspberry Pi to gather physiological signals from sensors strapped on the human body. Two main sensors were used: one for the myocardial heartbeat detection (ECG sensor) and another for the body temperature monitoring. The data read from each sensor was logged in a MySQL database, and subsequently presented on a web-based interface to facilitate real-time observation. This configuration provided constant monitoring of a patient's health condition by clinicians. Moreover, the system used GSM mobile communication technology, which automatically notified medical experts whenever a patient's pulse rate fell below or rose above specified levels. The application of this system assisted healthcare professionals in monitoring patients more effectively and reacting quickly to any irregularities (Figures 13.1 and 13.2).

Ghosh suggested an IoT-based remote patient health monitoring system that gathers information from the body of the patient as well as the environment. The physiological parameters being monitored include electrocardiogram (ECG) readings, body temperature, posture, and environmental data like room temperature and humidity. All this is stored in a central repository together with personal data such as the patient's age, gender, and admission time. The system entails three layers of access control: patient, medical practitioner, and family member. The multi-access architecture allows both the physician and the family members to see the patient's health status in real time through authenticated channels (Figure 13.3).

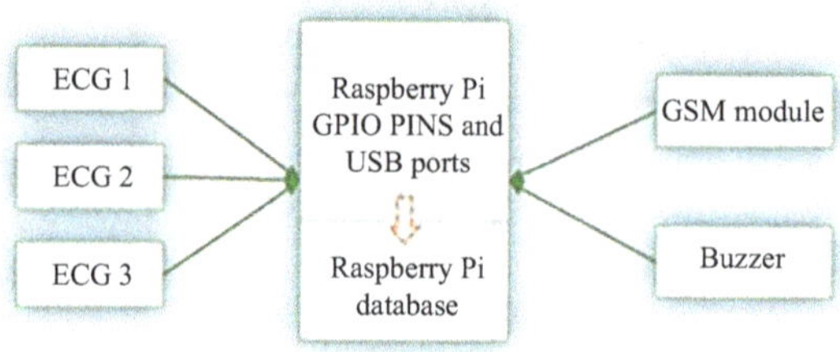

Figure 13.1 Function blocks of proposed system

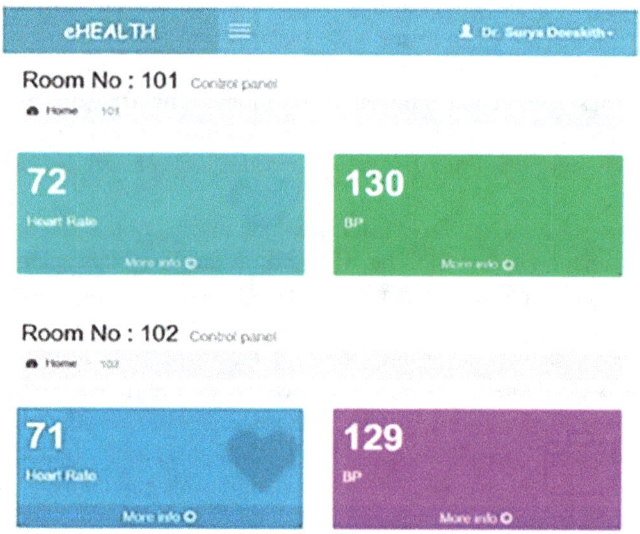

Figure 13.2 Website designed

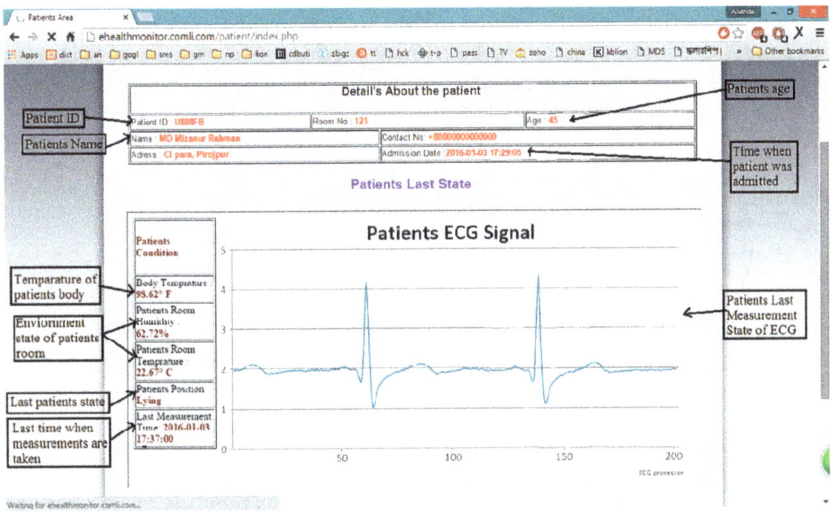

Figure 13.3 Website designed

Park *et al.* [13] brought the terms "Food-IoT" and "Health-IoT" into the overall context of IoT technologies for wellness and healthcare. The paper highlighted the difficulty of mapping conventional real-world organizational systems to IoT-based systems, with developers tending to find it hard to bring IoT capabilities into effective, scalable solutions. The research underscored the importance of more collaboration between commercial developers and IoT researchers in closing this gap. Pang estimated that within the next 10–15 years, IoT would deliver remarkable worldwide change, with remote healthcare (home health monitoring) being a main area of impact. The author envisioned an entirely connected system, with medical devices and health monitoring instruments speaking to each other seamlessly through IoT. Concurrently, Food-IoT targets the automation and digitization of food processing plants, where hygiene and nutritional quality are enhanced through IoT integration. The facilitating technologies supporting both applications—Food-IoT and Health-IoT—were investigated in depth, demonstrating how they help to support improved quality of life (Figure 13.4) [10,14].

Sowirelessc wrote about the adoption of IoT technologies in the Bosnian and Herzegovinian market, stressing the necessity to have harmonized technological frameworks and business models to enable the digitalization of the country. The research highlighted that the telecommunications sector within the region continues to make use of traditional business schemes and services. In order to successfully enter the IoT market, three main avenues were determined: IP-based connectivity solutions, manufacturers of IoT devices and hardware, and IoT data management platforms. Based on the analysis, IoT products in the region could be segmented into diverse consumer segments such as home automation, lifestyle, wireless systems, mobility, and others. The main aim of this research was to make clear the

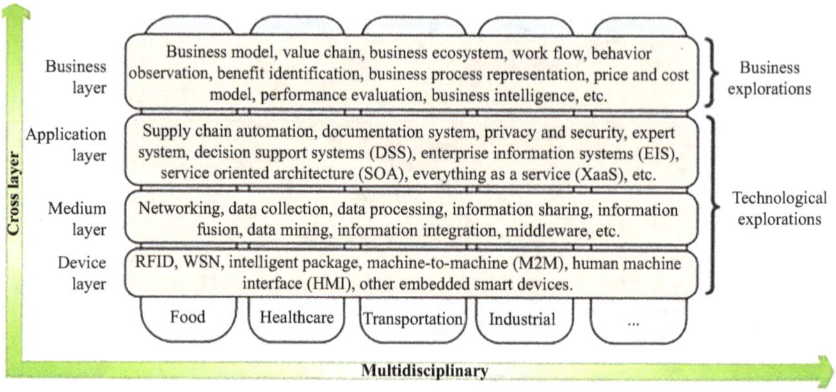

Figure 13.4 Gap b/w the business model and IoT

communication standards and architectural specifications required to ensure proper implementation of IoT in Bosnia and Herzegovina [15].

To facilitate secure communication in low-resource IoT settings, Maleh introduced a new link-layer interface designed to optimize Datagram Transport Layer Security (DTLS) protocols for low-power devices with the Constrained Application Protocol (CoAP). The creation of hybrid communication architectures was key to enhancing both the reliability and quality of DTLS-based transmissions over IoT networks. Based on this, additional research contributions by researchers such as Sikder *et al.*, and Maleh *et al.* highlighted the significance of lightweight yet secure encryption techniques optimized for resource-constrained medical IoT devices. Concurrently, Nasri *et al.* [16] discussed the application of WSNs and platforms like Samsung's IoT environment to facilitate autonomous healthcare operations. These systems proved to have the capability of unproblematic integration of operating systems and medical IoT infrastructure to improve data capture and automated medication support (Figure 13.5).

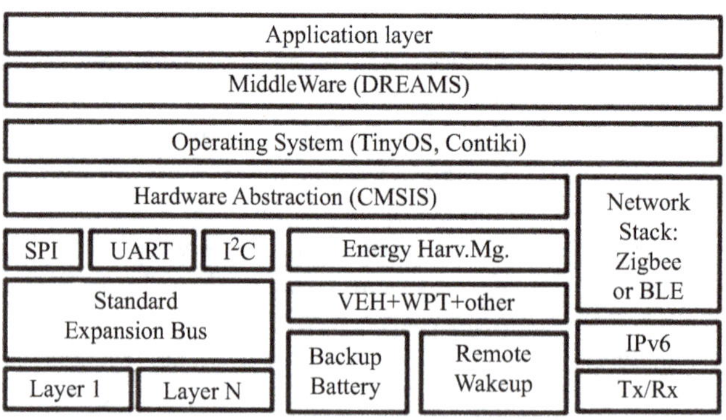

Figure 13.5 Sprout platforms for wireless sensors

A standardized protocol for brain data acquisition was proposed and implemented in mobile devices, enabling seamless integration into remote health monitoring systems. The framework supports the continuous tracking of physiological parameters such as pulse rate, oxygen saturation, and ECG signals. These biometric readings are transmitted to medical centers for centralized storage, maintenance, and diagnostic analysis. This approach facilitates real-time data flow between patients and clinicians, supporting improved care delivery and timely interventions (Figure 13.6) [16].

Ray gave a tutorial overview of recent cloud computing technologies in association with IoT devices. This research, coupled with the work of Maleh *et al.*, and Ray *et al.*, built upon the fundamental knowledge of IoT concepts, especially those that pertain to real-world applications. Principal areas of attention involved the detection of key research issues in IoT systems—notably, most prominently, security vulnerabilities and responsiveness of networked infrastructures. These concerns remain at the core of the current development and credible application of IoT in healthcare and other data-sensitive contexts. Accessibility of facts: Modern health systems employ a fixed number of sensors placed on patients. These sensors collect important biometric and environmental data, then transmit them wirelessly to a central processing unit. There are three types of sensor systems: body area sensors, area field sensors, and web server sensors. The connection between sensors and other devices occurs through short-range radio frequency such as Bluetooth or Zigbee technology.

1. A reference for minimum gadgets.
2. Safety and safety.
3. Merchandise and strategy.

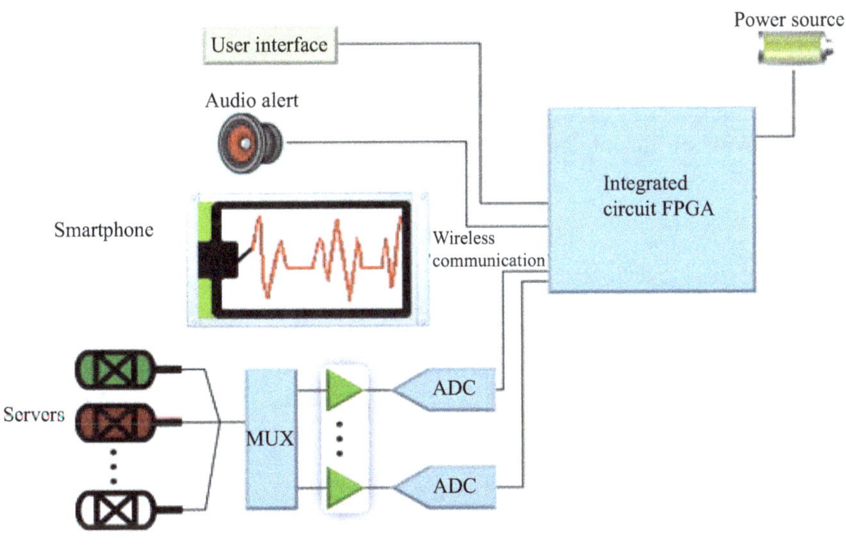

Figure 13.6 Healthcare systems with smartphone

4. Analytics.
5. Software.

Bedi discussed key issues concerning the deployment and management of IoT devices, especially in energy systems. Although IoT technologies offer a broad spectrum of opportunities, they also raise important concerns, notably in user intuitiveness, data privacy, and security. Secure acquisition, transmission, and storage of data are still among the most critical challenges. Bedi *et al.* provided some of the several limitations that must be overcome to facilitate the efficient and ethical deployment of IoT in massive applications.

In another related work, Luzuriaga *et al.* discussed using the MQTT protocol as a solution to device mobility management in IoT networks. MQTT and other light protocols such as CoAP and LWM2M have gained significant relevance in M2M and IoT communications. These protocols are backed by the TCP/IP suite and are optimally designed for applications in which low power, high efficiency, and real-time responsiveness are critical. The research highlighted the merit of MQTT's lightweight broadcast mechanism, which is public, economical, and easy to implement, making it ideally fit for dynamic and mobile IoT applications.

Ullah presented the successful integration of IoT technologies into clinical and sustainable healthcare systems. In this paper, the author proposed a system model known as S_n, which was specifically intended for clinical purposes. The C_d structure in this model is realized with a four-layered architecture, which supports systematic data acquisition, processing, and communication in medical IoT environments. This multi-level design provides efficient patient data handling and enables scale-out deployment for healthcare facilities (Figure 13.7).

1. A sensing layer
2. Machine achievement
3. The integrity layer of internet
4. The service layers

Based on Ullah *et al.*, as well as Stankovic insights, multi-tiered system architecture provides efficient and effective platforms for harvesting patient health

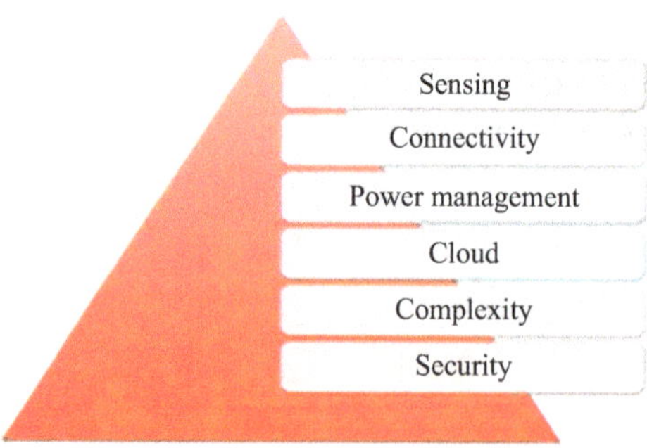

Figure 13.7 Challenges of IoT

information using mobile devices. Layered architecture provides structured data flow and improves the dependability of healthcare monitoring systems in real-world settings.

In another study, Rajavi *et al.* proposed a hyper-energy harvesting transceiver designed for Internet of Health Services. The system incorporated a high-efficiency energy digitizer to store energy for prolonged operation in medical devices. Through time-division duplexing, the transceiver enabled bilingual, full-duplex voice communication at 1.85 GHz. The measured data rates were 7.2 Mbps for Tx and 1.8 Mbps for Rx, while the analogous energy consumption rates were 55 μW for Tx and 9.4 μW for Rx. This innovation is a major improvement in low-power, high-performance healthcare-oriented IoT system communication technologies (Figure 13.8).

Stankovic mapped the increased scholarly interest in gaining insights into the root causes of smart environments, such as intelligent buildings, intelligent transportation systems, and intelligent cars, within the grander concept of a connected "smart planet." The research examined how wearable technologies, smart objects, WSNs, and novel information security systems interplayed with one another in the IoT platform. Some of the most important research challenges identified in this work are scalability of big networks, complexity of data structures, real-time response, openness of systems, robustness, security, privacy, and human-device interaction each of which is still a top priority area in IoT development.

Grounded in real-world applications, Marsico presented a working prototype that harmonizes IoT-based infrastructure with paradigms of public cloud computing. The platform was designed as a working commercial-grade model for monitoring electricity consumption and production, integrating IoT technologies with

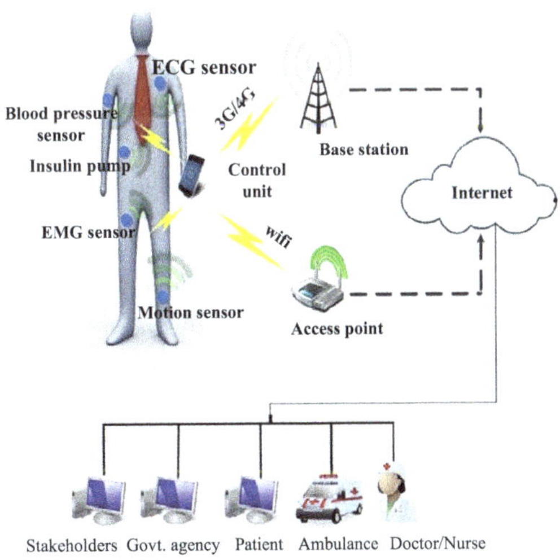

Figure 13.8 K-healthcare system

CC architecture. To make this happen, three key construction elements were necessary, providing real-time data visualization and fluent communication between physical sensors and cloud services.

1. Sensors and actuators: to get data from the real realm but then just initiate motion.
2. IoT entry point: It unites the core of IoT and distant applications.

Cloud platforms like Microsoft Azure have become robust solutions for computing and storage requirements in IoT-based healthcare systems. These platforms enable scalable infrastructure for data processing and management, facilitating effective integration of garage systems and computational frameworks in health monitoring applications [17].

In a similar contribution, Liu *et al.* [18] developed a remote adaptive medical monitoring system intended to gather physiological parameters from patients and send the data to a central monitoring center. The system allows for real-time, intelligent health monitoring through the use of smart IoT technologies to monitor patient conditions remotely. This method greatly increases access to healthcare services while decreasing the necessity for face-to-face consultations [18].

Rathore *et al.* [17] proposed an IoT-based emergency medical treatment framework developed on the Hadoop Distributed File System platform. The system, which is meant to operate in massive-scale medical emergencies, leverages data gathered from thousands—or even millions—of sensor-enabled devices that are strapped to patients. This data is dispersed throughout the network and processed by adaptive computing mechanisms to provide secure and scalable access. The architecture utilizes high-performance hardware in the form of quad-core Intel microcontrollers (HPUs) and dedicated devices for statistical computation and candidate data analysis. The software layer comprises Apache Kafka, which is run on a TMi5 tablet with Debian 14.04 LTS, providing effective message queuing and real-time data streaming in a mobile healthcare environment [17].

Wang proposed a wearable Internet mote designed specifically for ubiquitous healthcare applications. The device features a bare titanium housing integrated with a 6LoWPAN (IPv6 over low-power wireless personal area networks) interface, enabling low-power wireless connectivity in body-area networks. This design facilitates seamless, continuous monitoring of physiological data, allowing healthcare providers to access patient information remotely and in real time. The integration of 6LoWPAN ensures interoperability with modern IoT infrastructures while maintaining energy efficiency and compact form factors essential for wearable medical technology (Figure 13.9).

13.2.1.1 IoT implementation in the medical field

The use of the IoT in healthcare has shown enormous capacity to increase healthcare provision using high-tech techniques. During the COVID-19 pandemic, IoT is a revolutionary idea that provides enhanced patient care, such as distant monitoring and assistance for serious procedures. Research has emphasized IoT's role in

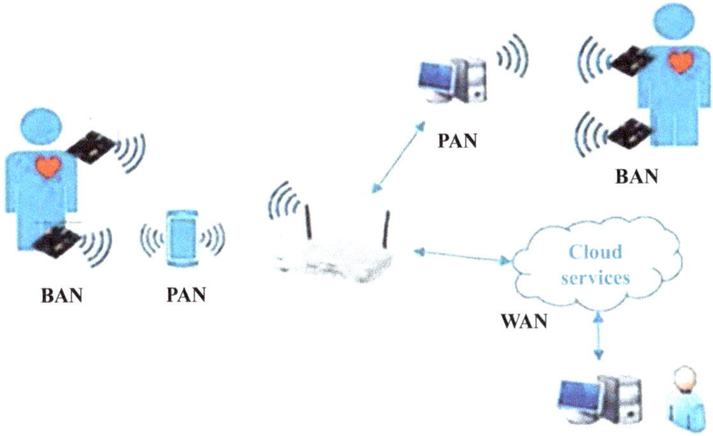

Figure 13.9 Overview of 6LoWPAN

enabling digital handling of complicated illnesses in the context of continuous healthcare crises [19,20,21]. Such applications have immense social and operational assistance for healthcare professionals, institutions, and patients by coupling new data science approaches. Plans for implementation are being carefully planned to be both efficient and scalable. As represented in Figure 13.1, IoT devices are utilized to track the health status of patients and obtain sensitive physiological information. All the electronic devices that are part of the system are internet-enabled, enabling ongoing system monitoring and real-time transfer of medical data to healthcare providers in real time directly, customized according to their unique diagnostic and treatment requirements.

13.2.1.2 Technologies of IoT for the healthcare during COVID-19 pandemic

IoT technologies couple machines, systems, and spectrometers together in order to form detailed data structures specific to each COVID-19 patient's medical needs. This is done following an interdisciplinary agenda with the intention of enhancing diagnostics, quality of treatment, and general knowledge of possible diseases. These systems can identify deviations in key patient data and define relevant information for clinical analysis [22]. Several IoT-based solutions have made a tremendous impact in improving healthcare services throughout the COVID-19 pandemic. Data from medical sensors can be retrieved automatically, stored, and processed using IoT platforms. All data for research purposes are kept electronically, and demographic data for patients can be relayed effectively through internet-based systems. This ability is particularly useful in times of crisis, enabling quick access to critical information and enabling healthcare professionals and researchers to react appropriately and in concert.

Sun *et al.* [23] emphasized the considerable promise of smart sensors for monitoring and managing essential clinical parameters like body temperature, insulin levels, and pulse rate, as well as monitoring changes in the functional status of those infected with COVID-19. Such technologies allow for real-time monitoring and support more precise patient evaluations. As per Li *et al.* [24], software is crucial in augmenting communication protocols and overseeing system operations effectively. The entire recorded medical data are encrypted to secure patient privacy while facilitating long-term care planning. Artificial intelligence (AI) integration also boosts the accuracy, reliability, and cost savings of medical decision-making, facilitating healthcare professionals to treat and diagnose patients more effectively. AI-based systems can minimize patient discomfort and enable early detection of diseases like bone degeneration, enabling timely interventions. Several actuators also allow physical control of medical devices, reacting to system commands to modify or activate parts as required. High-quality IoT solutions, especially those backed by computer simulation, are instrumental in enhancing both procedural precision and data fidelity in medical practice.

13.2.1.3 IoT enabled healthcare helpful during COVID-19 pandemic

The IoT has been profoundly and positively influential on clinical healthcare, enhancing the quality of life of many people. In addition to illness monitoring, IoT allows for steady data capturing from medical devices and facilitates personalized care using smart health management systems. Contemporary technologies can monitor day-to-day activities, schedule reminders, keep track of physical activity, log calorie consumption, measure pulse rate, and regulate disease states—functions that have proven particularly useful for tackling the COVID-19 pandemic [25,26]. Numerous high-priority IoT research projects were pursued during the pandemic to enhance responsive and accessible healthcare services by means of automation as well as remote monitoring.

IoT has brought a new era of software-driven innovation in the healthcare industry, especially post-COVID-19 pandemic. It is a useful technology for ensuring real-time coordination among patients and medical personnel, drastically decreasing waiting times and improving operational efficiency. With the provision of diverse intelligent healthcare solutions, IoT makes the environment more comfortable and assuring for patients. Inpatient treatment has been significantly enhanced by incorporating cutting-edge technologies like blood plasma analyzers, temperature control systems, intelligent pillows, glucose meters, MRI machines, and X-ray equipment. Furthermore, IoT is used in both the alteration of biological systems and the replacement of inorganic materials in medical interventions [27,28].

Some of the main characteristics of IoT systems are their application in surgical devices, networked diagnostic equipment, and facilitation of clinical research. The technologies allow medical professionals to provide best-in-class care through drug dispensing, patient monitoring, blood analysis, and primary healthcare meeting important needs that were particularly highlighted by the COVID-19 pandemic [29,30].

In healthcare facilities, IoT systems provide integrated analytics platforms that can record and analyze each patient's activities related to their health digitally. These systems utilize sophisticated data analytics to inform decision-making and potentially help manage potential future public health emergencies like the COVID-19 pandemic. Through the analysis of sophisticated datasets, such tools are able to identify best courses of action in life-threatening medical emergencies and report patient status in real time without a break [11,31,32,33]. Through the monitoring of patients' health status in real-time, the system can identify early symptoms of potential diseases and suggest preventive treatments. Furthermore, it facilitates diagnostic testing- e.g., for allergies- and serves as a good reminder system for taking medication.

13.3 Future work

In the very near future, IoT will be monitoring the vital signs of patients in real time and in smart healthcare settings. Early and precise data collection will facilitate prompt interventions, averting complications associated with prescriptions and treatments for COVID-19 patients. The use of advanced databases can greatly improve healthcare delivery by enabling clinicians to make more informed and optimized choices. Being an expanding digital platform, IoT provides enormous potential for rendering customized hospital care, and enabling sophisticated examination of medical data, diagnostics, and testing.

IoT technologies are also being used to streamline healthcare supply chains to provide the right resources at the right time and location. Smart systems are becoming increasingly autonomous in their ability to function, taking advantage of data storage capabilities through both public and private cloud infrastructures. These innovations enable disease identification and distributed health database integration, lessening dependency on manual record-keeping and making information more accessible.

This digital shift improves the efficiency of healthcare systems and facilitates evidence-based decision-making, especially in time-critical situations. Biomedical technology progressed at a faster pace during the COVID-19 pandemic, making care more sustainable and effective. These technologies are likely to influence the future of patient-centric healthcare, enabling people to live healthier lives and making systems better prepared to deal with future crises. These technologies also facilitate the wider objectives of the Med 4.0 framework, making healthcare ecosystems intelligent.

13.4 Discussion

The IoT has had a transformative impact on the medical field, enhancing infrastructure and strengthening security frameworks throughout the COVID-19 pandemic. It has contributed to improved medical governance and facilitated the digitization of elective surgical procedures. IoT applications in healthcare extend to real-time patient monitoring using wireless sensor systems. These devices enable clinicians to closely track patients' vital signs and provide timely interventions. Moreover, IoT-based platforms can assess environmental factors such as weather conditions that may influence public health risks, offering early warnings for population safety.

During the pandemic, IoT played a vital role in streamlining healthcare operations by transmitting verified medical data, thus supporting accurate supervision of pharmaceuticals and treatment plans. Intelligent systems powered by IoT have also improved the delivery of essential equipment and medications, ensuring they reach the correct patient efficiently. By supporting data-driven resource management, these technologies help reduce hospital waste and prevent potential medical errors or equipment misuse. Additionally, IoT systems aid in preventing theft of high-value medical devices through real-time tracking and alerts.

One of the major benefits of IoT in healthcare is its ability to provide relevant, timely, and secure information to medical professionals, thereby reducing the need for high-risk experimental procedures on patients. The integration of innovative technologies into complex medical environments has enhanced the response to crises such as the COVID-19 pandemic. IoT not only facilitates the development of advanced medical tools but also plays a life-saving role by supporting critical care delivery and emergency intervention. Ultimately, due to its efficiency, adaptability, and broad utility, IoT has driven substantial progress in modern medicine.

13.5 Conclusion

The promoted human fitness monitoring device can prove to be highly effective in emergency situations since it can be traced, stored, and saved as a daily database. In fate, the IOT device and cloud technology might actually be combined so that all hospitals share a database for intensive care and treatment.

Despite the internet age as it remains at its youthful stage, it is capable of a top-notch effect on that human care sector and various organizations. It is easy to pursue human beings and various objects due to fast internet and cutting sensing devices. Researchers are beginning to erstwhile enumerate technological advancements for the health machine. This paper provides deeper understandings of the internet of factors-based completely healthcare packages, allowing technology, present challenging situations and issues of healthcare.

The worldwide spread of COVID-19 continues to impact thousands of people daily, posing a serious public health issue. Early and proper treatment is essential in preventing mortality. In order to aid effective delivery of care, a number of preventive measures like ongoing temperature checking, pulse oximetry, and heart rate monitoring have been implemented. One of the major issues with COVID-19 patients is the sudden drop in oxygen levels, which, if not treated immediately, can be fatal.

To overcome this challenge, an intelligent health monitoring system using IoT technology has been created. The system works through a mobile app, allowing both patients and healthcare professionals to send and receive real-time notifications. Its portability and remote accessibility enable it to be used in almost any location. Being an IoT-based platform, the system is also scalable, with the possibility of adding more sophisticated features in the future.

It encompasses several hardware and software elements with each playing an individual role that contributes to its functionality. It also provides the practical implementation schemes, hence posing as a promising tool for dealing with healthcare problems. The system is meant to assist not only those with COVID-19 but also those patients with chronic respiratory diseases like COPD and the flu. Its usability, flexibility, and affordability allow it to track patient health across geographical distance, including in underserved or distant areas.

Through early health notifications to both the patient and medical personnel, the system allows intervention before critical circumstances can develop. General adoption of smart health monitoring technologies such as these may significantly curb hospitalizations, especially in countries like Bangladesh where medical care may be limited. Detection of impending health danger earlier empowers a person to initiate measures that will save his life in the end.

In summary, healthcare technology based on IoT can revolutionize clinical practice and increase human longevity worldwide. Its value in enhancing safety for patients, optimizing the efficiency of treatment, and aiding medical infrastructure particularly during pandemics is making it an essential part of future medicine.

References

[1] Nasajpour, M, S Pouriyeh, R M Parizi, M Dorodchi, M Valero, and H R Arabnia. 2020. "Internet of Things for current COVID-19 and future pandemics: An exploratory study." *Journal of Healthcare Informatics Research.* 4 (4): 325–364.

[2] Bai, L. 2020. "Chinese experts' consensus on the Internet of Things-aided diagnosis and treatment of coronavirus disease 2019 (COVID-19)." *Clinical eHealth* 3:7–15.

[3] Singh, R P, M Javaid, A Haleem, and R Suman. 2020. "Internet of things (IoT) applications to fight against COVID-19 pandemic." *Diabetes & Metabolic Syndrome* 14 (4): 521.

[4] Alqahtani, F H. 2018. "The application of the Internet of Things in healthcare." *International Journal of Computer Applications* 180 (18): 19–23.

[5] Vafea, M T, *et al.* 2020. "Emerging technologies for use in the study, diagnosis, and treatment of patients with COVID-19." *Cellular and Molecular Bioengineering* 13 (4): 249–257.

[6] Singh, S, A Bansal, R Sandhu, and J Sidhu. 2018. "Fog computing and IoT based health-care support service for dengue fever." *International Journal of Pervasive Computing and Communications* 14 (2): 197–207.

[7] Lai, Y L, Y H Chou, and L C Chang. 2018. "An intelligent IoT emergency vehicle warning system using RFID and Wi-Fi technologies for emergency medical services." *Technology and Health Care* 26 (1): 43–55.

[8] Rath, M, and B Pattanayak. 2019. "Technological improvement in modern health care applications using Internet of Things (IoT) and proposal of novel health care approach." *International Journal of Human Rights* 12 (2):148–162.

[9] Aggarwal, K, K R Joshi, and Y Rajavi, *et al.* 2017. "A millimeter-wave digital link for wireless MRI." *IEEE Transactions on Medical Imaging*. 36 (2): 574–583.

[10] Senthamilarasi, C, J J Rani, B Vidhya, *et al.* 2018. "Predicting students yearly performance using neural network: a case study of BSMRSTU." In *2016 5th International Conference on Informatics, Electronics and Vision (ICIEV)*, p. 2016.

[11] Haleem, A, and M. Javaid. 2019. "Industry 5.0 and its applications in orthopaedics." *Journal of Clinical Orthopaedics* 10 (4): 807–808.

[12] Suresh, M, C Ravisankar, Swayamsiddha, S, *et al.* 2018. "Integrated performance evaluation of the smart body area networks physical layer for future medical and healthcare IoT." *Sensors* 19 (1): 30.

[13] Park, A, H Chang, and K J Lee. 2018. "How to sustain smart connected hospital services: an experience from a pilot project on IoT-based healthcare services." *Healthcare Informatics Research* 24 (4): 387–393.

[14] Javaid, M, A Haleem, R Vaishya, S Bahl, R Suman, and A Vaish. "Industry 4.0 technologies and their applications in fighting COVID-19 pandemic." *Diabetes & Metabolic Syndrome*.

[15] Sofić, A, and J B Husić, and J A Stankovic. 2016. "Research directions for the Internet of Things." *IEEE Internet of Things Journal* 1 (1): 3–9.

[16] Nasri, F, N Moussa, A Mtibaa, *et al.* 2020. "IoT in the wake of COVID-19: A survey on contributions, challenges and evolution." *IEEE Access* 8:186821.

[17] Rathore, M M, A Ahmad, A Paul, Ray, S, Y Jin, and A Raychowdhury. 2016. "The changing computing paradigm with internet of things: a tutorial introduction." *IEEE Design & Test* 33 (2): 76–96.

[18] Liu, Y, B Dong, B Guo, J Yang, and W Peng. 2015. "Combination of cloud computing and internet of things (IOT) in medical monitoring systems." *International Journal of Hybrid Information Technology* 8(12): 367–376.

[19] Azizy, A, M Fayaz, and M Agirbasli. 2020. "Do not forget Afghanistan in times of COVID-19: telemedicine and the Internet of things to strengthen planetary health systems." *Omics: a Journal of Integrative Biology*, 24.

[20] Haleem, A, and M Javaid. 2019. "3D scanning applications in medical field: a literature-based review." *Clinical Epidemiology and Global Health* 7 (2): 199–210.

[21] Ruiz-Fernández, D, D Marcos-Jorquera, V Gilart-Iglesias, *et al.* 2017. "IoT driven healthcare system for remote monitoring of patients." *International Journal for Modern Trends in Science and Technology* 3 (6): 100–103.

[22] Končar, J, A Grubor, R Marić, S Vučenović, and G Vukmirović. 2020. "Setbacks to IoT implementation in the function of FMCG supply chain sustainability during COVID-19 pandemic." *Sustainability* 12(18): 7391.

[23] Sun, E, X Zhang, and Z Li. 2012. "The internet of things (IOT) and cloud computing (CC) based tailings dam monitoring and pre-alarm system in mines." *Safety Science* 50 (4): 811–815.

[24] Li, C T, T Y Wu, C L Chen, C C Lee, and C M Chen. 2017. "An efficient user authentication and user anonymity scheme with provably security for IoT-based medical care system." *Sensors* 17 (7): 1482.

[25] Janeh, O., O Fründt, B Schönwald, A Gulberti, C Buhmann, and C Gerloff. 2019. "Gait training in virtual reality: short-term effects of different virtual manipulation techniques in Parkinson's disease." *Cells* 8 (5): 419.

[26] Khan, R, J Plahouras, B C Johnston, M A Scaffidi, S C Grover, and C M Walsh. 2019. "Virtual reality simulation training in endoscopy: a Cochrane review and meta-analysis." *Endoscopy* 51 (7): 653–664.

[27] Farahani, B, F Firouzi, V Chang, M Badaroglu, N Constant, and K Mankodiya. 2018. "Towards fog-driven IoT eHealth: Promises and challenges of IoT in medicine and healthcare." *Future Generation Computer Systems* 78 (2): 659–676.

[28] Ma, Y, C Wu, K Ping, *et al.* 2019. "IoT in health-care: Achieving interoperability of high-quality data acquired by IoT medical devices." *Sensors* 19 (9): 1978.

[29] Deng, Y Y, C L Chen, W J Tsaur, Y W Tang, and J H Chen. 2017. "Internet of Things (IoT) based design of a secure and lightweight body area network (BAN) healthcare system." *Sensors* 17 (12): 2919.

[30] Sivathanu, B. 2018. "Adoption of internet of things (IOT) based wearables for healthcare of older adults a behavioural reasoning theory (BRT) approach." *Journal of Enabling Technologies* 12 (4): 169–185.

[31] Hossam, A, A Magdy, A Fawzy, *et al.* 2017. "Action research on development and application of Internet of Things services in hospital." *Healthcare Informatics Research* 23 (1): 25–34.

[32] Janet, B, P Raj, Javaid, M, and A Haleem. 2019. "Industry 4.0 applications in medical field: a brief review." *Current Medicine Research and Practice* 9 (3): 102–109.

[33] Yang, Z, Q Zhou, L Lei, K Zheng, and W Xiang. 2016. "An IoT-cloud based wearable ECG monitoring system for smart healthcare." *Journal of Medical Systems* 40 (12): 1–11.

Chapter 14

The potential and challenges of ChatGPT in medical applications: a comprehensive review

V. Aravindan[1], P. Anirudh[1], Rajkanwar Singh[1], Patil Krishna Reddy[1], Chandan Vishwas[2] and Sukanta Nayak[3]

Abstract

This chapter delves into the fascinating realm of ChatGPT's medical applications, exploring its potential to revolutionize diagnosis, treatment, and patient interaction. Through this study, we embark on a journey to explore the current landscape and analyze the tools being a step toward future where AI fosters a healthier world. By finding gaps and navigating ethical considerations, it is illuminated the path forward, ensuring ChatGPT's contributions are both impactful and responsible. ChatGPT is a language model developed by OpenAI that uses deep learning techniques to generate human-like responses to natural language inputs. Here, it is discussed the history of ChatGPT, its applications in various fields, and importance in the medical domain. Further, the potential impact of using ChatGPT in medical research, education, and clinical care, as well as the challenges and limitations associated with its use, is highlighted.

Natural language models like ChatGPT, developed by OpenAI, hold immense potential to transform the future of healthcare. This research delves into the exciting possibilities of medical applications for ChatGPT, exploring its potential to change diagnosis, treatment, and patient interaction. The aim of this chapter is to show the use of ChatGPT in medical fields where it could improvise efficiency and potentially perform better at some case scenarios highlighting applications in medical diagnosis, mental health support system, monitoring administrative tasks, aiding the patients in communication with their practitioner and provide a patient a better way for second opinion, live monitoring for people suffering from mental illness, thus exploring the possible benefits and some downsizes of the same.

This chapter extensively examines the advancement of ChatGPT and its use in the healthcare sector, while also highlighting important gaps and ethical issues in its current implementation in medicine. It introduces a suggested framework for

[1]School of Computer Science and Engineering, VIT-AP University, India
[2]School of Social Sciences and Humanities, VIT-AP University, India
[3]Department of Mathematics, School of Advanced Sciences, VIT-AP University, India

incorporating ChatGPT into healthcare settings, backed by unique visuals, and juxtaposes current research to highlight progress and persistent obstacles. It is structured as follows: Section 14.1 introduces the fundamentals of subject line; Section 14.2 explains the mechanism and function of ChatGPT; Section 14.3 describes the research methods; Section 14.4 provides a review of literatures; Section 14.5 showcases the proposed system; Section 14.6 talks about the outcomes; Section 14.7 outlines key contributions and organization; Section 14.8 discusses limitations and future directions; and finally, Section 14.9 includes the concluding remarks.

Keywords: ChatGPT; medical applications; AI in healthcare; diagnosis; patient interaction; ethical considerations; AI research

14.1 Introduction

ChatGPT, an iteration of OpenAI's renowned language model, has seen remarkable evolution in performance over the years. Beginning with earlier versions like GPT-2 and progressing through GPT-3 and beyond, each iteration has brought significant improvements in language understanding, coherence, and response generation.

GPT-2 is introduced in 2019. It marked a significant leap in natural language processing (NLP) capabilities. It highlighted the potential of large-scale language models by generating coherent and contextually relevant text across a wide range of topics. However, limitations in model size and training data led to occasional inconsistencies and lack of deep understanding. Whereas GPT-3 is launched in 2020. It brought a breakthrough in language generation. With 175 billion parameters, it surpassed its predecessors in both scale and performance. GPT-3 proved enhanced contextual understanding, improved coherence, and the ability to generate more human-like responses. Its versatility allowed for various applications, from Chabot's to content generation and even code completion. While specific details about future iterations like GPT-4 may not be available, it is reasonable to expect further advancements in performance. With continued research in NLP, future versions of ChatGPT are likely to show even greater language understanding, contextual reasoning, and generation capabilities.

The evolution of ChatGPT has paved the way for its integration into various industrial and current applications across different sectors. Some notable developments include:

- Customer support chatbots: ChatGPT powers customer support chatbots, enabling businesses to provide instant responses to customer queries. Its ability to understand natural language allows for personalized interactions, improving customer satisfaction and reducing response times and human intervention.
- Content creation: Content creators utilize ChatGPT to generate articles, blog posts, product descriptions, and more. By providing prompts or outlines, users

can leverage ChatGPT to produce coherent and relevant content, saving time and effort in the content creation process.

- Language translation: ChatGPT facilitates language translation by providing real-time translation services. Its ability to understand and generate text in multiple languages make it valuable for overcoming language barriers in communication.
- Personal assistants: ChatGPT serves as the backbone for personal assistants, providing users with aid in tasks such as scheduling appointments, setting reminders, answering questions, and providing recommendations based on user preferences.
- Educational tools: In the education sector, ChatGPT is used as an educational tool for language learning, essay writing help, and tutoring. It can provide explanations, answer questions, and offer learning resources tailored to individual student needs.
- Healthcare applications: In healthcare, ChatGPT is employed for patient engagement, virtual health assistants, and answering medical inquiries. It helps healthcare professionals by providing relevant information and guidance to patients.
- Creative writing and storytelling: Writers and storytellers use ChatGPT for creative inspiration, character development, and plot generation. By interacting with ChatGPT, writers can overcome creative blocks and explore new narrative possibilities.

14.2 ChatGPT, its mechanism and utility

ChatGPT, developed by OpenAI, stands for a big leap in NLP. As an artificial intelligence (AI)-powered language model, it combines the sophistication of machine learning with the nuances of human conversation. Let us delve into what makes ChatGPT remarkable.

- The genesis of ChatGPT: At its core, ChatGPT is a transformer-based neural network. It builds upon the foundation laid by its predecessor, GPT-3.5, which itself was a marvel in language modeling. GPT stands for Generative Pre-trained Transformer, emphasizing its ability to generate coherent and contextually relevant text.
- Fine-tuning and adaptability: ChatGPT's journey begins with pre-training on vast amounts of text data. During this phase, it learns grammar, semantics, and world knowledge. But what sets ChatGPT apart is its fine-tuning process. Unlike a static model, ChatGPT adapts to specific tasks and contexts.
- Reinforcement learning with human feedback (RLHF): To refine ChatGPT's conversational abilities, OpenAI employs RLHF. It works with the following steps. Step 1: ChatGPT learns from examples provided by human experts. These demonstrations guide it toward desirable behavior. Step 2: Users rank different responses, allowing ChatGPT to learn from their preferences.

- Conversational flexibility: ChatGPT is not confined to rigid scripts or templates. It can engage in open-ended conversations across diverse topics. Whether you seek trivia, explanations, or creative content, ChatGPT adapts.
- Use cases and applications: The major utility includes (i) its conversational flow feels remarkably human, (ii) ChatGPT gives concise answers, (iii) it is a versatile assistant, and (iv) students can use ChatGPT for explanations, expanding their understanding of complex concepts.
- Future of AI interaction: ChatGPT is a step toward more natural and intuitive interactions with AI. As models evolve, we will see even more sophisticated language understanding and generation.

14.3 Methodology

In the following, the general framework of method is discussed. In Figure 14.1, a flow diagram of the involved procedure is presented.

The implementation of ChatGPT in the medical field presents a promising approach to enhance supply chain operations. A structured review method is needed to provide a comprehensive overview of the current state of knowledge on ChatGPT in the medical field. This research method aims to conduct a systematic literature review to examine the existing applications of ChatGPT in the medical field, name gaps and limitations, and propose future research directions.

14.3.1 Research questions

This study seeks to investigate the complex role of ChatGPT in the medical area by posing three fundamental research issues. First, it aims to show and investigate ChatGPT's present applications in medical practice and research, focusing on diagnostics, patient engagement, and medical education. Second, the study will investigate the major problems and constraints of using ChatGPT in medical settings, such as accuracy, ethical considerations, and data protection. Finally, the study aims to find possible areas for future inquiry and development and propose future research approaches that could improve the efficacy and integration of ChatGPT in medical situations. The study's goal is to help people comprehend ChatGPT through a thorough investigation.

14.3.2 Literature survey

The literature survey will be conducted using the Scopus database, which is one of the largest databases for peer-reviewed scientific literature and has significant

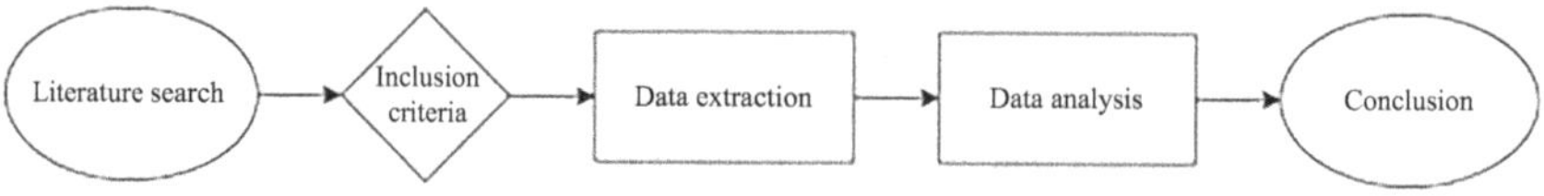

Figure 14.1 Schematic diagram of the method

overlaps with other databases such as Web of Science and Google Scholar. The search terms "ChatGPT's" "Medical ChatGPT," and "Generative Pre-trained Transformers" will be used to show relevant papers published between 2017 and 2022.

14.3.3 Inclusion and exclusion criteria

The following inclusion criteria are used here.

(a) The publication should be in English.
(b) The publication should be related to ChatGPT in the medical field.
(c) The publication should be published between 2017 and 2022.

The following exclusion criteria are used here.

(a) The publication is not related to ChatGPT or ChatGPT in the medical field.
(b) The publication is not in English.
(c) The publication is not available in full text.

14.3.4 Data extraction

After applying the inclusion and exclusion criteria, relevant publications will be downloaded, and the data extraction process will begin. The following data points will be extracted from each publication.

(a) Title of the publication
(b) Author name(s)
(c) Year of publication
(d) Research method used
(e) Key findings
(f) Limitations
(g) Future research directions

14.3.5 Data analysis

The data analysis approach will form both quantitative and qualitative methods. Initially, a descriptive analysis will be conducted to examine the frequency of publications, research methodologies, and primary discoveries. Then, a thematic analysis will be performed to find common themes and patterns across all articles. The main limitations and constraints of medical ChatGPT and potential areas for further investigation will be decided based on the outcomes of the data analysis.

14.3.6 Concluding remarks

This research method aims to provide a comprehensive understanding of the current applications of ChatGPT in the medical field, name gaps and limitations, and propose a research agenda for the future. The approach involves a systematic literature review, with clear inclusion and exclusion criteria, data extraction, and a combination of quantitative and qualitative methods for data analysis. By conducting this study, the findings can be used to gain insights into the current state of knowledge on medical application of ChatGPT, find major challenges and limitations, and provide guidance for future research in this area.

14.4 Literature review

Based on the method of the comprehensive survey the following important articles are gathered for study.

- Wang *et al.* [1] discuss the integration of large-scale AI models, like ChatGPT, into biomedical research and healthcare. They review the potential of AI to enhance healthcare quality, including diagnosis, patient monitoring, and personalized care and examines large AI models, their complexity, and the advancements in GPU programming that may make them more accessible to healthcare. They explore how some large-scale AI models show emergent abilities, improving performance on tasks as model size increases. Potential applications in healthcare are highlighted, such as medical record abstraction and AI-assisted diagnosis while acknowledging the challenges in implementing these models, emphasizing the need for alignment with human values and goals. Finally, they suggest that future research should focus on deploying these AI models in healthcare to improve patient outcomes and reduce workloads.
- Yanagita *et al.* [2] evaluate the accuracy of ChatGPT, in answering medical questions from the National Medical Licensing Examination in Japan by including all 400 questions from the 2022 examination, excluding those with figures, tables, and testing ChatGPT versions GPT-3.5 and GPT-4 by inputting the questions in Japanese and assessing the correctness of the responses by two general practice physicians. GPT-4 achieved an 81.5% correct response rate, surpassing the passing standard for the examination while GPT-3.5 had a lower correct response rate of 42.8% thus suggesting that as AI models continue to learn, they could become valuable decision support systems for medical professionals.
- Haupt and Marks [3] explore how AI tools, particularly in radiology and dermatology, may outperform physicians in specific tasks, raising questions about the future of certain medical specialties. They also address concerns about the ethical use of AI, such as whether it is conscionable for physicians to use AI without fully understanding it, and the potential for AI to be considered a standard of care, highlight the limitations of AI, including the inability to replace the human touch in medicine, and the importance of human judgment in clinical decisions and suggest that future AI tools could improve by filtering out low-quality information sources, enhancing the accuracy and reliability of medical advice generated by AI.
- Roos *et al.* [4] discuss the use of AI in medical education, specifically comparing the performance of ChatGPT, Bing, and medical students in Germany on the German Medical State Examinations of 2022. Ginson *et al.* assess and

compare the performance of three large language models (LLMs) and medical students by testing them on 630 questions from the 2022 exams [26], evaluating their ability to answer correctly using statistical methods like ANOVA and t-tests were used for comparison.

- Cheng *et al.* [5] discuss the role of AI, specifically GPT-4, in sports medicine emphasizing that GPT-4 would not replace human doctors but will serve as a valuable assistant. The authors conducted an online survey and reviewed literature to explore GPT-4's applications in sports medicine and named potential uses in diagnostic imaging, exercise prescription, medical supervision, surgery treatment, sports nutrition, and scientific research. GPT-4 is seen as an indispensable tool for future sports medicine, aiding but not replacing human ability.
- Johnson *et al.* [6] discuss the evaluation of AI-generated medical responses, specifically assessing the Chat-GPT model's accuracy and reliability. 33 physicians across 17 specialties answer 284 medical questions of varying difficulty (easy, medium, hard). Physicians rate Chat-GPT's answers using a 6-point Likert scale for accuracy and a 3-point Likert scale for completeness. Accuracy median score was 5.5, showing answers were mostly correct. Completeness median score was 3, suggesting answers were complete and comprehensive. Questions requested showed significant improvement in accuracy over time.
- Caruccio *et al.* [7] aim to define a new medical diagnostic bot by analyzing the benefits, limitations, and implications of using AI tools like ChatGPT in healthcare. They discuss the use of AI in medical diagnostics, comparing ChatGPT with traditional machine learning models and other LLMs for diagnosing diseases based on symptoms. They evaluate ChatGPT's performance using different engines and a new prompt engineering method tailored for exact diagnostics and compare these results with traditional predictive models, Google BARD, and two domain-specific NLP models. Experiments were conducted using two medical datasets for disease prediction, which included over 100 symptoms associated with various diagnoses and a new interactive bot was proposed, based on the best-performing models from the study.
- Nori *et al.* [8] present an evaluation of GPT-4's capabilities in medical contexts. They observe GPT-4 exceeds the passing score for the United States Medical Licensing Examination (USMLE), a medical licensure exam in the USA. It outperforms both general-purpose models like GPT-3.5 and specialized medical models. GPT-4 shows improved probability calibration, predicting the likelihood of its answers being correct, which is crucial for medical applications. They discuss potential uses of GPT-4 in medical education, assessment, and clinical practice, emphasizing accuracy and safety challenges.
- Wang *et al.* [1] used datasets from the China National Medical Licensing Examination and the China National Entrance Examination for Postgraduate Clinical Medicine Comprehensive Ability to assess ChatGPT's performance. It also tested ChatGPT's ability to generate discharge summaries and ease group learning in a problem-based learning context. GPT-4 showed significant

improvements over GPT-3.5, particularly in understanding Chinese medical tasks. GPT-4 achieved high accuracy rates on medical exams and proved promising potential in discharge summarization and group learning.

- Wang *et al.* [1] saw that ChatGPT achieved an impressive 81.5% correct response rate on these medical questions. This study suggests that while AI systems like ChatGPT have potential in medical applications, there are still hurdles to overcome, particularly in ensuring accuracy and addressing ethical concerns.
- Balas *et al.* [9] concluded that GPT-4 shows promise in addressing medical ethical issues but requires further development to handle nuanced dilemmas effectively before practical implementation in clinical settings. A survey with six ethicists assessed GPT-4's responses to eight ethical vignettes. GPT-4 performed well in technical and non-technical clarity but lacked depth and acceptability. Although it found key ethical issues, it struggled with nuanced aspects of dilemmas, suggesting the need for further refinement before clinical use. Use of ChatGPT could potentially accelerate the pace of diagnosis but all results must be cross verified by a practitioner.
- Garg *et al.* [10] examine ChatGPT's role in patient care and medical research, following PRISMA guidelines for systematic review. Results suggest ChatGPT aids in patient inquiries, note writing, decision-making, trial enrollment, and research support but raises concerns about accuracy, bias, and authorship legitimacy. It concludes that while ChatGPT is valuable in clinical and research contexts, human judgment stays essential, and ethical challenges persist.
- Liu *et al.* [11] found ChatGPT aids in generating differential diagnosis lists, supporting clinical decision-making, and perfecting clinical decision support systems. It shows high diagnostic accuracy for common complaints and helps in cancer screening decisions. ChatGPT is effective in creating patient clinical letters, radiology reports, medical notes, and discharge summaries, enhancing efficiency and accuracy for healthcare providers. It provides reliable information about diseases and medical queries, although it is important to note that ChatGPT's responses may change over time and biases.
- Waisberg *et al.* [12] observed GPT-4's enhanced problem-solving abilities and broader knowledge base, particularly in the medical field. The method involved testing various functions of GPT-4, such as generating discharge summaries and summarizing clinical trials, to evaluate its effectiveness. In terms of summarization tasks, GPT-4 proved success in generating complete discharge summaries and providing relevant information on clinical trials, specifically for interstitial lung disease. However, when it came to image analysis, a new feature of GPT-4, limitations were seen. GPT-4 incorrectly found fundus photographs, mistaking them for a schematic of the CRISPR-Cas system, showing room for improvement in this area.
- Gilson *et al.* [13] discuss ChatGPT's performance on USMLE questions, comparing it with GPT-3 and instruct. Using AMBOSS and NBME questions, ChatGPT achieved accuracies of 42–64.4%, surpassing Instruct and

performing better than random chance against GPT-3. However, its accuracy declined with increasing question difficulty. Qualitative assessments praised ChatGPT's logic and information delivery. The findings propose ChatGPT as a promising tool for medical education and assessment, highlighting its potential utility in enhancing knowledge acquisition and evaluation in the medical field.

- Ali *et al.* [14] mentioned the use of AI, specifically OpenAI's ChatGPT, to improve the efficiency and quality of clinical letters to patients. An evaluation of ChatGPT-generated clinical letters showed elevated levels of readability, factual correctness, and humanness, suggesting potential for real-world application. This study aimed to produce clinical letters at a reading level suitable for a wide audience, adhering to the recommended sixth grade level in the USA.
- Tangadulrat *et al.* [15] showed that medical students generally have a positive feeling of ChatGPT for treatment guidance and education, while graduated doctors are more cautious, though both groups see value in using ChatGPT for creating patient educational material.
- Eysenbach [16] explores ChatGPT's potential in medical education. Notably, it performs at a level comparable to a third-year medical student. ChatGPT's ability to generate lifelike patient scenarios and personalized learning materials enhances medical textbooks and research summaries. Additionally, the paper addresses the importance of detecting machine-generated text to keep academic integrity in medical education.
- Ashraf and Ashfaq [17] critically examine the capabilities and constraints of AI language models like ChatGPT in the medical field. It highlights that while ChatGPT can offer basic guidance and clarify concepts, it lacks the ability to access the latest scientific information and original medical databases, which is crucial for up-to-date medical research. The authors caution against over-reliance on ChatGPT for complex tasks, emphasizing the irreplaceable value of the human element and real-world experiences in science. They advocate for consulting experts for reliable insights, as AI-generated content cannot substitute personal clinical experiences.
- Haze *et al.* [18] assess ChatGPT models (GPT-3.5 and GPT-4) on medical examination questions, using the Japanese National Medical Examination. GPT-4 proves enhanced accuracy (81%) and consistency (88.8%) compared to GPT-3.5. It is noted that accuracy correlates positively with available information per medical field. However, questions needing multiple answers or calculations pose accuracy risks. Despite GPT-4's advancements, caution is urged about its clinical use, especially in less studied medical domains, due to inconsistencies and potential inaccuracies.
- Chintagunta *et al.* [19] introduces an inventive strategy for summarizing medical dialogues employing GPT-3. Key components include a novel algorithm for synthetic data generation focusing on medically pertinent information, and the use of GPT-3 as the foundational element, effectively scaling human-labeled examples. By incorporating low shot learning and an ensemble method, the approach enhances the quality of synthetic training data. Results

prove that when combined with human-labeled data, this method yields summaries showing superior medical accuracy and coherency compared to those trained solely on human data. Consequently, the research suggests a substantial reduction in the requirement for extensive labeled data in medical summarization endeavors.

- Tian *et al.* [20] investigated LLMs present promising prospects in revolutionizing biomedicine by enhancing tasks like information retrieval and text summarization. This literature review conducts an extensive survey, revealing significant advancements in text generation. While LLMs hold potential in accelerating discoveries and improving healthcare, challenges such as misinformation dissemination and privacy concerns persist. Despite not yet fully transforming the field, ongoing research is crucial for refining LLMs' benefits and mitigating associated risks, ensuring their effective integration into biomedical applications.
- Wang *et al.* [1] evaluated ChatGPT's performance in Chinese medical contexts, comparing GPT-3.5 and GPT-4. Utilizing datasets from China's National Medical Licensing Examination and postgraduate clinical medicine, it assesses accuracy, verbal fluency, and hallucination classification. GPT-4 shows significant accuracy improvement over GPT-3.5 across all datasets, achieving 84%, 86%, and 82% accuracy rates, respectively. Verbal fluency exceeds 95%, showing high readability. GPT-4 also reduces hallucinations, particularly in open-domain errors. The findings suggest GPT-4's potential to aid in medical tasks and education, marking advancements in AI's healthcare role.
- Arif *et al.* [21] used ChatGPT in academic settings raises concerns about originality and critical thinking, as students may rely on it to cheat in exams and assignments, potentially losing their ability to produce original ideas and present proper arguments. Ethical issues, such as accountability for content and potential misuse, are also highlighted, alongside limitations like outdated training data and restricted database access, which question its credibility in tasks like medical literature review. Experts suggest that ChatGPT should be used as an aid for constructive writing and reviewing material, rather than as a source of original content, emphasizing the need for human oversight and policy control to ensure its responsible use.
- Biswas [22] studied ChatGPT holds promise for public health by providing accessible information on health issues and strategies for disease prevention and health promotion. It can elucidate the roles of community health workers and health educators, as well as the influence of social and environmental factors on health outcomes. However, its application in public health comes with challenges such as limited accuracy, potential biases, and the absence of direct interaction with healthcare professionals. The integration of ChatGPT into public health efforts must be approached with caution to complement and not replace professional medical advice.
- Vaishya *et al.* [23] noted that the current version of ChatGPT offers rapid responses to medical queries but primarily relies on general information from existing literature up to September 2021. While it admits its limitations in the

medical field and learns from past interactions, errors in responses have been noted. As such, it may serve as a narrative AI chatbot for medical professionals with caution advised, recommending fact-checking due to its limitations.

- Li *et al.* [24] provide one of the first systematic review and taxonomy of ChatGPT in healthcare, offering a comprehensive classification system for related publications. It critiques the current state of ChatGPT in medical applications, noting its moderate performance in various tests and its unsuitability for clinical deployment due to its original design not being intended for such purposes. The review highlights the necessity for specialized NLP models trained on biomedical datasets for critical clinical applications and warns against the uncritical adoption of ChatGPT in healthcare despite its potential. The authors stress the importance of prudence amidst the AI hype, advocating for human oversight and the development of more specialized tools tailored for healthcare applications.
- Souza *et al.* [25] discuss the potential of ChatGPT in transforming the field of medicine. They discuss that ChatGPT can analyze clinical trials and medical studies to find effective treatments. It offers personalized learning experiences for students and professionals. ChatGPT aids in diagnosis, treatment decisions, and provides quick access to medical knowledge. The model aims to reduce diagnostic errors and improve communication, enhancing treatment plans.

14.5 Medical ChatGPT

ChatGPT is an advanced language model developed by OpenAI. Leveraging deep learning techniques, it produces human-like responses to natural language inputs. It belongs to the family of generative pre-training transformer (GPT) models and is currently one of the largest publicly available language models. The advantages to using such models is the remembrance of context of the situation. It can recollect and understand references about topics within the conversation effortlessly which can be difficult when an individual has a vast history of traumatizing events, and his/her case must be dealt with in immediate situations. It can vastly improve the efficiency and working of medical systems and can further aid doctors for possibilities. Further a centralized collection of vast amounts of cases can further help these models get better over time and provide even more correct advice and suggestions to the patients as well as communicate the relevant cases of a particular patient with the doctors/caregivers to help them in their diagnosis.

14.5.1 Applications in medicine

- Clinical aid: ChatGPT [27,28] can help healthcare professionals in clinical and laboratory diagnosis. By analyzing patient data, it can provide insights and recommendations. Such diagnosis can ensure faster and well addressed suggestions without having dependence. This ensures better care and can improvise recovery times of the patient. Additionally, it becomes easier to find better solutions, hence leading to better medical practices in the future.

- Research topics: It helps to name potential research topics by processing vast amounts of medical literature. It can also be used to develop new methodologies which involve techniques combining different medical procedures from various cultures. This can bring new techniques, further improvising medical practices.
- Health updates: Medical students, doctors, nurses, and other healthcare professionals can stay informed about updates and new developments in their respective fields. Additionally, it can be used to prescribe different drugs if in case patients experience allergy due to medications.
- Virtual assistants: Developing virtual assistants powered by ChatGPT aids patients in managing their health.
- Therapeutic chatbots: Using ChatGPT, one can develop conversational agents that are programmed to provide therapeutic support for affected individuals dealing with medical conditions such as depression. These chatbots can be deployed and continuously trained upon from vast data to provide better treatments.
- Mood tracking and analysis: By interacting with individuals regularly, medical GPT can help track mood patterns and find potential triggers for medical health concerns. This analysis can be crucial in patients dealing with bipolar disorders whose moods need to be closely watched for peculiar analysis.
- Virtual therapy sessions: By using ChatGPT patients can have extended therapy sessions in scenarios where their assigned mentors can be trained upon to emulate their respective mentors so that patients can have continuous access even at odd times making their treatments more effective.
- Personalized mental health education: ChatGPT's can be used to solve issues which include people who are under stress from various causes such as exams or the passing away of a family member. It can further be aided with programs such as recreational center run by NGO's for helping and answering the tough questions which is hard to convey to others as it is a personal emotional feeling.
- Second opinion for medical decisions: Patients can use ChatGPT for further consultation and understanding their problems it can also be used to be trained upon to clarify some of the frequent questions and practices which could not have been answered by the practitioner/doctor during consultation.
- Crisis intervention and suicide prevention: Mental health support hotlines or crisis intervention services, chatbots can offer immediate help to individuals in distress. It can provide emotional support and guide individuals toward proper resources and support networks such as suicide prevention helplines.

The advantage of using such models over traditional phone calls can be dealing with vast amounts of scenarios and immediate action based on the sentiments of the statements to decide the severity of the scenario.

14.5.2 Advantages

- Nuanced responses: ChatGPT captures the intricacies of human language, generating contextually relevant answers.
- Efficiency: It can handle a broad spectrum of prompts, making it versatile for various medical tasks.
- Education: Medical students can learn from its responses, enhancing their knowledge base.
- Availability: ChatGPT can aid around the clock offering support to healthcare professionals and patients. This ensures prompt access to information in emergency situations.
- Multilingual support: ChatGPT's can be trained on diverse language dataset to provide support in multiple languages.
- Privacy and confidentiality: when integrating such models, we can redact some personal information so that the privacy of the individual stays intact. This can be further enhanced by communicating with the patient what information is specifically being used to train upon enhancing the trust around such systems.

14.5.3 Ethical considerations

- Copyright laws: Using ChatGPT for medical writing must consider copyright infringement.
- Transparency: AI-generated content should be transparent to keep trust.
- Medico-legal concerns: Legal implications need careful handling.

14.5.4 Prospects

- Improved applications: Despite limitations, further improvements can enhance ChatGPT's utility in medicine.
- Collaboration: Researchers, developers, and healthcare professionals should collaborate to explore its full potential.

14.5.5 Impact of this research

A new client can be added to the system through the client maintenance section available in the left menu of the application. With the advent of GPT-3 (Generative Pre-trained Transformer 3), developed by OpenAI, we see a potential revolution—one that could redefine patient care, research, and diagnostics. Let us delve into the impact of GPT-3 in medical contexts.

14.5.5.1 Automating medical tasks

GPT-3's advanced NLP capabilities empower it to understand and interpret medical information. This ability opens doors to automating tasks that were previously human-dependent. Impact can be brought upon by utilizing GPT as a preliminary diagnosis measure for faster and correct assessments. Imagine GPT-3 handling medical record keeping, data analysis, and even diagnosis. Automating routine tasks helps streamline administrative tasks increase efficiency and precision of medical care, healthcare professionals can focus on critical decision-making, leading to faster and more correct diagnoses and treatments. For example, with the help of GPT doctors can test their procedure sheet after a surgery has been completed.

14.5.5.2 Personalized healthcare

The sheer volume of medical knowledge grows exponentially. By analyzing and interpreting medical information, it can provide personalized healthcare solutions, treatments, and consultations. These solutions include:

- Genomic information: Based on certain factors there can be early diagnosis of certain diseases, preventive measures, potential response to medications and overall health risks.
- Lifestyle factors: Consideration of people's diet, exercise and stress levels can help aid in tailoring recommendations and medications based on the ailment.
- Patient preferences: Does the individual have certain preferences will the medication be in the form of a syrup or tablet. Building a plan and adjusting for the patient based on their lifestyle goals.
- Predictive analytics: Using data driven insights, forecasts on potential health risks and recommend proactive measures for prevention or early intervention can be done.
- Continuous monitoring: Watching the patients' health regularly helps diagnose the root cause of the ailment and change the treatment plans accordingly.

14.5.5.3 Diagnoses and patient outcomes

GPT-3's impact on medical diagnosis cannot be overstated. It can analyze patient data, correlate symptoms, and make predictions. Physicians armed with GPT-3 can make informed decisions, potentially saving lives. Moreover, better diagnoses lead to better patient outcomes—reducing suffering and enhancing quality of life. It acts as a validator for the diagnosis, can track whether the patient has a steady improvement in health.

14.5.5.4 Medical research and hypothesis generation

In the realm of medical research, GPT-3 has become a powerful ally. It can generate hypotheses based on patient data, suggesting new avenues for exploration. Assuming the medication that the patient is provided with is not improving the individual's health, change the medication offered and help it with a better one and consider these findings into the research. Researchers can leverage GPT to accelerate discoveries, improve resource use, and advance medical science. Whether it is drug development, disease modeling, or epidemiological studies, GPT-3 offers fresh perspectives to try and include over the research period.

14.5.5.5 Challenges and ethical considerations

Implementing GPT-3 in healthcare is not without hurdles. Accuracy is still paramount. Mistakes in medical advice can have life-or-death consequences.

Additionally, GPT-3's "unsupported use" status for diagnosing conditions underscores the need for caution. Balancing automation with human oversight is crucial. A doctor requires around 4.5 years of study, 1-year internship and 3 years of residency to be a qualified doctor. Here, GPT has no credible source of information and not as many trials as that of a doctor, therefore, there needs to be a qualified person to use it so that no harm can be done to the patients' health.

14.5.5.6 Beyond diagnosis: education and research

Beyond clinical settings, GPT-3 serves as an interactive encyclopedia for medical education. It can simulate patient interactions, helping students hone history-taking skills. In research, GPT-3 aids scientists in formulating questions, developing study protocols, summarizing data. Its potential extends beyond the clinic.

In [13], authors tested ChatGPT's performance on USMLE Step 1 and Step 2 exams. The NLP models executed tasks with greater than 60% threshold on the datasets presented. These were evaluated based on logical and informational context of the answers presented. The research also points out that the models were on par with a third-year medical student. This proves that ChatGPT could be used for educational and research purposes.

14.5.5.7 The road ahead

As GPT-3 evolves, its impact will ripple across healthcare. We must tread carefully, ensuring rigorous validation, ethical guidelines, and transparency. Collaborations between AI and healthcare professionals will shape this transformative journey.

14.5.6 Challenges

The challenges are mentioned below.

14.5.6.1 Data privacy and security

GPT-3 requires access to vast amounts of patient data for exact predictions. However, handling sensitive health information raises concerns about data breaches and unauthorized use. Striking a balance between data availability and privacy protection is crucial.

14.5.6.2 Biases and fairness

AI algorithms, including GPT-3, can perpetuate existing biases present in the training data. In healthcare, this could lead to unequal treatment for certain patient populations. Ensuring fairness and addressing biases is essential to avoid worsening health disparities.

14.5.6.3 Clinical validation and accuracy

GPT-3's predictions must undergo rigorous clinical validation. Mistakes in medical advice can have profound consequences. Achieving high accuracy and reliability is paramount before widespread adoption.

14.5.6.4 Dynamic interactions and error correction

GPT-3's inability to correct itself upon an error pose challenge. In dynamically changing healthcare interactions, real-time adjustments are crucial. Handling unexpected scenarios and gracefully recovering from errors is essential.

14.5.6.5 Human-AI collaboration

GPT-3 should complement human ability rather than replace it entirely. Striking the right balance between automation and human oversight is complex. Collaborative models that combine AI insights with clinical judgment are promising.

14.5.6.6 High-stakes situations

GPT-3 cannot replace healthcare providers in critical situations like emergencies. Its limitations in coherence over long conversations and potential contradictions must be acknowledged. Knowing when to rely on GPT-3 and when to involve human experts is crucial.

14.5.6.7 Ethical considerations

Transparency, accountability, and informed consent are vital. Patients and healthcare professionals need to understand GPT-3's role and limitations. Balancing innovation with ethical safeguards is an ongoing challenge.

- Identification of gaps and ethical considerations: Imagine ChatGPT in medicine like a treasure hunt, looking for where we do not know enough. It is like a friend in the medical world, but we need to check if it is good at handling real situations, not just on paper. We are figuring out where it might struggle.
- Now, the ethical question is important, like making sure it respects privacy and does not have unfair biases. Think of it like ChatGPT joining the healthcare team, but everyone needs to be on the same page.
- In the talks between doctors and patients, ChatGPT might change things up. It is like an adventure, seeing where ChatGPT is useful and where it might have a bit of trouble in the medical world.

14.5.6.8 Insights and future directions

AI in its current state is very impressive. AI is used for typically being an intermediary and not replacing professional medical experts and cannot be trusted all the time as it might generate false positives in certain scenarios. But deploying AI would certainly help in future medical diagnosis and can help bring affordable healthcare to the masses and severely reduce the time for communication. It can also enable better transparency for medical institutions who are involved in nefarious practices and can be trusted as a valid alternative. The government can deploy these for public welfare as well so that it can be more trusted.

Right now, AI is doing some cool stuff, it is like a helper, not taking over from human medical experts. But here is the catch, one cannot always trust it completely. It might goof up, especially in tricky situations, giving incorrect information. So, even though AI is nifty, having human experts around is utmost important to make sure everything is on the right track.

We are on a mission to understand the usefulness and compatibility of ChatGPT in the medical world. We want to know how well ChatGPT plays its role in today's tech scene and how it shakes up the game in taking care of patients. Think of it as trying to figure out if ChatGPT is going to disrupt the healthcare sector.

14.5.6.9 Guidance for relevant parties

Once we unravel the mysteries, we are not keeping it to ourselves. We are turning our discoveries into down-to-earth tips for doctors, the rule-makers, and the brainy researchers. We are not just pointing out the good stuff—we are also shining a light on the tricky parts and giving practical advice to doctors and rule-makers to make things run smoother.

14.5.6.10 Significance

This review is crucial for evaluating the impact of ChatGPT in the field of medical. Our goal is to provide insights into both the advantages and potential challenges of integrating ChatGPT into practices. By doing we aim to contribute to discussions on effective use of AI in healthcare, which can shape policies, enhance patient care and advance research.

14.6 Future scope

14.6.1 Evolution of ChatGPT

Currently, the version of ChatGPT used under this paper is GPT3.5, free to public as of March 2024. This model comes with set limitations which can be subjected to change. The upcoming GPT4.o (o standing for omni) model is said to already bring in a lot of changes which includes real time access to vision, audio and text. This can change significantly and further affect the implementation of virtual assistants making the process more seamless and free flowing. As we evolve toward more sophisticated models the implementation would get much easier and thus push the public to adapt and get more comfortable with the technology.

14.6.2 Interaction via speech

While ChatGPT may be able to answer the patients' questions and relay its answers using text. The prospect of it having interactivity with the patient using voice in a pleasant or eager tone may help to better understand the severity of the situation and take actions accordingly. There are high chances where a person may take a hasty decision based on the information they are given and ChatGPT being a chatbot cannot reply at once unless it is prompted to. But, with the voice enabled there could be quick feedback to remain calm or take an immediate step to save the patient.

14.6.3 Easier documentation of cases

ChatGPT can fasten the pace of administrative tasks such as documentation, Distribution of tasks in medical institutions and actively aid in keeping track of important stats which significantly can enhance the chances of errors in critical institutions. The implementation of such systems is subjected to change and thus decides upon the models and their evolution. With upcoming models which do not require users to be specific and use certain wake words the process of administration can change significantly, and more smaller jobs can be easily taken upon by such models, thus potentially leading to better inclusion and better managed and well-informed administrative tasks.

14.6.4 Identifying medicinal drugs accurately

Traditional drugs are made after a lot of trials, errors and testing. This takes time and effort which can be used effectively if there was a tool that would the trials and errors at a much faster pace than a human being. ChatGPT can be that solution where it investigates the Documentation given to it which has past results and combining it with the present knowledge to make a unique drug that can cure the disease. So, the accuracy at which we could use that drug and successfully run tests would be much higher and can be scaled down to patients swiftly.

14.6.5 Advanced software-driven diagnostic tools

Current diagnostic tools are often outdated, and a lot of tasks require an assisted physician to address and analyze the cases which is time consuming and tedious task which requires years of experience. Sometimes certain elements are often overlooked resulting in inaccurate diagnostics. With Models like GPT Vision such tasks can be revamped and integrated into medical software to help the physician to diagnose at a much faster pace than before. Such changes are more often expected to be seen in future as more such models get cheaper and easier to be deployed, we can expect better results in correctly classifying and aiding the physician for much seamless and time efficient diagnosis.

14.7 Limitations

14.7.1 Precision and verification in clinical settings

It is crucial to confirm thoroughly GPT-3's predictions in clinical settings to guarantee real world use case scenarios and confirm upon reliability and errors. When deploying such solutions it must be done in limited phase under careful guidance and such systems cannot be deployable unless tested for many cases.

14.7.2 Lack of large case study of active deployments

As technology has recently appeared, a lack of real-world driven tests is still yet to be seen and studied upon. As these AI models continue to evolve and mature the room for errors is significantly expected to reduce, however sometimes they tend to hallucinate thus leading to incorrect results. Such cases must be thoroughly studied.

14.7.3 Lack of studies of biased dataset

The datasets often used to train these models tend to have societal biases, resulting in biased predictions and recommendations. If the training data mostly includes specific demographics, the model might not work well for minority groups, leading to differences in health results. To reduce this risk, it is essential to conduct thorough research on the datasets used in the training of these models, pinpointing any biases and evaluating how they affect the predictions.

14.7.4 Deployment in critical scenario

Implementing AI models such as GPT-3 in crucial medical situations necessitates careful preparation and thorough verification because of the minimal room for mistakes in urgent settings like emergency healthcare or intensive care, where inaccurate forecasts might result in life-threatening outcomes. Ensuring safety and effectiveness includes a gradual process, beginning with small-scale implementation in monitored environments with human supervision to confirm suggestions and enable needed actions. Thorough testing in various real-life situations is essential prior to wider adoption.

14.7.5 Closed source technology

Although OpenAI's ChatGPT's allows for third-party extensions the technology driving such systems stays closed source as of now thus impossible to evaluate some edge case applications in certain scenarios. It is hard to understand the functioning of the model and thus the consequent training that would be needed to produce correct chatbots.

14.8 Concluding remarks

This chapter provides a systematic approach to how ChatGPT can be included in applications, especially in the medical domain. We identify gaps, ethical considerations, and some ideas for future directions. In the discussion, we show that, though ChatGPT is promising for enhancing medical practices through applications in diagnostics, patient engagement, and medical education, several issues need to be addressed for its efficient implementation.

To address challenges, key points that should be looked into are data privacy and security, since GPT-3 and the like require a large volume of patient data, hence making this data vulnerable to leakage and unauthorized use. Biases and fairness are an important issue, as AI algorithms carry the risk of continuing the same type of biases that were on the source data, potentially leading to unequal treatment of patient populations. Clinical validation and accuracy are paramount since mistakes in medical advice can be grave. Dynamic interaction and error correction are also a huge hurdle, with GPT-3 lacking the ability to self-correct, possibly creating grave errors in real-time scenarios of healthcare.

The potential of AI to work with rather than for human capabilities is very promising, but it is necessary to do so in a balanced way. Human-AI collaboration may enhance the decision-making of medical practice, but the right level of automation and human oversight must be achieved. Ethical issues of transparency, accountability, and informed consent are central in integrating AI into healthcare.

References

[1] Wang, H., Wu, W., Dou, Z., He, L., and Yang, L. (2023). Performance and exploration of ChatGPT in medical examination, records and education in Chinese: Pave the way for medical AI. *International Journal of Medical Informatics*, 177, 105173. https://doi.org/10.1016/j.ijmedinf.2023.105173.

[2] Yanagita, Y., Yokokawa, D., Uchida, S., Tawara, J., and Ikusaka, M. (2023). Accuracy of ChatGPT on medical questions in the National Medical Licensing Examination in Japan: Evaluation study. *JMIR Formative Research*, 7, e48023. https://doi.org/10.2196/48023.

[3] Haupt, C. E., and Marks, M. (2023). AI-generated medical advice—GPT and beyond. *JAMA*, 329(16), 1349–1350. https://doi.org/10.1001/jama.2023.5321.

[4] Roos, J., Kasapovic, A., Jansen, T., and Kaczmarczyk, R. (2023). Artificial intelligence in medical education: Comparative analysis of ChatGPT, Bing, and medical students in Germany. *JMIR Medical Education*, 9(1), e46482. https://doi.org/10.2196/46482.

[5] Cheng, K., Guo, Q., He, Y., *et al.* (2023). Artificial intelligence in sports medicine: Could GPT-4 make human doctors obsolete? *Annals of Biomedical Engineering*, 51(8), 1658–1662. https://doi.org/10.1007/s10439-023-03213-1.

[6] Johnson, D., Goodman, R., Patrinely, J., *et al.* (2023). Assessing the accuracy and reliability of AI-generated medical responses: An evaluation of the ChatGPT model. *Research Square, rs.3.rs-2566942*. https://doi.org/10.21203/rs.3.rs-2566942/v1.

[7] Caruccio, L., Cirillo, S., Polese, G., Solimando, G., Sundaramurthy, S., and Tortora, G. (2024). Can ChatGPT provide intelligent diagnoses? A comparative study between predictive models and ChatGPT to define a new medical diagnostic bot. *Expert Systems with Applications*, 235, 121186. https://doi.org/10.1016/j.eswa.2023.121186.

[8] Nori, H., King, N., McKinney, S. M., Carignan, D., and Horvitz, E. (2023). Capabilities of GPT-4 on medical challenge problems (arXiv:2303.13375). arXiv. https://doi.org/10.48550/arXiv.2303.13375.

[9] Balas, M., Wadden, J. J., Hébert, P. C., *et al.* (2024). Exploring the potential utility of AI large language models for medical ethics: An expert panel

evaluation of GPT-4. *Journal of Medical Ethics*, 50(2), 90–96. https://doi.org/10.1136/jme-2023-109549.

[10] Garg, R. K., Urs, V. L., Agarwal, A. A., Chaudhary, S. K., Paliwal, V., and Kar, S. K. (2023). Exploring the role of ChatGPT in patient care (diagnosis and treatment) and medical research: A systematic review. *Health Promotion Perspectives*, 13(3), 183–191. https://doi.org/10.34172/hpp.2023.22.

[11] Liu, J., Wang, C., and Liu, S. (2023). Utility of ChatGPT in clinical practice. *Journal of Medical Internet Research*, 25(1), e48568. https://doi.org/10.2196/48568.

[12] Waisberg, E., Ong, J., Masalkhi, M., *et al.* (2023). GPT-4: A new era of artificial intelligence in medicine. *Irish Journal of Medical Science (1971-)*, 192(6), 3197–3200. https://doi.org/10.1007/s11845-023-03377-8.

[13] Gilson, A., Safranek, C. W., Huang, T., *et al.* (2023). How does ChatGPT perform on the United States Medical Licensing Examination? The implications of large language models for medical education and knowledge assessment. *JMIR Medical Education*, *9*(1), e45312.

[14] Ali, S. R., Dobbs, T. D., Hutchings, H. A., and Whitaker, I. S. (2023). Using ChatGPT to write patient clinic letters. *The Lancet Digital Health*, 5(4), e179–e181. https://doi.org/10.1016/S2589-7500(23)00048-1.

[15] Tangadulrat, P., Sono, S., and Tangtrakulwanich, B. (2023). Using ChatGPT for clinical practice and medical education: Cross-sectional survey of medical students' and physicians' perceptions. *JMIR Medical Education*, 9(1), e50658. https://doi.org/10.2196/50658.

[16] Eysenbach, G. (2023). The role of ChatGPT, generative language models, and artificial intelligence in medical education: A conversation with ChatGPT and a call for papers. *JMIR Medical Education*, 9(1), e46885. https://doi.org/10.2196/46885.

[17] Ashraf, H., and Ashfaq, H. (2024). The role of ChatGPT in medical research: Progress and limitations. *Annals of Biomedical Engineering*, 52(3), 458–461. https://doi.org/10.1007/s10439-023-03311-0.

[18] Haze, T., Kawano, R., Takase, H., Suzuki, S., Hirawa, N., and Tamura, K. (2023). Influence on the accuracy in ChatGPT: Differences for information per medical field. *International Journal of Medical Informatics*, 180, 105283. https://doi.org/10.1016/j.ijmedinf.2023.105283.

[19] Chintagunta, B., Katariya, N., Amatriain, X., and Kannan, A. (2021). Medically aware GPT-3 as a data generator for medical dialogue summarization. *Proceedings of the 6th Machine Learning for Healthcare Conference*, 354–372. https://proceedings.mlr.press/v149/chintagunta21a.html.

[20] Tian, S., Jin, Q., Yeganova, L., *et al.* (2024). Opportunities and challenges for ChatGPT and large language models in biomedicine and health. *Briefings in Bioinformatics*, 25(1), bbad493. https://doi.org/10.1093/bib/bbad493.

[21] Arif, T. B., Munaf, U., and Ul-Haque, I. (2023). The future of medical education and research: Is ChatGPT a blessing or blight in disguise? *Medical Education Online*, 28(1), 2181052. https://doi.org/10.1080/10872981.2023.2181052.

[22] Biswas, S. S. (2023). Role of Chat GPT in Public Health. *Annals of Biomedical Engineering*, 51(5), 868–869. https://doi.org/10.1007/s10439-023-03172-7.

[23] Vaishya, R., Misra, A., and Vaish, A. (2023). ChatGPT: Is this version good for healthcare and research? *Diabetes & Metabolic Syndrome: Clinical Research & Reviews*, 17(4), 102744. https://doi.org/10.1016/j.dsx.2023.102744.

[24] Li, J., Dada, A., Puladi, B., Kleesiek, J., and Egger, J. (2024). ChatGPT in healthcare: A taxonomy and systematic review. *Computer Methods and Programs in Biomedicine*, 245, 108013. https://doi.org/10.1016/j.cmpb.2024.108013.

[25] Souza, L. L. de, Fonseca, F. P., Martins, M. D., *et al.* (2023). ChatGPT and medicine: A potential threat to science or a step towards the future? *Journal of Medical Artificial Intelligence*, 6(9). https://doi.org/10.21037/jmai-23-70.

[26] Gilson, A., Safranek, C. W., Huang, T., *et al.* (2023). How does ChatGPT perform on the United States Medical Licensing Examination (USMLE)? The implications of large language models for medical education and knowledge assessment. *JMIR Medical Education*, 9(1), e45312. https://doi.org/10.2196/45312.

[27] Muftić, F., Kadunić, M., Mušinbegović, A., and Almisreb, A. A. (2023). Exploring medical breakthroughs: A systematic review of ChatGPT applications in healthcare. *Southeast Europe Journal of Soft Computing*, 12(1), Article 1. https://doi.org/10.21533/scjournal.v12i1.252.

[28] Wang, D.-Q., Feng, L.-Y., Ye, J.-G., Zou, J.-G., and Zheng, Y.-F. (2023). Accelerating the integration of ChatGPT and other large-scale AI models into biomedical research and healthcare. *MedComm – Future Medicine*, 2(2), e43. https://doi.org/10.1002/mef2.43.

Chapter 15

The integration of the Internet of Things (IoT) in healthcare analytics: a transformative force

Shafi Shereef[1] *and Nisha Varghese*[2]

Abstract

The healthcare landscape is undergoing a dynamic transformation driven by a confluence of factors. Patient expectations for personalized and accessible care are rising, fueled by rapid technological advancements and demographic shifts. To meet these demands, the healthcare sector is actively integrating emerging technologies such as big data analytics, electronic health records (EHRs), telemedicine, and remote patient monitoring (RPM)—all contributing to a value-based care model. This paradigm shift prioritizes preventative care and patient-centered approaches, leveraging technological innovations to fundamentally alter how healthcare is delivered and experienced. Advances in artificial intelligence (AI) and machine learning (ML) algorithms empower doctors with early disease diagnosis and prompt decision-making, potentially preventing illnesses before their onset. However, a relatively new development with significant transformative potential is the integration of the Internet of Things (IoT) into healthcare analytics systems.

The core concept of IoT in healthcare revolves around facilitating seamless data sharing, networking, and communication between various entities. This encompasses patients, medical devices, sensors, and healthcare professionals, creating a fully interconnected ecosystem. However, the true power of IoT lies in its data generation. This data fuels sophisticated analytics systems that utilize ML algorithms, predictive modeling, and data visualization techniques to uncover hidden patterns and relationships within the vast information pool. These analytical methods empower healthcare professionals with early detection of abnormalities, accurate diagnoses, and robust disease monitoring capabilities. The resulting connectivity fostered by IoT translates into numerous benefits for the healthcare industry- increased operational efficiency, improved patient care, and advancements in medical research. This convergence of technologies is redefining how healthcare data is collected, exchanged, and analyzed, ultimately providing crucial insights to support clinical decision-making and evidence-based guidance for healthcare practitioners.

[1]Department of CS&IT, Jain University, India
[2]Department of Computer Science, Christ University, India

This chapter delves into the multifaceted integration of IoT into healthcare analytics systems, highlighting its transformative potential for patient outcomes, data-driven decision-making, and healthcare delivery itself. We explore the diverse applications of IoT technology in healthcare analytics, encompassing population health management, remote diagnostics, real-time patient monitoring, and clinical research. Furthermore, we investigate the role of IoT gadgets such as wearables, sensors, and smart medical instruments in data collection. These devices capture a comprehensive picture of a patient's health through information on behavior, environmental factors, and physiological parameters, providing healthcare professionals with a holistic and continuous view. Additionally, the chapter addresses critical challenges associated with IoT integration, including data interoperability, security, and scalability. We examine how technologies like edge computing, blockchain, and cloud computing play a vital role in safeguarding patient privacy and ensuring data integrity.

Keywords: IoT; healthcare analytics; EHR; telemedicine; sensors; digital therapeutics; blockchain; cloud, and edge computing

15.1 Introduction

The term "Internet of Things" (IoT) refers to a network of networked items, devices, or "things" that are equipped with sensors, actuators, and software to facilitate autonomous data collecting, sharing, and analysis [1,2]. Through internet-based communication, these gadgets create a seamless network of physical things that can share data and interact in real time with one another and with centralized systems [3]. The IoT seeks to improve automation, efficiency, and connection in a variety of industries, including manufacturing, transportation, home automation, healthcare, and agriculture [4,5]. By leveraging IoT technologies, both individuals and organizations may monitor, regulate, and optimize operations. This can lead to improved decision-making abilities and the development of creative solutions to address challenging problems and improve overall quality of life.

Considering the architecture (Figure 15.1), generally IoT follows four-layer architecture [4,6]. The first layer is sensing or device layer which is a data gathering layer. This layer equipped with sensors and actuator to detect changes in the environment or device state. The second layer is network layer. This layer deals with data transmission. It consists of gateways and connection medias, which built a connection to exchange data collected from sensors with central system. Next layer is data processing layer, in which data analysis and decision-making is carried out. Through the final layer, application layer, the processed information received from the data processing layer is communicated to the users through various interfaces such as mobile apps and dashboard.

Healthcare is just one of the many areas that the IoT has quickly turned from futuristic concept to impacting reality. Its reach extends deep into hospital operations, medical research, and patient care, paving the way for a new era of connected health [7]. Significant changes have resulted from this transformation, including

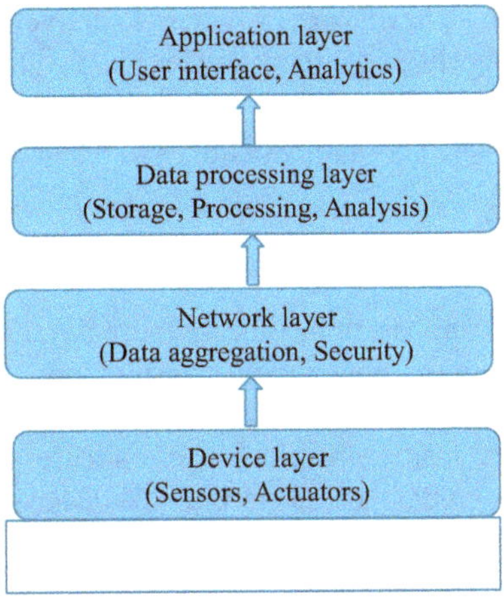

Figure 15.1 Architecture of IoT

faster research, streamlined procedures, and patient empowerment through data-driven insights.

Wearable technology and remote monitoring tools were increasingly popular in the 2010s, giving people the ability to track their health and provide ongoing care for long-term illnesses. This significantly increased connectivity and data interchange by paving the way for seamless integration with electronic health records (EHRs) and telemedicine [8]. the convergence of IoT data with advanced analytics and machine learning has opened the door to predictive analytics and population health management. The future looks even more promising. Blockchain, 5G networks, and edge computing are examples of emerging technologies that could expand the capabilities and uses of IoT in healthcare [9].

15.2 Significance of IoT in healthcare

The integration of IoT technology in healthcare represents a pivotal advancement with profound implications for patient care, healthcare delivery, and medical research [10]. In this section, we will cover how IoT enhances health industry.

15.2.1 Improving patient outcomes

The healthcare professional continuously monitors health state of patients remotely by using sensors, wearable and medical monitors and quickly detect any anomalies or changes promptly. Healthcare teams can prevent complications and improve patient outcomes by regularly monitoring their patients and taking appropriate action early on.

15.2.2 Tailored attention

IoT enables to collect vast amounts of patient data, which can be analyzed to gain insights into individual health profiles and preferences of a particular patients. This data-driven approach allows healthcare providers to deliver personalized care tailed to each patient's unique needs and circumstances. So that healthcare person can fix personalized treatment to targeted patient. For examples some medicine has some side effect for some patient. So, that medicine will be substituted with some-other depending on these data.

15.2.3 Remote patient management

Telehealth plays a significant part in healthcare these days. Healthcare professionals can monitor patients' progress, hold virtual consultations, and make prompt interventions when necessary, with the use of remote monitoring devices and telehealth platforms. Remote patient management not only increases patient convenience but also saves healthcare expenses and lowers the risk of hospital-acquired infections. This will be discussed in detail in applications of IoT in healthcare.

15.2.4 Early screening and action

Nowadays, IoT-driven analytics and machine learning algorithms used to analyze patient data to identify patterns, trends, and risk factors associated with undesirable effects on wellness. By leveraging predictive analytics, healthcare providers can address potential health risks, such as disease exacerbation or complications, and intervene proactively to prevent them. Early detection and intervention can lead to better management of chronic conditions, reduced hospitalizations, and improved overall patient outcomes.

15.2.5 Patient involvement and adherence

The way people interact with their healthcare is changing as a result of the development of IoT technologies. People now have unparalleled access to personal health data, health related educational resources, and direct channels of communication with healthcare practitioners through the use of internet-connected gadgets, mobile applications, and online patient portals [11]. Patients now have the ability to take control of their health, which encourages a sense of accountability and ownership. Engaging actively in their care process helps patients follow treatment plans more closely, track their progress, and make wise health decisions.

15.2.6 Enhancing operational efficiency

Sensors and radio frequency identification (RFID) tags provide real-time auditing into pharmaceuticals, medical supplies, and equipment. This reduces stockouts, optimizes purchasing, and prevents expired or misplaced items. IoT-based systems automatically trigger restocking orders for essential supplies, streamlining procurement processes and ensuring availability [12]. These devices predict equipment failures in advance or in real time, allowing for preventative maintenance instead of costly downtime and repairs [13].

Tracking equipment location and utilization improves staff efficiency by locating items like wheelchairs and beds quickly, maximizing their use, and freeing up resources. Smart sensors monitor and adjust lighting, heating, and cooling based on real-time needs, leading to significant energy cost savings. Automation eliminates manual data entry for vital signs and medication administration, reducing errors and freeing up staff time for patient care. IoT platforms facilitate secure communication and data sharing between departments, enhancing coordination and improving patient care continuity. By implementing IoT in healthcare, we can improve the staff productivity, enhance patient's experience, and reduce running cost.

15.2.7 Advancing innovation in the provision of healthcare

The healthcare industry faces constant pressure to improve patient outcomes, experiences, and service optimization. Fortunately, numerous avenues for fostering innovation are emerging, leading to a future of more efficient, effective, and patient-centered care [14].

One key area of innovation lies in leveraging technology to provide remote care options. Telemedicine and virtual care solutions allow healthcare organizations to reach patients in underserved areas, reduce wait times, and offer greater convenience. This increased accessibility fosters a more inclusive healthcare landscape.

Furthermore, integrating artificial intelligence (AI) and machine learning into clinical workflows holds immense potential. By analyzing vast amounts of healthcare data, AI can identify patterns, predict potential outcomes, and personalize treatment plans. This translates to improved diagnostic accuracy, tailored treatment approaches, and enhanced patient safety.

Innovation also extends to improving communication and collaboration within the healthcare ecosystem. Secure messaging platforms, EHRs, and interoperable health information exchanges enable seamless information sharing between healthcare providers, patients, and caregivers. This reduces duplicated efforts, fosters care coordination, and ultimately leads to improved continuity of care.

Beyond improving care delivery processes, fostering patient-centered design and engagement is crucial for enhancing the overall care experience. User-friendly mobile apps, patient portals, and interactive health education tools empower patients to actively participate in managing their health and wellness, leading to increased satisfaction and better health outcomes.

Finally, promoting preventive care and wellness through innovative solutions like population health management programs and predictive analytics tools allows healthcare professionals to identify and address potential risks before they develop into full-blown illnesses. This focus on proactive prevention can significantly reduce the burden of preventable chronic conditions, leading to a healthier population overall.

By embracing these diverse avenues for innovation, the healthcare industry can create a future where technology seamlessly augments care delivery, empowering patients and healthcare professionals alike to achieve improved health outcomes and a more positive healthcare experience.

15.3 Key components of IoT in healthcare

The healthcare industry is being transformed by the IoT through the establishment of an interconnected network of devices, sensors, and software. This network serves to gather, analyze, and exchange data with the aim of enhancing patient care, medical research, and hospital operations. The subsequent section delineates the underlying factors that facilitate this revolutionary phenomenon.

When IoT is used in the healthcare sector, its core is sensors and IoT-enabled devices. These work like tireless assistants. Vital signs of diseases or patients, drug use, activity levels, and environmental conditions are continuously monitored and collected. This real-time data stream provides healthcare providers with a deeper understanding of individual and collective health.

The invisible threads that weave together this ecosystem are connectivity solutions. Consider cellular technologies, wireless networks, and Bluetooth. They guarantee smooth data transfer and communication between central systems, healthcare providers, and devices. This enables the timely delivery of this vital information to the appropriate individuals at the appropriate time, as well as prompt interventions and decision-making.

Yet, data on its own is insufficient. Platforms for data analytics and insight serve as the organization's brains. Large volumes of data are analyzed by them using cutting-edge methods like AI and machine learning. These discoveries open the door for population health management, personalized medicine, and predictive analytics by highlighting patterns, trends, and possible issues. The infrastructure for these large-scale data operations is provided by cloud and edge computing. Strong computational resources and centralized storage are provided by cloud-based solutions. Edge computing allows for real-time analysis by transferring this data to the processing device in the interim. It makes sure that data is handled and analyzed effectively to satisfy the various demands of the healthcare ecosystem.

EHR integration is essential. As a result, patient data from multiple sources is accessible without interruption within current workflows. Unified views of patient health are produced by open data vaults and their deductions. Improved coordination, care, and sound decision-making are facilitated by this. It is critical to safeguard sensitive data. Access limits, authentication, and encryption are examples of crucial security and privacy protection techniques. Ensuring the privacy, accuracy, and accessibility of medical records while adhering to national regulations.

Ultimately, user interfaces like wearables, web portals, and mobile apps let patients and caregivers take an active role in their health journey. These intuitive interfaces give users direct control over information stewardship by facilitating communication, self-management, and access to health information.

In summary, IoT facilitates a connected healthcare system that empowers patients and healthcare professionals with data-driven insights. This results in improved decision-making, enhanced patient care, and ultimately a healthier future for everyone.

15.4 Applications of IoT in healthcare

15.4.1 Remote patient management

Remote patient monitoring (RPM) is a rapidly advancing field in healthcare. It does the mission that allows medical professionals to track a patient's health data remotely, outside of a traditional clinical setting. IoT is utilized for this. Primary data collection is done through wearable devices that contain sensors. These tools collect real-time vital health sign such as blood sugar levels, sleep patterns activity levels and are transferred to the cloud platform through mobile apps. These apps not only transmit health data securely, but also offer additional features such as educational resources, medication reminders, and secure messaging with healthcare providers. The cloud platform serves as a secure central hub that enables healthcare professionals to store patient data. Ultimately, healthcare providers use these data and web interfaces or mobile apps to remotely monitor patient health, identify concerns early, tailor treatment plans, and reduce the need for in-person visits. Gateways can be used for added protection in environments with multiple sensors that collect and transmit data to a cloud platform.

RPM is revolutionizing healthcare by offering a four-pronged attack on improving patient well-being and healthcare efficiency. First of all, RPM makes early health change detection possible. Through the continuous collection of real-time health data, healthcare providers can potentially reduce complications via early identification of potential abnormalities and timely intervention. Second, RPM addresses the growing expense of healthcare. Enabling remote monitoring reduces avoidable emergency room visits and hospital admissions, which saves the healthcare system a substantial amount of money. Third, RPM encourages individuals to take a more proactive approach to managing their health. Patients are empowered to take charge of their own health by having access to real-time data and insights about their condition. Lastly, RPM improves the coordination of care. A more all-encompassing and cohesive approach to patient care is ensured by better communication and cooperation between patients, physicians, and other healthcare professionals. All things considered, RPM offers a potent strategy for revolutionizing healthcare delivery, benefiting both individuals and the system at large.

15.4.2 Hospital asset management

Hospitals juggle a vast amount of critical equipment, and keeping track of it all can be a logistical nightmare. Here is where the IoT steps in to revolutionize hospital asset management. Implementing IoT technology in the healthcare industry makes equipment location awareness and real-time tracking easier. IoT tags and sensors attached to devices and equipment like wheelchairs or infusion pumps can transmit real-time location data, eliminating the need for manual searches and ensuring efficient utilization of resources [8,12,13]. Additionally, by offering predictive insights into equipment performance, IoT facilitates improved maintenance methods. Proactive maintenance reduces downtime and guarantees continuous operations. Automated maintenance alarms, including those for low battery levels or expired calibrations, enable quick action in response, minimizing disruptions and maximizing equipment performance. For example, managing temperature-sensitive items such as those stored in blood banks or vaccine refrigerators is critical to maintaining their efficacy and safety. Healthcare institutions may ensure that these storage units are maintained within the acceptable range for optimal storage conditions by using real-time temperature monitoring provided by IoT-based hospital asset management systems [14,15].

There are a number of noteworthy advantages to using IoT-based hospital asset management systems. First off, it increases productivity by cutting down on the amount of time employees must spend looking for equipment, freeing them up to concentrate more on providing patient care. By enhancing equipment utilization, reducing loss or theft, and facilitating preventative maintenance procedures, it also helps to cut costs. Additionally, IoT-based asset management systems improve patient care by ensuring the ready availability of equipment, facilitating timely delivery of care, and reducing patient waiting times. Furthermore, by providing real-time tracking of vital equipment in an emergency and guaranteeing proper equipment function through preventative maintenance procedures, these systems improve safety and ultimately contribute well-being of patients.

15.4.3 Patient safety and rehabilitation

IoT technology offers multifaceted solutions in healthcare and some of them are fall detection and prevention, medication management, wandering prevention, and infection control. Wearables equipped with fall detection capabilities serve as life-saving tools for elderly patients or those prone to falls by sensing sudden changes in motion and promptly alerting caregivers or healthcare providers, thereby minimizing fall-related injuries. Smart pill bottles and dispensers aid in medication management by reminding patients to take their medication, tracking adherence, and issuing alerts if doses are missed, thereby enhancing medication compliance and reducing the risk of errors. For patients with dementia or Alzheimer's, wearable tracking devices with geo-fencing capabilities prevent wandering incidents by triggering alerts if a patient strays outside a designated safe zone, facilitating quick location by caregivers [16,17]. Additionally, IoT sensors embedded in hospital equipment and environmental monitoring systems contribute to infection control by monitoring cleanliness levels and detecting pathogens, ensuring a hygienic environment and lowering the risk of healthcare-associated infections.

IoT technology in rehabilitation has many advantages for both patients and medical professionals. Through the use of wearables and sensors, physical therapists may remotely monitor their patients' progress in rehabilitation [18]. By analyzing data on range of motion, exercise performance, and pain levels, they can customize rehabilitation plans and offer direction. When IoT sensors are coupled with virtual reality (VR) therapy, immersive rehabilitation settings are created where patients may virtually execute exercises while receiving feedback on form and progress. This is especially beneficial for the restoration of motor skills following accidents or strokes. Moreover, IoT-enabled pill dispensers guarantee medication compliance in conjunction with physical therapy exercises, fostering the best possible course of treatment. Additionally, a data-driven rehabilitation strategy is made possible by the information gathered by wearables and sensors, which enables medical personnel to objectively monitor patients' progress and tailor rehabilitation programs.

15.4.4 Telemedicine and virtual care

The integration of IoT technology into telemedicine and virtual care has revolutionized remote communication and consultation in healthcare. Telemedicine platforms utilize video conferencing technology to facilitate virtual consultations between patients and healthcare providers, enabling doctors to access real-time health data stored in the cloud platform during consultations for informed decision-making [19]. The benefits of IoT-powered telemedicine and virtual care include improved accessibility, enhanced convenience by eliminating the need for travel, data-driven decisions supported by real-time health data, early intervention through continuous monitoring, and effective management of chronic diseases like diabetes or heart disease. For instance, in post-surgical care, wearables equipped with IoT sensors enable remote monitoring of patients' recovery progress, allowing healthcare providers to track healing and detect complications early on, ensuring optimal patient outcomes.

15.4.5 Clinical trials and research

The backbone of medical progress is clinical trials that test the safety and effectiveness of new drugs and treatments. Traditionally, data collection in these trials has relied on self-reported information and sparse clinical information. However, in this new era the IoT is revolutionizing clinical research by enabling more comprehensive and real-time data collection [20]. This leads to faster and more insightful experiments. It also provides an insight into the effectiveness of medications and the decisions to be made thereon.

Let us discuss how IoT is empowering clinical trials in a variety of ways. For the first time, wearable devices and sensors are used to collect continuous data from multiple individuals. This comprehensive data includes each individual's heart rate, blood pressure, sleep patterns, activity levels, their medication use and side effects, and disease-specific biomarkers such as blood sugar levels for diabetes trials.

Additionally, IoT fosters enhanced patient engagement by empowering participants to actively contribute to their trials through wearable feedback mechanisms and mobile apps offering medication reminders and educational resources, thus improving compliance and involvement. Moreover, IoT facilitates the utilization of real-world data (RWD), derived from wearables, reflecting participants' daily lives and providing valuable insights into treatment effectiveness in real-world scenarios.

Leveraging RWD enables faster trial completion by allowing researchers to identify trends and potential issues in real time, potentially expediting trial timelines. Furthermore, the automation of data collection through IoT devices ensures improved data quality by minimizing human error, thereby enhancing the accuracy and reliability of trial data. Overall, IoT-driven advancements in clinical trials hold promise for revolutionizing the research landscape by enabling more efficient, insightful, and patient-centric trials [20–22].

15.5 Challenges

The IoT has the potential to completely transform healthcare, but before it can be widely used, a number of important issues must be resolved [23,24]. Now we are going to discuss about it.

15.5.1 Security and data privacy

Security and data governance are the main areas of difficulty in IoT healthcare. Sensitive patient health information may be exposed to hackers and data breaches due to weak security measures on vulnerable devices. In addition, the ownership and management of patient data gathered from diverse sources present regulatory and compliance issues, necessitating the development of clear guidelines to manage data privacy laws [25]. It is imperative that these issues be resolved in order to protect patient privacy and preserve faith in IoT-enabled healthcare systems.

15.5.2 Interoperability and integration

The lack of standardization in IoT healthcare poses a major challenge as it affects interoperability between devices and healthcare systems. Without standardized communication protocols and data formats, integrating devices from different manufacturers becomes cumbersome, making it difficult to seamlessly exchange data and develop a comprehensive view of a patient's health [26]. Solving this problem is critical to enabling the seamless integration of IoT devices into healthcare ecosystems, enabling efficient data exchange and improving patient care.

15.5.3 Scalability and infrastructure

The proliferation of IoT devices in healthcare brings challenges in data management and network connectivity [27]. As large amounts of data are generated, healthcare IT infrastructure is faced with the task of robustly storing, managing, and analyzing this information in order to use it effectively. Additionally, RPM relies heavily on a stable internet connection, but the presence of limited or unreliable internet access in certain areas poses a barrier to the effectiveness of IoT-based healthcare solutions. Addressing these challenges is critical to ensure the smooth operation and effectiveness of IoT applications in healthcare.

15.5.4 Cost and reimbursement

Healthcare organizations face challenges related to the investment costs and reimbursement models associated with implementing IoT infrastructure. Costs associated with device procurement, data storage, and platform integration present significant barriers to adoption [28]. Additionally, uncertain reimbursement models within healthcare systems may not fully cover the costs of IoT-based remote monitoring and interventions.

15.5.5 Ethical considerations

In the field of IoT healthcare solutions, maintaining patient autonomy and consent to data collection is paramount. Transparency about how data is used and protected is critical to building trust [29]. Additionally, there is an urgent need to address algorithmic biases in AI algorithms for analyzing IoT data in healthcare. Measures to avoid bias are crucial to maintaining fairness and ethical standards in healthcare.

15.5.6 Workforce training and adoption

Two critical aspects of integrating IoT into healthcare systems are upskilling the workforce and ensuring user adoption and adoption. Healthcare professionals need to be trained to effectively interpret and use data from IoT devices to make clinical decisions. This includes understanding how to analyze and utilize the wealth of information provided by these devices. Additionally, both patients and healthcare providers need to be familiar with IoT technologies and have confidence that they can be seamlessly integrated into healthcare workflows. Patient acceptance and provider trust are critical to the successful implementation of IoT solutions in healthcare [30].

15.6 Advancements in healthcare IoT

The world of healthcare IoT is brimming with innovation, constantly pushing the boundaries of what is possible. Here's a glimpse into some exciting new trends and emerging technologies in healthcare IoT.

15.6.1 Artificial intelligence (AI) integration

Patient care and administration are being completely transformed in the healthcare industry by the fusion of IoT and AI [31]. IoT devices are gathering enormous amounts of data, which are being processed in real time using AI-powered analytics. Through trend detection, risk assessment, and individualized treatment suggestions, these algorithms evaluate data to improve treatment results and diagnostic precision. AI-powered chatbots can also be used as virtual health assistants; they can answer common medical questions, diagnose symptoms, make appointments, and even provide conversational analysis to support mental health. This AI and IoT integration has the potential to improve patient engagement, streamline healthcare delivery, and improve overall healthcare outcomes. Proactive prevention is replacing reactive healthcare as the trend. AI and IoT data can be used to identify people who are at risk of different illnesses, enabling early intervention and individualized preventive measures.

15.6.2 Advanced wearables and sensors

The next frontier in wearable technology involves advanced sensors that offer minimally invasive or implantable solutions [32]. Biocompatible sensors are being developed for implantation under the skin, enabling continuous monitoring of vital signs, blood sugar levels, and even brain activity. Additionally, smart clothing embedded with integrated sensors is emerging as a novel approach to healthcare monitoring. These fabrics track key health metrics such as heart rate, respiration, and activity levels, providing a seamless and non-intrusive method for collecting valuable health data. These advancements in wearables and sensors hold promise for enhancing healthcare monitoring, enabling early detection of health issues, and promoting proactive management of chronic conditions.

15.6.3 Blockchain for secure health data management

Healthcare companies can make sure that patient data is securely stored and shared among approved healthcare providers by utilizing blockchain technology. Data security and integrity are improved by this decentralized ledger system's strong defenses against illegal access and manipulation [9]. Patients have more control over their health information because they can handle and securely access their medical records from various healthcare providers. Because of blockchain's immutability and transparency, healthcare data management is made more trustworthy and accountable, which encourages cooperation and interoperability amongst a variety of stakeholders.

15.6.4 Telepresence robots and digital therapeutics

The cameras and microphone-equipped robots enable doctors to perform physical examinations from a distance, virtually visit patients in their homes, and even engage in real-time emotional support sessions.

Digital therapeutics (DTx) represents a burgeoning field in healthcare that harnesses the power of mobile apps, wearables, and sensors to provide therapeutic interventions [33]. These digital tools enable innovative approaches to treatment delivery, such as using VR apps for physical therapy exercises or gamified apps to manage chronic pain. By integrating technology into treatment protocols, DTx offers novel avenues for patient engagement and enhanced treatment outcomes. Patients can actively participate in their therapy through interactive experiences tailored to their needs, leading to increased adherence and better results. Overall, DTx holds great promise for revolutionizing healthcare delivery by leveraging digital solutions to optimize patient care and well-being.

15.7 Future of IoT in healthcare

There is a plethora of fascinating opportunities for IoT in healthcare in the future, which could completely transform how we provide and receive healthcare [34]. Here are a few things that could really change the game.

15.7.1 Hyper-personalized medicine

Envision a day when medical care is completely customized to meet your individual needs. Cutting-edge wearables and sensors along with analytics driven by AI can create a complete picture of your health. This information can be used to tailor treatment regimens depending on your unique response, anticipate possible health problems, and suggest personalized preventative measures [35].

15.7.2 AI-powered diagnostics and treatment decisions

The future of healthcare will include not only the integration of AI algorithms as potent diagnostic tools but also the utilization of massive patient data from wearables, sensors, and EHRs. AI's ability to evaluate this data makes it possible to identify diseases early, depending on the person to person make a recommendation for treatment options, and forecast the possible results of those options. By improving patient outcomes, treatment decisions, and diagnostic accuracy, this approach transforms the healthcare industry.

15.7.3 Smart homes for holistic health management

Smart homes offer more than just convenience; By attaching sensors to walls or furniture, they can act as comprehensive health centers, monitoring medication adherence, activity levels, and sleep patterns, especially for elderly and bedridden patients [36]. Combined with information from wearables, this data provides a comprehensive picture of a patient's health status, enabling improved RPM and early intervention. Providing proactive environments that promote well-being and improve overall health outcomes, households are being transformed by this cutting-edge approach to home health care.

15.7.4 Bioprinting and implantable sensors

The field of regenerative medicine, bioprinting, represents a promising path to treating injuries and organ failure through the creation of bioprinted tissues or organs. Healthcare could see a shift towards the innovative treatments that this technology enables. In addition, with the development of biocompatible implantable sensors, a new aspect is added that allows continuous monitoring of the body's vital functions and health indicators [37]. Real-time data from these sensors can be used to devise proactive health management strategies.

15.7.5 Focus on mental and behavioral health

The field of mental health is becoming increasingly important and wearables and AI are essential for early detection and monitoring. By analyzing physiological responses, activity levels and sleep patterns, these technologies make it possible to detect possible signs of stress, anxiety or depression. With the help of digital therapeutic interventions and telemedical consultations, mental disorders can be identified early and measures to improve mental health can be initiated regularly [38,39].

15.8 Conclusion

To sum up, IoT is transforming healthcare by connecting devices, sensors, and software to improve patient care, improve medical research and improve hospital operations. IoT allows for real-time monitoring, continuous data collection and advanced analytics to deliver personalized medicine, remotely manage patients, optimize hospital asset management and revolutionize clinical trials. There are many challenges to overcome in the healthcare IoT space, such as security, data privacy, interoperability and scalability issues, infrastructure scalability and limitations, cost/reimbursement model, ethical considerations, and workforce adoption. However, advances in healthcare IoT provide promising solutions to address these challenges. The integration of AI, advanced wearables, sensors, blockchain technology, telemedicine, DTx opens new doors to improve healthcare delivery, improve patient outcomes, improve patient engagement and promote proactive approaches to healthcare management. Looking ahead, IoT in healthcare promises to create a more interconnected, efficient and patient-centric healthcare system that will ultimately lead to improved health outcomes for everyone. With continued innovation and collaboration, IoT has the potential to transform healthcare delivery and improve the quality of life for individuals worldwide.

References

[1] M. Afzal, K. Fatima, P. Khalid, *et al.* (2018). Internet of things its environmental applications and challenges. *Environmental Contaminants Reviews (ECR)* 1(2), 1–3.

[2] Z. H. Ali, H. A. Ali, and M. M. Badawy (2015). Internet of Things (IoT): Definitions, challenges and recent research directions. *International Journal of Computer Applications*, 128(1), 37–47.

[3] "How IoT Works?", https://techvidvan.com/tutorials/how-iot-works/ (assessed Oct. 21, 2021).

[4] P. Gokhale, O. Bhat, and S. Bhat (2018). Introduction to IOT. *International Advanced Research Journal in Science, Engineering and Technology*, 5(1), 41–44.

[5] R. A. Rayan, C. Tsagkaris and R. B. Iryna (2021). The Internet of Things for healthcare: Applications, selected cases and challenges. In G. Marques, A. K. Bhoi, V. H. C. de Albuquerque, K.S. Hareesha Rayan, C. Tsagkaris and R. B. Iryna (eds.), IoT in Healthcare and Ambient Assisted Living (pp. 1–15). Singapore: Springer.https://doi.org/10.1007/978-981-15-9897-5_1.

[6] "IoT Architecture – Detailed Explanation", https://www.interviewbit.com/blog/iot-architecture/ (assessed Jun. 3, 2022).

[7] K. H. Almotairi (2023). Application of Internet of Things in healthcare domain. *Journal of Umm Al-Qura University for Engineering and Architecture*, 14, 1–12.

[8] M. Kumar, A. Kumar, S. Verma, *et al.* (2023). Healthcare Internet of Things (H-IoT): Current trends, future prospects, applications, challenges, and security issues. *Electronics*, 12, 2050. https://doi.org/10.3390/electronics12092050.

[9] B. Farahani, F. Firouzi, V. Chang, M. Badaroglu, N. Constant, and K. Mankodiya (2018). Towards fog-driven IoT eHealth: Promises and challenges of IoT in medicine and healthcare. *Future Generation Computer Systems*, 78, 659–676.

[10] H. Bhatia, S. N. Panda and D. Nagpal (2020). Internet of Things and its applications in healthcare: A survey. In *2020 8th International Conference on Reliability, Infocom Technologies and Optimization (Trends and Future Directions) (ICRITO)* (pp. 305–310). Noida, India. https://doi.org/10.1109/ICRITO48877.2020.9197816.

[11] D. N. Nasser, (2018). Augmented reality in education learning and training. In *2018 JCCO Joint International Conference on ICT in Education and Training, International Conference on Computing in Arabic, and International Conference on Geocomputing (JCCO: TICET-ICCA-GECO)* (pp. 1–7). Tunisia/Hammamet, Tunisia.

[12] P. S. Saarika, K. Sandhya and T. Sudha (2017). Smart transportation system using IoT. In *2017 International Conference on Smart Technologies for Smart Nation (SmartTechCon)* (pp. 1104–1107). Bengaluru, India.

[13] R. Sarmah, M. Bhuyan and M. H. Bhuyan (2019). SURE-H: A secure IoT enabled smart home system. In *2019 IEEE 5th World Forum on Internet of Things (WF-IoT)* (pp. 59–63). Limerick, Ireland.

[14] J. Xiao, J. Xie, X. Chen, K. Yu, Z. Chen and Z. Li (2017). Energy cost reduction robust optimization for meeting scheduling in smart commercial buildings. In *2017 IEEE Conference on Energy Internet and Energy System Integration (EI2)* (pp. 1–5). Beijing, China.

[15] M. Challa, K. Shreya Reddy, M. Jacob and N. S. Varna (2023). Smart energy management in classroom using IoT. In *2023 International Conference on Intelligent Data Communication Technologies and Internet of Things (IDCIoT)* (pp. 81–83). Bengaluru, India.

[16] T. Ueda and Y. Ikeda (2017). Assisting student's health consciousness by the use of wearable device. In *2017 IEEE Region 10 Conference – TENCON 2017* (pp. 2083–2087). Penang, Malaysia, 2017.

[17] S. Neogi, R. Mukherjee, S. Mukherjee, U. Chaudhuri and T. K. Rana (2021). Personal Health Monitoring System with Notification Alert. In *2021 5th International Conference on Electronics, Materials Engineering & Nano-Technology (IEMENTech)* (pp. 1–4). Kolkata, India.

[18] S. Joshi and S. Joshi (2019). Research review on IoT in healthcare. *International Journal of Engineering Applied Sciences and Technology*, 4(1), 49–53.

[19] "What can IoT do for healthcare?", https://www.wipro.com/business-process/what-can-iot-do-for-healthcare-/ (assessed Jul. 2018).

[20] S. Agnihotri and K. R. Ramkumar (2021). IoT and healthcare: A review. In *International Conference on Emerging Technologies: AI, IoT, and CPS for Science & Technology Applications*.

[21] S. Badugu, K. Srikanth, and L. N. Inampudi (2016). IoT for healthcare. *International Journal of Science and Research*, 5(2), 2319–7064.

[22] K. Aryan Nakhale, A. Tiwari, C. Choudhary, and V. Garg (2023). Healthcare monitoring system using IoT. *International Journal for Research in Applied Science & Engineering Technology (IJRASET)*, 11(1), 482–488. Available: https://doi.org/10.22214/ijraset.2023.48608.

[23] S. Selvaraj, and S. Sundaravaradhan (2020). Challenges and opportunities in IoT healthcare systems: A systematic review. *SN Applied Sciences*, 2(1), 139.

[24] M. S. Sharbaf (2022). IoT driving new business model, and IoT security, privacy, and awareness challenges. In *2022 IEEE 8th World Forum on Internet of Things (WF-IoT)* (pp. 1–4). Yokohama, Japan.

[25] R. Sivapriyan, S. V. Sushmitha, K. Pooja and N. Sakshi (2021). Analysis of security challenges and issues in IoT enabled smart homes. In *2021 IEEE International Conference on Computation System and Information Technology for Sustainable Solutions (CSITSS)* (pp. 1–6). Bangalore, India.

[26] R. Somasundaram, and M. Thirugnanam (2021). Review of security challenges in healthcare internet of things. *Wireless Networks*, 27(8), 5503–5509.

[27] M. Shamila, K. Vinuthna, and A. K. Tyagi (2019). A review on several critical issues and challenges in IoT based e-healthcare system. In *2019 International Conference on Intelligent Computing and Control Systems (ICCS)* (pp. 1036–1043). IEEE.

[28] The integration of the Internet of Things in healthcare analytics "The impact of IoT on healthcare costs and resource utilization", https://iotbusinessnews.com/2023/02/17/26547-the-impact-of-iot-on-healthcare-costs-and-resource-utilization/.

[29] N. Zainuddin, M. Daud, S. Ahmad, M. Maslizan and S. A. L. Abdullah (2021). A study on privacy issues in Internet of Things (IoT). In *2021 IEEE 5th International Conference on Cryptography, Security and Privacy (CSP)* (pp. 96–100). Zhuhai, China.

[30] S. Anand, and S. K. Routray (2017). Issues and challenges in healthcare narrowband IoT. In *2017 International Conference on Inventive Communication and Computational Technologies (ICICCT)* (pp. 486–489). IEEE.

[31] M. Mamun-Ibn-Abdullah, and M. H. Kabir (2021). A healthcare system for Internet of Things (IoT) application: Machine learning based approach. *Journal of Computer and Communications*, 9(7), 21–30. https://doi.org/10.4236/jcc.2021.97003.

[32] S. Kolarkar (2020). Modelling of Internet of Things (IoT) for Healthcare. M.S. Theses, University of Wisconsin-Milwaukee, 2020. [Online]. Available: https://dc.uwm.edu/etd/2540.

[33] I. Muzyleva, L. Yazykova, A. Gorlach and Y. Gorlach (2021). Augmented and virtual reality technologies in education. In *2021 1st International Conference on Technology Enhanced Learning in Higher Education (TELE)* (pp. 99–103). Lipetsk, Russia.

[34] G. J. Joyia, R. M. Liaqat, A. Farooq, and S. Rehman (2017). Internet of Medical Things (IoMT): Applications, benefits and future challenges in healthcare domain. *Journal of Communication*, 12(4), 240–247.

[35] D. Stefanicka-Wojtas and D. Kurpas (2023). Barriers and facilitators to the implementation of personalised medicine across Europe. *Journal of Personalized Medicine*, 13, 203. https://doi.org/10.3390/jpm13020203.

[36] M. S. Al-Kahtani, F. Khan, and W. Taekeun (2022). Application of Internet of Things and sensors in healthcare. *Sensors (Basel)*, 22(15), 5738.

[37] S. Tripathi, S.S. Mandal, S. Bauri and P. Maiti (2022). 3D bioprinting and its innovative approach for biomedical applications. *MedComm (2020)*, 4(1), e194. https://doi.org/10.1002/mco2.194.

[38] A. Tyagi, A. Baliyan and R. Kumar (2023). Impact of Internet of Things (IoT) in healthcare industry. *International Journal of Emerging Technologies and Innovative Research*, 10(6), a689–a701. Available: http://www.jetir.org/papers/JETIR2306095.pdf.

[39] N. K. Iyortsuun, S.-H. Kim, M. Jhon, H.-J. Yang and S. Pant (2023). A review of machine learning and deep learning approaches on mental health diagnosis. *Healthcare*, 11, 285. https://doi.org/10.3390/healthcare11030285.

Chapter 16
Overview of lung cancer detection: a short survey

Bhumika Choksi[1], *Priyanka Roy*[1], *Pawan Hingane*[2], *Snehal Dnyane*[2] *and Tejas Nehete*[2]

Abstract

Lung cancer is among the most fatal cancers globally, claiming millions of lives annually. Perfecting survival rates and reducing the global complaint burden rely on early and accurate discovery. Traditional individual styles, like imaging and necropsies, face limitations related to perceptivity, availability, and cost. Recent advancements in medical technologies, particularly artificial intelligence, machine learning, and radiomics, have revolutionized lung cancer discovery, offering more precise and cost-effective results. This chapter explores the biology and epidemiology of lung cancer, reviews traditional and arising individual ways, and highlights how inventions are being integrated into clinical practices. Crucial advancements, including image processing styles like noise reduction, point birth, and comparison with literal medical data, have enhanced the capability to identify cancer-affected regions in the lung. This chapter also discusses the challenges and ethical counter accusations of these inventions, emphasizing the necessity of global collaboration to maximize the benefits of high-tech results. By addressing these challenges, medical professionals can achieve better issues and give better care for cases worldwide. This work underscores the critical significance of using advanced technologies and fostering cooperation to combat one of the most significant global health challenges.

Keywords: Lung cancer detection; diagnostic advancements; imaging techniques; artificial intelligence (AI) in healthcare; machine learning models

16.1 Introduction

Lung cancer is one of the largest public health enterprises, being the leading cause of death due to cancer globally. It is quite aggressive and fatal, having a high rate of

[1]Mathematics Division, School of Advanced Sciences and Languages, VIT Bhopal University, India
[2]School of Computing Science and Engineering (SCOPE), VIT Bhopal University, India

death, which is primarily ascribed to its late diagnosis. Opinion beforehand is indispensable in perfecting survival rates, as the complaint is more manageable and treatable in its original stages. Still, traditional individual styles such as necropsies, X-rays, and computed tomography (CT) reviews have essential limitations, including perceptivity, particularity, and availability. These challenges have paved the way for integrating advanced technologies, like machine learning (ML), into the field of medical diagnostics.

ML, a division of artificial intelligence (AI), has revolutionized various disciplines, and its mode of functioning in healthcare, especially, is a transformative one.

ML algorithms excel in the processing of large, sophisticated datasets, as well as drawing patterns, even if they were minuscule enough to remain inconsiderable before the mortal viewer. For the discovery of lung cancer, there is a great potential for improving the delicacy and efficacy at individual levels. Using imaging data and other inputs that are applicable, ML models will improve early discovery capabilities to a potential extent that the mortality rate associated with the complaint might be reduced. This chapter talks about the types of lung cancer, the challenges associated with traditional discovery styles, and the revolutionary role. ML has played a role in solving such challenges while discussing unborn directions and the broader counter accusations of this technology in clinical settings. Understanding lung cancer and its challenges, lung cancer is astronomically divided into two primary types: small cell lung cancer (SCLC) and non-small cell lung cancer (NSCLC).

Each type has distinct natural characteristics and treatment approaches. SCLC is more aggressive and constitutes a smaller percentage of cases, whereas NSCLC, including adenocarcinoma, scaled-cell melanoma, and large-cell melanoma, is more common but different in terms of progression and response to treatment.

Regardless of the type, early diagnosis is the cornerstone of effective treatment, as the prognosis worsens dramatically with complaint progression. The main imaging techniques for diagnosing lung cancer include casket X-rays, CT, and positron emission tomography (PET) scans, besides invasive procedures such as necropsies. Although these techniques are effective up to a certain extent, they have many drawbacks. Imaging techniques often provide vague results, which require expert interpretation, leading to detentions or crimes. Also, invasive styles like necropsies carry pitfalls, are resource-ferocious, and may not be doable for all cases.

These challenges emphasize the need for innovative individual results that are both accurate and accessible.

The part of ML in lung cancer detection learning among machines has emerged as an important tool in ultra-modern drugs, and one such area is oncology. The ability to reuse and dissect large volumes of data coupled with its prophetic capabilities makes it well-suited for addressing the complications of lung cancer. ML can decompose vivid forms of data, such as imaging reviews, clinical records, and genomic data, to detect patterns representative of cancer. Some of the most impactful operations of ML in lung cancer discovery involve medical imaging ways like CT and PET reviews making detailed images of lung cancer, which can be anatomy using ML algorithms to describe abnormalities, classify nodes, and separate between benign and nasty growths.

Deep learning models, especially convolutional neural networks (CNNs) have shown magnificent performance in image analysis, including the automatic discovery of lung nodes in high perceptivity and particularity. The quality of the data used for training is very essential to the effectiveness of these ML models. Intimately available datasets like LUNA16 and NSCLC Radiomics have played a pivotal part in advancing exploration in this sphere. These datasets give labeled imaging data, which ML models use to learn and ameliorate their individual capabilities. Still, data preprocessing is vital for guaranteeing the effectiveness of these models in a way similar to addition, normalization, and segmentation are used to prepare data for analysis about challenges similar to class imbalance and noise.

16.2 Background and related works

Lung cancer is a leading cause of cancer-related deaths globally, causing millions of fatalities annually. The disease is often difficult to detect in its early stages due to the lack of clear symptoms or the presence of symptoms that mimic less severe conditions, resulting in delayed diagnosis. It is generally categorized into two types: SCLC and NSCLC, with NSCLC being more common. Smoking, genetic factors, and exposure to harmful substances such as radon and asbestos are major risk factors. Early detection is essential to improving survival rates, as treatment effectiveness decreases in later stages. Advances in imaging methods, such as CT scans and X-rays, have improved the ability to detect lung abnormalities. However, the manual analysis of these images is often slow and error-prone, especially in detecting subtle or early stage abnormalities. To address these challenges, ML and image processing technologies are being increasingly applied to enhance the accuracy and efficiency of lung cancer detection.

Without explicit programming, computers can now analyze data and make judgements or predictions based on patterns thanks to ML, a subfield of AI. The analysis and modification of images to improve their quality or extract useful information is the main emphasis of image processing. When combined, these technologies have revolutionized diagnostic techniques by providing automated, accurate, and effective lung cancer detection options. This study highlights developments, methods, and difficulties in the field as it examines the use of ML and image processing in the diagnosis of lung cancer.

It kills millions of people around the world every year. To prevent that, early detection is a must because it enhances survival and improves the chances of a cure for this disease. Researchers show that patients with an early stage of lung cancer have more positive responses to the therapy than those diagnosed with late-stage diseases. Such diagnostic techniques have been significantly developed through advancements in ML and medical imaging. It is becoming a very precious resource for the support given to medical practitioners through these computer-aided diagnosis (CAD) systems that automatically identify and categorize lung cancer. Such systems analyze chest CT scans, which are commonly employed to detect lung cancer, using powerful techniques from both image processing and ML. Image processing techniques are fundamental to the functioning of CAD systems. The process begins with preprocessing, which enhances the quality of medical images by minimizing noise and emphasizing important features. Techniques such as the geometric mean filter are frequently applied to CT images to reduce noise, resulting in clearer and more detailed visuals of lung structures. Following preprocessing, segmentation is performed to identify specific regions of interest within the image. Methods like K-means clustering and fuzzy C-means clustering are utilized to distinguish cancerous tissues from healthy ones, effectively segmenting images based on texture and intensity. To improve the accuracy of nodule detection, advanced segmentation techniques, including active contour models and selective binary Gaussian filtering, have been developed. These approaches help to outline lung nodules and other abnormalities, which are critical for accurate and reliable diagnosis.

There are numerous applications for ML algorithms in the discovery and classification of lung cancer. Commonly used are artificial neural networks (ANN), K-nearest neighbors (KNN), and random forests (RF). ANNs are found to work very effectively for predicting lung cancer by simulating the functions of the human brain. Through the learning approach from labeled datasets, the CT images can be categorized as either cancerous or not. KNN is a supervised learning algorithm, valued for simplicity and effectiveness in classification tasks because it predicts based on the proximity of data points in the feature space. KNN, however, is computationally expensive and less efficient with high-dimensional data. Random Forest is an ensemble method that uses multiple decision trees to make more reliable and accurate predictions. The studies indicate that ANNs usually outperform KNN and RF in accuracy, sensitivity, and specificity for the classification of lung cancer. Researchers have made good progress using computational techniques to develop innovative methods for improving accuracy and reliability in the detection system for lung cancer. To be specific, Palani and Venkatalakshmi [1] introduced a predictive modelling system. They combined health monitoring based on IoT, fuzzy clustering, and ML algorithms with classification. Their procedure utilizes fuzzy C-means to perform segmentation upon medical images and then classifies the cancer stages utilizing various ML algorithms. This system enhances the accuracy of segmentation as well as real-time monitoring of health. Therefore, it is also helpful in early detection and continuous management of patients.

Sood *et al.* [2] also developed a deep learning pipeline involving the combination of UNet and ResNet models to draw features from CT scans. This ensemble system combines classifiers like XGBoost as well as RF that enhance the accuracy of identification of malignant nodules. Compared to the traditional methods, their method produces better results and shows that it is possible to use several models to have robust detection. Talukdar and Sarma [3] also contributed by focusing on the issues of false positives and negatives in manual radiology. They presented automated systems using CNNs to minimize the reliance on the input of the radiologist and maximize the efficiency of detection.

A notable contribution here is the optimal deep neural network proposed by Lakshmanaprabu *et al.* [4], which determines the optimal number of image features that need to be processed so that the classification accuracy can be improved. Their model reached 96.2% accuracy, indicating that feature optimization in deep learning models will be a great asset in lung cancer detection. Also, Joon *et al.* [5] used an active spline model for segmenting the X-ray images for lung cancer regions. Their approach combines K-means and fuzzy C-means feature extraction techniques and employs SVMs for the classification. The multi-stage methodology ensures proper segmentation and classification of the lesion area. Nithila and Kumar [6] employed an active contour model with a variational level set function to outline the borders of the lung parenchyma, which is very critical for diagnosis in the case of lung diseases.

Advances in ML have also aided in histopathological analysis. Yu *et al.* [7] applied whole-slide images of lung cancer stained with hematoxylin and eosin to predict patient prognosis. Thousands of quantitative features were extracted from these images, and then ML algorithms were used to split patients into groups with significantly different survival outcomes. This integration of pathology and computational techniques makes clear the scope for developing personalized treatment plans based on predictive analytics.

Novoy methodologies have been developed, for which rapid progress has resulted in automated detection systems. Lung nodules from CT images were identified by Kurkure and Thakare [8] using a system based on a genetic algorithm. Its approach was a combination of Naive Bayes classifiers as well as genetic algorithms, to which up to 80% of accuracy in cancer stage classifications came at their end. Chaudhary and Singh [9] have stressed the need for an overall multi-step approach involving preprocessing, segmentation, and feature extraction to identify the phases of lung cancer with better accuracy.

Although all these improvements are present, there are challenges with enhancing the sensitivity and specificity of detection systems, especially when dealing with lesions that are smaller than 10 mm in diameter. Reducing false positives and negatives is essential for the enhancement of the reliability of CAD systems. There has been a great emphasis on such approaches as radiomics that involve extracting high-dimensional data from medical images. For example, Westaway *et al.* [10] utilized a radiomic approach to analyze the 3D properties of CT scans and predict lung cancer outcomes. Their findings highlight the potential of imaging phenotypes in improving diagnostic accuracy and understanding disease progression.

Another integration that has reduced dependency on manual input from the radiologist is ML and image processing. In this way, automated systems with minimal human intervention can reduce noise, extract features, and classify. Advanced neural ensemble-based detection methods, which combine sophisticated feature extraction techniques and robust classification models, have been developed to enhance biopsy result accuracy. These provide a much stronger identification of cancerous tissues and find their place in the modern lung cancer detection system.

In conclusion, in a nutshell, with advanced computational techniques such as deep learning, fuzzy clustering, genetic algorithms, and radiomics, the detection of lung cancer has been revolutionized. These systems are opening doors to early diagnosis, personalized treatment, and improved patient outcomes because of challenges such as segmentation accuracy, small lesion detection, and false classification. Further research is therefore important in fine-tuning these methods for their proper adaptation into clinical settings.

16.3 Literature survey

Javed *et al.* [11] in “Deep Learning for Lungs Cancer Detection: A Review” (2024), discuss the use of deep convolutional neural networks (DCNN) in lung cancer diagnostics, emphasizing imaging modalities and systematic studies. It brings out the precision, accuracy, and AUC of DCNN, making it a game-changer in the field using multiple imaging modalities. Another paper is, “Deep Machine Learning for Medical Diagnosis: Application to Lung Cancer Detection: A Review” by Gayap and Akhloufi [12] in 2024. It aims for advanced models, including 2D/3D CNNs, dual-path networks, and vision transformers. Challenges cited include the dependency on the data used and interpretability, but it is highly sensitive on a dataset such as LIDC and LUNA16.

The work by Wan *et al.* [13], published in 2024, integrates explainable AI (XAI) with deep learning to increase model transparency and encourage trust in medical AI systems. In the same year, Tan *et al.* [14] examined AI architectures and methodologies for CAD systems based on PRISMA guidelines. This research reviewed 119 studies, classifying AI techniques and identifying future research avenues, with studies indexed in Scopus and WoS.

Another notable contribution by Quasar *et al.* [15] employs ensemble techniques like Boosting and Weighted Box Fusion by combining BEiT, DenseNet, and Sequential CNN models. This method achieved 98% accuracy, outperforming individual models using a chest CT-scan dataset. Similarly, Obayya *et al.* [16] in their work applied Gabor filtering, GhostNet for feature extraction and tuna swarm algorithm (TSA) optimized with an echo state network (ESN) classifier achieving a maximum accuracy of 99.33%, which is effective in early cancer detection.

An effective method for lung cancer diagnosis using deep learning-based support vector network, Shafi *et al.* [17] combine deep learning with SVM in analyzing pulmonary nodules to reach an accuracy of 94% with its reliability for early diagnosis using the LIDC/IDRI dataset. Similarly, Sori *et al.* [18] in 2020 introduced DFD-Net, which combines DR-Net for noise reduction with two-path CNN for local and global feature extraction, achieving balanced and competitive results on the CT scan datasets.

In 2022, Shimazaki *et al.* [19] published five-fold cross-validation on chest radiographs, achieving a sensitivity of 0.73 and low mFPI of 0.13, outperforming competitors in the non-overlapping category. Last but not least, Rehman *et al.* [20] in 2021 employed LBP and DCT to extract features using SVM and KNN classifiers with accuracy at 93% and 91%, respectively, surpassing the accuracy of other state-of-the-art techniques based on chest CT scan images.

Sl. no.	Year	Authors	Key techniques approach	Dataset used
1	2024	Javed *et al.* [11]	Explored deep learning with an emphasis on Deep Convolutional Neural Networks (DCNN); reviewed imaging modalities and systematic studies.	Multiple Imaging Modalities
2	2024	Gayap and Akhloufi [12]	Focused on advanced models like 2D/3D CNNs, dual-path networks, and vision transformers	LIDC, LUNA16
3	2024	Wani *et al.* [13]	Integrated Explainable AI with deep learning for better interpretability in diagnostics	N/A
4	2024	Tan *et al.* [14]	Analyzed AI architectures and methodologies for CAD systems using PRISMA guidelines	Studies Indexed (Scopus, WoS)
5	2024	Quasar *et al.* [15]	Combined BEiT, DenseNet, and Sequential CNN using ensemble methods like boosting and Weighted Box Fusion	Chest CT-Scan Dataset
6	2021	Obayya *et al.* [16]	Gabor filtering for preprocessing, GhostNet for feature extraction, TSA optimized with ESN classifier	N/A
7	2022	Shafi *et al.* [17]	Combined deep learning with SVM for detecting pulmonary nodules based on physiological changes	LIDC/IDRI Dataset
8	2020	Sori *et al.* [18]	Introduced DFD-Net, combining DR-Net for denoising and two-path CNN for local and global feature extraction	CT scan datasets
9	2022	Shimazaki *et al.* [19]	Developed a segmentation-based DL model using five-fold cross-validation for chest radiographs	Chest radiographs from 2006–2018
10	2021	Rehman *et al.* [20]	Feature extraction using LBP and DCT, classification via SVM and KNN	Chest CT scan images

16.4 Research gap

Despite the promising development in deep learning for lung cancer detection, several research gaps are yet to be addressed. One of the challenges is the dependency on big, well-labeled datasets, such as LIDC and LUNA16, that fail to represent the diversity encountered in clinical settings [11,12]. These powerful models often lack generalizability across diverse datasets, reducing their utility in a clinical setting. Another issue is the interpretability of the deep learning models used. Many approaches, such as DCNN, perform well in terms of accuracy but operate as "black-box" systems, making it difficult for clinicians to trust or understand the model's decision-making process [13]. This lack of transparency is a significant bottleneck in their clinical adoption. Finally, computational complexity is another bottleneck. While accuracy is touted as the goal, deep learning models are often computationally prohibitive for real-time applications in clinical settings [20,21]. Moreover, while ensemble models and hybrid approaches have been proven useful in enhancing detection accuracy, they introduce new complexity problems and training times that have to be optimized for practical use. Other aspects that have not been dealt with appropriately in current research are XAI and real-time model deployment. These are gaps that will be essential to bridge for more robust, efficient, and clinically viable solutions for lung cancer detection.

16.5 Overview of model

The model for lung cancer detection involves a systematic pipeline consisting of multiple stages, each designed to prepare, process, and analyze the input data to achieve an accurate prediction [21–23].

Automation: This model automates the lung cancer detection process, reducing reliance on manual interpretation and improving efficiency.

Feature extraction: Advanced preprocessing techniques ensure that the most critical features are effectively captured for accurate prediction.

Scalability: The model is scalable and can handle large datasets, making it suitable for real-world medical applications.

Customizability: The model's architecture can be customized with advanced techniques like deep learning, ensemble learning, or radiomic feature analysis to improve performance.

This approach addresses common challenges in medical image analysis, such as noise reduction, small lesion detection, and minimizing false positive and false negative rates, paving the way for earlier and more accurate lung cancer diagnoses Figure 16.1).

16.5.1 Lung image dataset

16.5.1.1 Image collection

Medical imaging data, such as CT scans or X-rays, are gathered from a dataset. These images provide crucial visual details about the lung structure and any possible abnormalities [24].

16.5.1.2 Labeling

Along with the images, labels are included to indicate whether the images show cancerous or non-cancerous conditions. These labels are essential for supervised learning and for assessing the model's performance.

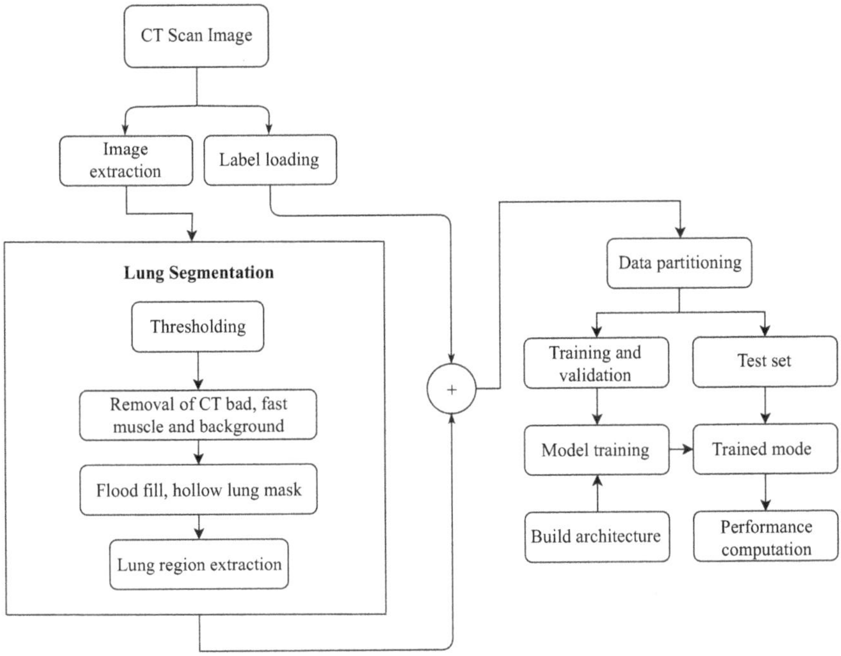

Figure 16.1 Diagrammatic representation of the model

16.5.2 Lung segmentation

This section outlines the preprocessing steps to isolate and segment the lung regions from the CT images:

16.5.2.1 Thresholding

A threshold is applied to identify regions of interest (lungs) based on pixel intensity.

16.5.2.2 Removal of CT artifacts

Unwanted elements like fat, muscle, and background noise are removed to clean the image.

16.5.2.3 Flood fill and hollow lung mask creation

A flood-fill algorithm is applied to create a mask that represents the hollow lung regions.

16.5.2.4 Lung region extraction

The final step isolates the lungs from the CT image, ready for further processing.

16.5.3 Data partitioning

The dataset is divided into three subsets:

16.5.3.1 Training set

This set is used to train the ML model, allowing it to learn from the labeled data.

16.5.3.2 Validation set

This set is used during training to adjust hyperparameters and prevent overfitting.

16.5.3.3 Test set

This set is kept aside for evaluating the model's performance on new, unseen data to assess its generalization ability.

16.5.4 Model architecture

16.5.4.1 Model design

A ML or deep learning model is developed for lung cancer detection. This typically involves constructing neural networks like CNNs, which are well-suited for image processing tasks. The architecture typically includes convolutional and pooling layers for feature extraction, followed by fully connected layers for classification.

16.5.5 Training

16.5.5.1 Model learning

The model is trained by inputting the labeled data, allowing it to learn to differentiate between cancerous and non-cancerous lung images. The training process aims to reduce errors by applying optimization algorithms, such as stochastic gradient descent or Adam. During this phase, feature extraction is key as the model learns to identify important patterns like the size, shape, and texture of lung nodules.

16.5.6 Finalized model

After completing the training process, the model is ready for evaluation. It should be capable of making accurate predictions based on the features it has learned.

16.5.7 Performance evaluation

The model's performance is tested on the validation set to compute key metrics like accuracy, sensitivity, specificity, and F1-score. These metrics are used to assess the model's effectiveness and determine its potential for practical use in clinical environments.

16.6 Epidemiology of lung cancer

Lung cancer is the second most common cancer globally, following breast cancer. It leads to the highest number of cancer-related deaths worldwide, primarily due to its aggressive nature. The diagnosis of lung cancer often occurs at later stages, making it highly fatal. The disease is influenced by a complex interplay of environmental, genetic, and socio-economic factors.

In 2020, the World Health Organization reported approximately 2.2 million new cases of lung cancer, accounting for 11.4% of all cancer diagnoses globally. Lung cancer was responsible for about 1.8 million deaths in the same year, representing 18% of all cancer-related deaths. Historically, lung cancer has been more prevalent in males due to higher smoking rates, but as smoking rates in females have increased and environmental exposure has increased, the gender disparity in prevalence is narrowing.

The incidence of lung cancer is higher in high-income countries, such as the United States and Europe, largely due to smoking and the aging population. In contrast, developing countries, particularly in Asia and Africa, face rising rates of lung cancer due to factors such as increased air pollution and occupational exposures.

Lung cancer is classified into two main types based on histological characteristics:

16.6.1 Non-small cell lung cancer

This type accounts for approximately 85% of lung cancer cases. It includes:

16.6.1.1 Adenocarcinoma

Often found in non-smokers and females, originating from glandular cells.

16.6.1.2 Squamous cell carcinoma

Typically linked to a history of smoking, originating from the bronchial epithelium.

16.6.1.3 Large cell carcinoma

A rare and rapidly developing form of cancer.

16.6.2 Small cell lung cancer

Making up about 15% of lung cancer cases, this type is highly aggressive, with early metastasis and rapid growth.

16.7 Risk factors and causes

Lung cancer has several known causes, with smoking being the most significant. It is responsible for about 85% of all cases. Tobacco smoke contains over 7,000 harmful chemicals, many of which can cause cancer. Even people who do not smoke are at risk if they are exposed to second-hand smoke, often called passive smoking. Environmental factors and workplace hazards also contribute to lung cancer. Long-term exposure to air pollution, such as fine particles (PM2.5) and industrial emissions, has become a growing problem. Certain jobs, like those in mining, construction, and manufacturing, expose workers to harmful substances like asbestos, silica, and radon gas, which can greatly increase their risk.

Genetics also play a role, as some individuals are more prone to lung cancer due to mutations in specific genes like epidermal growth factor receptor (EGFR) and anaplastic lymphoma kinase (ALK). These mutations can make even non-smokers susceptible to the disease. Age and gender also matter. Lung cancer is most commonly diagnosed in people over 65, and while it has historically been more common in men, the number of cases in women is rising due to changing smoking habits and more exposure to environmental risks. Other factors, such as certain viruses like human papillomavirus, can also increase the risk. Additionally, chronic lung conditions such as COPD and pulmonary fibrosis make individuals more likely to develop lung cancer.

16.8 Socioeconomic impact

Lung cancer has a profound socioeconomic impact, affecting both individuals and society as a whole. The financial burden is substantial, with high costs associated with diagnosis, treatment, and palliative care placing immense strain on patients and healthcare systems. Beyond these direct expenses, the economic impact is further amplified by the loss of productivity caused by illness and premature death, particularly among individuals of working age. Additionally, disparities in access to healthcare services are a significant concern. In resource-limited areas, the lack of early detection programs and advanced treatment options often results in worse outcomes for patients, highlighting the need for equitable healthcare solutions.

16.9 Current diagnostic methods

Lung cancer diagnosis relies on a combination of imaging techniques, laboratory tests, and pathological examinations. These methods aim to identify lung abnormalities, confirm malignancy, and determine the stage and type of cancer. While traditional approaches have been instrumental in lung cancer management, their limitations in early detection and accuracy necessitate the development of advanced diagnostic tools.

16.9.1 Imaging techniques

Imaging techniques are essential in diagnosing lung cancer, aiding in biopsies, and tracking treatment effectiveness.

16.9.1.1 Chest X-rays

Often used as the first diagnostic step, chest X-rays help detect lung abnormalities. However, they are less effective at identifying small nodules or early stages of lung cancer despite being cost-effective and widely accessible.

16.9.1.2 CT scans

Low-dose CT scans are especially beneficial for high-risk groups like heavy smokers, as they can reveal smaller nodules that might not be visible on X-rays. These scans produce detailed cross-sectional images of the lung, helping in tumor detection and staging, although they sometimes lead to false positives, prompting unnecessary follow-ups.

16.9.1.3 Positron emission tomography scans

PET scans use radioactive tracers to identify areas of active cancer growth, which helps in assessing the spread of cancer and determining its stage. When combined with CT scans (PET-CT), they offer both functional and structural information, leading to more precise diagnostic results.

16.9.1.4 Magnetic resonance imaging

While magnetic resonance imaging (MRI) is not commonly used to detect lung cancer directly, it is highly valuable in evaluating whether the cancer has spread to the brain or spinal cord, providing crucial information for treatment planning.

16.9.2 Biopsy and histopathology

A definitive diagnosis of lung cancer typically requires a biopsy, which involves collecting a tissue sample for microscopic examination to identify cancer cells. This process is essential in determining the type and stage of cancer, aiding in the selection of the most effective treatment plan.

16.9.2.1 Bronchoscopy

Bronchoscopy is a method used to obtain tissue samples for biopsy. During this procedure, a flexible tube with a camera and light is inserted through the nose or mouth and guided into the airways. It allows physicians to directly examine the lung and collect tissue or fluid samples from suspicious areas for further testing.

16.9.2.2 Needle biopsy

A needle biopsy involves inserting a thin needle through the chest wall into the tumor to obtain a tissue sample for examination. Imaging techniques, such as CT or ultrasound, are used to guide the procedure and ensure precision. This approach is particularly effective for accessing tumors located in the outer areas of the lung.

16.9.2.3 Endobronchial ultrasound

Endobronchial ultrasound is a technique that combines bronchoscopy with ultrasound to locate and biopsy lymph node tumors near the lung. It offers a minimally invasive alternative to traditional surgical methods like mediastinoscopy, making it a preferred option for sampling tissue in these areas.

16.9.2.4 Thoracoscopy and video-assisted thoracic surgery

Thoracoscopy and video-assisted thoracic surgery are procedures used when less invasive methods cannot provide adequate lung or pleural biopsies. A small camera is inserted into the chest cavity, allowing direct visualization of the area and enabling precise tissue collection. These methods are particularly useful for challenging cases.

The histological examination of biopsy samples allows pathologists to classify lung cancer into its major subtypes (NSCLC and SCLC) and to further characterize tumor markers, which can guide treatment decisions.

16.9.3 Blood tests and biomarkers

Blood tests are emerging as non-invasive methods for detecting and monitoring lung cancer. While they are not currently a standalone diagnostic tool, they provide useful information about tumor biology and progression.

16.9.4 Circulating tumor cells

Circulating tumor cells are cancer cells that detach from the primary tumor and enter the bloodstream. Their detection can indicate the presence of lung cancer and provide information about metastasis, aiding in disease management.

16.9.5 Biomarkers for screening and diagnosis

16.9.5.1 EGFR mutations

EGFR mutations are frequently found in lung adenocarcinomas, particularly in non-smokers. Testing for these mutations helps identify patients who may benefit from targeted therapies [25].

16.9.5.2 ALK rearrangements

This genetic alteration is seen in NSCLC and can be treated with targeted therapies like crizotinib.

16.9.5.3 PD-L1 expression

Levels of programmed death-ligand 1 (PD-L1) are used to assess eligibility for immunotherapy treatments such as pembrolizumab, which enhance the immune system's ability to target cancer cells.

16.9.6 Plasma tumor DNA (liquid biopsy)

Liquid biopsy is a technique for detecting circulating tumor DNA (ctDNA) in the bloodstream. It provides real-time information about genetic mutations, tumor changes, and treatment responses. This method is used for screening, monitoring disease recurrence, and assessing minimal residual disease.

16.9.7 Other biomarkers

Biomarkers such as CYFRA 21-1, CEA, and NSE are being studied for their ability to support imaging and biopsy methods. However, their routine clinical application is limited by issues with sensitivity and specificity.

16.9.8 Genomic and molecular testing

Understanding the genetic characteristics of lung cancer is essential for developing targeted therapies and personalized treatments, which is crucial for developing targeted therapies and personalized treatments.

16.9.8.1 Next-generation sequencing

Next-generation sequencing (NGS) allows for a detailed analysis of tumor genomes, identifying mutations, copy number changes, and gene fusions. It is used to classify tumors and guide treatment decisions.

16.9.8.2 Mutation profiling

Profiling mutations such as EGFR, KRAS, and BRAF is used to customize treatments for lung adenocarcinomas. Targeted therapies for these mutations have improved outcomes, with promising new therapies for KRAS mutations.

16.9.8.3 Gene expression profiling

This method evaluates patterns of gene activity in cancer cells to predict prognosis and treatment responses. It also helps differentiate tumor types and identify potential treatment targets.

16.10 Comparative table

Sr. no	Author/ year	Key findings	Advantages	Disadvantages
1	Javed *et al.* (2024)	Utilized Deep Convolutional Neural Networks (DCNN) for detection of lung cancer. Attained high accuracy, precision, and AUC in various imaging modalities.	Extremely effective for various imaging methods, enhances detection rates.	Needs large datasets and high computation.
2	Gayap and Akhloufi (2024)	Used 2D/3D CNNs, Dual-Path Networks, and Vision Transformers for lung cancer diagnosis. Had high sensitivity, especially on LIDC and LUNA16 datasets.	Good feature extraction, enhances the accuracy of early detection.	Accurate, but depends on dataset quality and is hard to interpret.
3	Wani *et al.* (2024)	Applied Explainable AI (XAI) with Deep Learning to enhance transparency and trust in AI-powered medical diagnostics.	Enhances clinical use and enhances decision-making.	Hard to achieve high accuracy with explainability.
4	Tan *et al.* (2024)	Carried out a review of 119 AI architecture studies for CAD systems and determined gaps in diagnostic AI research.	Gives a complete overview of AI trends and the future.	Experimental verification for real-world use not available.

(Continues)

(*Continued*)

Sr. no	Author/ year	Key findings	Advantages	Disadvantages
5	Quasar *et al.* (2024)	Applied Ensemble Learning (Boosting, Weighted Box Fusion) with BEiT, DenseNet, and Sequential CNN with an accuracy of 98% on lung cancer classification.	better than single models, enhances classification performance.	High computational complexity, making real-time usage restrictive.
6	Obayya *et al.* (2023)	Applied Gabor Filtering, GhostNet, TSA, and ESN Classifier, which achieved a 99.33% accuracy for detection in early stage lung cancer. Integrated Deep Learning with SVM for pulmonary nodule classification with 94% accuracy.	Very accurate, suitable for early detection.	Need for high-quality input images, which can restrict its accessibility.
7	Shafi *et al.* (2022)	Integrated Deep Learning with SVM for pulmonary nodule classification with 94% accuracy.	Effective for the detection of small nodules, enhancing early diagnosis.	Is based on the dataset chosen and extraction of features.
8	Sori *et al.* (2020)	Created DFD-Net (DR-Net + Two-Path CNN) for segmentation and noise reduction in lung cancer CT scans.	Reduces the noise of images, enhances the accuracy of detection.	High-cost computation and need for expensive hardware.
9	Shimazaki *et al.* (2022)	Applied Segmentation-Based Deep Learning with five-fold cross-validation to chest radiographs, which gave 0.73 sensitivity and fewer false positives.	Reliable for large lung lesions, improving detection accuracy.	Has a low accuracy rate in identifying small, overlapping lesions.
10	Rehman *et al.* (2021)	Utilized Feature Extraction (LBP, DCT) with Classification (SVM, KNN), with 93% (SVM) and 91% (KNN) accuracy.	Competitive performance, beneficial for structured CT scan analysis.	Ineffective for real-time detection because it takes a lot of time.

The above table presents a comparative study of various deep learning methodologies applied to lung cancer detection, analyzing research conducted between 2020 and 2024. It highlights key techniques such as DCNNs, 2D/3D CNNs, dual-path networks, vision transformers, and ensemble learning approaches like boosting and weighted box fusion. Several studies integrate XAI to enhance model transparency and clinical trust, while others explore hybrid models combining deep learning with traditional classifiers like SVM and KNN.

The analysis spans multiple datasets, including LIDC, LUNA16, chest CT scans, and radiographs, assessing model performance in terms of accuracy, sensitivity, and precision. While some models, such as those using BEiT and DenseNet, achieve classification accuracies above 98%, the research also highlights key limitations. These include high computational costs, dependence on large and high-quality datasets, and challenges in achieving real-time performance. Additionally, studies focusing on CAD systems using AI frameworks identify gaps in diagnostic AI research, calling for further experimental validation and improvements in model interpretability. The findings underscore the importance of balancing model accuracy with explainability and real-world applicability to advance AI-driven lung cancer diagnostics.

16.11 Discussion

AI and ML are revolutionizing lung cancer diagnosis by enhancing accuracy and minimizing dependency on human analysis. Conventional techniques, like CT scans and X-rays, involve physicians interpreting images, which sometimes results in misdiagnosis or delays. AI-based methods, specifically deep learning models such as CNNs and Ensemble Learning, have proven high accuracy, at times over 98%. These models allow for early detection, which is essential for successful treatment and improved patient outcomes. Furthermore, AI accelerates the diagnostic process and delivers reproducible results, reducing human error. While these advantages are present, AI-based lung cancer detection is challenged by several concerns.

The biggest challenge is the reliance on certain datasets, that is, LIDC and LUNA16, that might not accurately reflect varied patient populations. This can result in skewed outcomes when used in various clinical settings. Another issue is the unexplainability—most AI models give predictions without explaining how they came to those conclusions, so doctors cannot always have complete faith in the results. Deep learning methods also need a lot of computational power, which makes them less feasible for resource-constrained hospitals. Privacy and ethical issues related to the use of AI in medical decision-making also have to be addressed. To render AI more useful in lung cancer detection, future advancements should aim toward developing models that can generalize across various datasets, making AI choices more explicit for medical practitioners, and maximizing computational performance to make it available. Solving these challenges will enable AI implementation within practical medical environments, rendering lung cancer diagnosis stronger and more efficient.

16.12 Future research direction

Future research in lung cancer detection should focus on the continuous development and refinement of advanced ML techniques to improve accuracy, interpretability, and clinical usability. XAI will be critical in ensuring that clinicians and patients understand how diagnostic decisions are made, while deep learning models, particularly CNNs, should be further enhanced to increase sensitivity and specificity in detecting early stage lung cancer. Radiomics, which extracts quantitative features from medical images, holds significant potential for improving early detection and predicting tumor behavior. Integrating radiomics with other forms of data—such as genomic profiles, electronic health records, and patient lifestyle factors—will enable more precise diagnostics and personalized treatment planning. Liquid biopsy technologies also offer promising avenues for research, particularly in enhancing the sensitivity of detecting ctDNA and other biomarkers in bodily fluids. Efforts should focus on standardizing these techniques for consistent clinical application, as well as exploring their potential for monitoring treatment effectiveness and detecting residual disease. Similarly, molecular and genomic profiling, facilitated by NGS, should be expanded to identify rare genetic mutations and guide precision medicine. This approach will help tailor treatments to individual patients, particularly through targeted therapies and immunotherapy strategies. Cost-effective diagnostic solutions are essential to make these advancements accessible to low-resource settings. Developing affordable technologies for AI-based diagnostics, liquid biopsies, and molecular testing can enable broader implementation of population-wide screening programs, particularly in underserved regions. Addressing ethical challenges and biases is equally important, necessitating diverse datasets for training ML models and ensuring transparency in AI decision-making processes. Further research should also prioritize the detection of small lesions, which are often missed in early diagnostic stages. Enhanced imaging algorithms and hybrid imaging techniques, such as combining CT with PET or MRI, can improve sensitivity and accuracy. Automation of diagnostic workflows, integrating AI and IoT-based systems for real-time monitoring, will not only streamline processes but also reduce human error. Finally, fostering global collaboration through international data sharing, standardization of diagnostic protocols, and equitable access to advanced technologies is vital. Such efforts will help address disparities in healthcare delivery and contribute to reducing the global burden of lung cancer through earlier detection and improved treatment outcomes.

16.13 Conclusion

The chapter on lung cancer detection concludes that significant advancements in technology, such as AI, ML, and radiomics, are transforming the landscape of early lung cancer diagnosis. These innovations address the limitations of traditional methods, including sensitivity, specificity, accessibility, and cost, enabling earlier and more accurate detection. The integration of imaging techniques, data analysis, and ML models has significantly enhanced diagnostic precision, automated workflows, and personalized treatment plans. Despite these achievements, challenges remain, such as ensuring data quality, addressing false positives and negatives, overcoming the high costs of advanced tools, and managing ethical concerns associated with AI in healthcare. The chapter emphasizes the critical need for global collaboration and equitable access to these technologies to maximize their benefits and improve patient outcomes. By leveraging technological advancements and fostering international cooperation, healthcare systems can better tackle lung cancer, one of the most pressing global health challenges.

References

[1] Palani, D., and Venkatalakshmi, K. (2019). An IoT based predictive modelling for predicting lung cancer using fuzzy cluster based segmentation and classification. *Journal of Medical Systems*, *43*(2), 21.

[2] Sood, T., Bhatia, R., and Khandnor, P. (2023). Cancer detection based on medical image analysis with the help of machine learning and deep learning techniques: a systematic literature review. *Current Medical Imaging*, *19*(13), 1487–1522.

[3] Talukdar, J., and Sarma, P. (2018). A survey on lung cancer detection in CT scans images using image processing techniques. *International Journal of Current Trends in Science and Technology*, *8*(3), 20181–20186.

[4] Lakshmanaprabu, S. K., Mohanty, S. N., Shankar, K., Arunkumar, N., and Ramirez, G. (2019). Optimal deep learning model for classification of lung cancer on CT images. *Future Generation Computer Systems*, *92*, 374–382.

[5] Joon, P., Bajaj, S. B., and Jatain, A. (2019). Segmentation and detection of lung cancer using image processing and clustering techniques. In *Progress in Advanced Computing and Intelligent Engineering: Proceedings of ICACIE 2017, Volume 1* (pp. 13–23). Springer Singapore.

[6] Nithila, E. E., and Kumar, S. S. (2019). Segmentation of lung from CT using various active contour models. *Biomedical Signal Processing and Control*, *47*, 57–62.

[7] Yu, K. H., Zhang, C., Berry, G. J., *et al.* (2016). Predicting non-small cell lung cancer prognosis by fully automated microscopic pathology image features. *Nature Communications*, *7*(1), 12474.

[8] Kurkure, M., and Thakare, A. (2016). Classification of stages of lung cancer using genetic candidate group search approach. *IOSR Journal of Computer Engineering*, *18*(05), 7–13.

[9] Chaudhary, A., and Singh, S. S. (2012, September). Lung cancer detection on CT images by using image processing. In *2012 International Conference on Computing Sciences* (pp. 142–146). IEEE.

[10] Westaway, D. D., Toon, C. W., Farzin, M., *et al.* (2013). The International Association for the Study of Lung Cancer/American Thoracic Society/ European Respiratory Society grading system has limited prognostic significance in advanced resected pulmonary adenocarcinoma. *Pathology*, *45* (6), 553–558.

[11] Javed, R., Abbas, T., Khan, A. H., Daud, A., Bukhari, A., and Alharbey, R. (2024). Deep learning for lung cancer detection: a review. *Artificial Intelligence Review*, *57*(8), 197.

[12] Gayap, H. T., and Akhloufi, M. A. (2024). Deep machine learning for medical diagnosis, application to lung cancer detection: a review. *BioMedInformatics*, *4*(1), 236–284.

[13] Wani, N. A., Kumar, R., and Bedi, J. (2024). DeepXplainer: An interpretable deep learning-based approach for lung cancer detection using explainable artificial intelligence. *Computer Methods and Programs in Biomedicine*, *243*, 107879.

[14] Tan, S. L., Selvachandran, G., Paramesran, R., and Ding, W. (2024). Lung cancer detection systems applied to medical images: a state-of-the-art survey. *Archives of Computational Methods in Engineering*, *32*, 1–38.

[15] Quasar, S. R., Sharma, R., Mittal, A., Sharma, M., Agarwal, D., and de La Torre Díez, I. (2024). Ensemble methods for computed tomography scan images to improve lung cancer detection and classification. *Multimedia Tools and Applications*, *83*(17), 52867–52897.

[16] Obayya, M., Arasi, M. A., Alruwais, N., Alsini, R., Mohamed, A., and Yaseen, I. (2023). Biomedical image analysis for colon and lung cancer detection using the Tuna Swarm algorithm with a deep learning model. *IEEE Access*, *11*, 94705–94712.

[17] Shafi, I., Din, S., Khan, A., *et al.* (2022). An effective method for lung cancer diagnosis from CT scan using deep learning-based support vector network. *Cancers*, *14*(21), 5457.

[18] Sori, W. J., Feng, J., Godana, A. W., Liu, S., and Gelmecha, D. J. (2021). DFD-Net: lung cancer detection from denoised CT scan image using deep learning. *Frontiers of Computer Science*, *15*, 1–13.

[19] Shimazaki, A., Ueda, D., Choppin, A., *et al.* (2022). Deep learning-based algorithm for lung cancer detection on chest radiographs using the segmentation method. *Scientific Reports*, *12*(1), 727.

[20] Rehman, A., Kashif, M., Abunadi, I., and Ayesha, N. (2021, April). Lung cancer detection and classification from chest CT scans using machine learning techniques. In *2021 1st International Conference on Artificial Intelligence and Data Analytics (CAIDA)* (pp. 101–104). IEEE.

[21] Reddy, N. S., and Khanaa, V. (2023). Intelligent deep learning algorithm for lung cancer detection and classification. *Bulletin of Electrical Engineering and Informatics*, *12*(3), 1747–1754.

[22] Abdullah, D. M., Abdulazeez, A. M., and Sallow, A. B. (2021). Lung cancer prediction and classification based on correlation selection method using machine learning techniques. *Qubahan Academic Journal*, *1*(2), 141–149.

[23] Pacurari, A. C., Bhattarai, S., Muhammad, A., *et al.* (2023). Diagnostic accuracy of machine learning AI architectures in detection and classification of lung cancer: a systematic review. *Diagnostics*, *13*(13), 2145.

[24] Kaur, C., and Garg, U. (2023). Artificial intelligence techniques for cancer detection in medical image processing: a review. *Materials Today: Proceedings*, *81*, 806–809.

[25] Lin, L. P., and Tan, M. T. T. (2023). Biosensors for the detection of lung cancer biomarkers: a review on biomarkers, transducing techniques, and recent graphene-based implementations. *Biosensors and Bioelectronics*, *237*, 115492.

Index

www.ingramcontent.com/pod-product-compliance
Lightning Source LLC
LaVergne TN
LVHW011836200926
848484LV00002B/361